Symbiosis

SYMBIOSIS

An Introduction to Biological Associations

SECOND EDITION

Surindar Paracer
Vernon Ahmadjian

UNIVERSITY PRESS

2000

OXFORD

Oxford   New York

Athens   Auckland   Bangkok   Bogotá   Buenos Aires   Calcutta
Cape Town   Chennai   Dar es Salaam   Delhi   Florence   Hong Kong   Istanbul
Karachi   Kuala Lumpur   Madrid   Melbourne   Mexico City   Mumbai
Nairobi   Paris   São Paulo   Singapore   Taipei   Tokyo   Toronto   Warsaw

and associated companies in
Berlin   Ibadan

Copyright © 2000 by Oxford University Press, Inc.

Published by Oxford University Press, Inc.
198 Madison Avenue, New York, New York 10016

Oxford is a registered trademark of Oxford University Press

Library of Congress Cataloging-in-Publication Data
Paracer, Surindar.
    Symbiosis : an introduction to biological associations / by
Surindar Paracer, Vernon Ahmadjian. — 2nd ed.
      p.   cm.
    Includes bibliographical references and index.
    ISBN 0-19-511806-5; 0-19-511807-3 (pbk.)
      1. Symbiosis.   I. Ahmadjian, Vernon. Symbiosis.   II. Title.
      [DNLM: 1. Symbiosis.   QH 548 P221s 1999]
    QH548.P27   1999
    577.8'5—dc21
    DNLM/DLC
    for Library of Congress   99-26963

9 8 7 6 5 4 3 2 1

Printed in the United States of America
on acid-free paper

To my wife Kathleen and our
children Amy and Mark,
in love and gratitude

Surindar Paracer

To Janice, for her love and patience

Vernon Ahmadjian

PREFACE

In an age of specialization, when scientists are learning more and more about less and less, we have taken a different approach in writing this book. We have broadly examined many disciplines of modern biology for examples of symbiosis, the living together of different species of organisms. Comparisons of such examples have revealed common themes among the different kingdoms of life and have emphasized the unity and relatedness of living organisms.

Life is complex and often involves a delicate balancing act between hosts and symbionts, in associations that range from parasitism to mutualism. In the long history of life on earth, symbionts in their attempts to overcome a host's defenses have evolved many protective strategies, including molecular camouflage, deception, mimicry, and subversion. An example of how a host defends itself against infection is pathogen-induced apoptosis. Advances in DNA biology have provided new insights into molecular mechanisms that are involved in symbioses. These insights have helped us understand the intricate details of microbial pathogenesis and how diseases have evolved. The recent emergence of fields such as cellular microbiology, immunoparasitology, and endocytobiology have revealed new aspects of symbiosis that are helping scientists develop innovative approaches to the synthesis of drugs and vaccines to combat some of the most widespread and persistent scourges of humanity. By studying symbiosis one gains a wide evolutionary perspective on the extent and nature of interspecific biological interactions.

The impact of symbiosis on modern biological thinking has been revolutionary due to the serial endosymbiosis theory (SET) of the origin of eukaryotic cells. What started out as a historical oddity evolved into an off-beat, curious hypothesis, and finally ended up as mainstream biology. This tortuous path has been shepherded by Lynn Margulis of the University of Massachusetts, a tireless champion of SET. The idea that mitochondria and chloroplasts are transformed symbiotic bacteria provides a common thread to the biological world and raises the hope of finding other symbiotic wonders among life's diversity. The recent findings of plastid genes in malarial parasites, DNA sequence similarities between *Rickettsia prowazekii*, the bacterium that causes typhus, and mitochondria and the symbiotic life forms around deep-sea hydrothermal vents have awakened our curiousity about the ancient origins of modern life.

This book is intended for students and faculty to help them learn about molecular and evolutionary aspects of symbioses. We believe that an understanding of symbiosis should be part of the education of all biologists. The molecular revolution in biology has also embraced the symbiotic world and has raised as many new questions as it has answered old ones. We have included molecular findings throughout the text in the overall context of organismal ecology and evolution.

This book can be used as a reference source and also as a text for undergraduate and graduate courses in biology. We have tried to maintain the diversity of information which was appreciated by users of the first edition. Our reference list is selective rather than exhaustive and designed to provide entry into particular topics. Many chapters have been revised, some extensively, and there is a new chapter on the molecular biology of infectious diseases. One of our most difficult tasks was the choice of topics and examples to be included in the text. We are sure that there are some investigators whose work should have been included. We invite our colleagues to view our work for its overall coverage.

We thank the following individuals and institutions for their help in preparing this book. Librarians at the Marine Biological Laboratory, Woods Hole, Massachusetts, Goddard and Science Libraries at Clark University, Worcester State College, and the University of Massachusetts Medical School helped us gather research materials. Professors Justin Thackeray, Denis Larochelle, Jerry Brink, and Thomas Leonard, all of the Department of Biology, Clark University, Lynn Margulis, University of Massachusetts, and Ann Hirsch, University of California at Los Angles, read various chapters and made helpful comments

and suggestions. Professors Peter Bradley, Department of Biology, and Ruth Haber, Department of Languages and Literature, both from Worcester State College, reviewed the entire text and made many helpful suggestions. Thomas White, Computer Services, Worcester State College, helped us when we encountered computer software problems. Kathryn Delisle rendered our ideas and rough sketches into a fine set of illustrations for this book. Figure 8.3 was drawn by Anne T. Sullivan. We deeply appreciate the skill and persistence of Rene Baril, Department of Biology, Clark University, for organizing the manuscript. The index was prepared by Cristin Hulslander. We thank Lisa Stallings and Kirk Jensen, Oxford University Press, for their encouragement and assistance in the production of this book. We gratefully acknowledge the scientists who provided us with photographs and illustrations.

Worcester, Massachusetts
January, 2000

S.P.
V.A.

CONTENTS

Symbiosis

1

INTRODUCTION

Symbiosis is an association between two or more different species of organisms. The association may be permanent, the organisms never being separated, or it may be long lasting. This definition excludes populations, which are associations between individuals of the same species. Organisms that are involved in a symbiosis may benefit from, be harmed by, or not be affected by the association. Symbiotic associations are common in nature, from bacteria and fungi that form close alliances with the roots of terrestrial plants to those between giant tube worms and sulfur-oxidizing bacteria that live together in the deepest depths of the oceans. No organism is an island—each one has a relationship to other organisms, directly or indirectly. Even humans bear a reminder of an ancient symbiosis—their cells contain mitochondria, organelles which once were symbiotic bacteria. In addition, each of us harbors several types of viruses and bacteria in our skin and intestinal tract. Similarly, chloroplasts in plant cells are organelles which have evolved from ancient symbiotic photosynthetic bacteria. Bacteria which form symbioses with higher forms of life are themselves hosts to symbiotic viruses. Satellite viruses depend on other viruses for their expression. It is difficult to imagine life and its evolutionary history without symbioses (Khakhina, 1992; Sapp, 1994a,b).

This book focuses on interspecific relationships, their importance in the evolution of living organisms, and their significance in the earth's economy. We examine in detail symbiotic associations that represent the different kingdoms of the living world (Rennie, 1992).

1.1 SYMBIOSIS AND ITS SIGNIFICANCE IN MODERN BIOLOGY

Many biologists study organisms and their natural history. Organisms have been collected, hunted, stuffed, preserved, classified, and dissected, and their life histories have been subjected to microscopic details, yet we are hard pressed to define an organism. After decades of cellular and molecular emphasis in biology, the organismal concept is once again drawing attention due to renewed interest in biodiversity and conservation biology. Organisms do not occur axenically in nature. Does the physical boundary contain and define an organism? At the genetic level we learn that an organism's genes are "selfish," genomic conflicts abound, and transposons and viruses play critical roles in genetic variability. All symbioses involve interspecific genetic interactions (Dyer, 1989).

Endocytosis is central to all cellular symbioses. The process involves the uptake of extracellular material into a cell in a membrane bound vacuole. Endocytosis also plays a significant role in: (1) antigen presentation, (2) nutrient acquisition, (3) clearance of apoptic cells, (4) receptor regulation, (5) hypertension, and (6) synaptic transmission. The process involves several distinct morphological and biochemical mechanisms and understanding it through a high-resolution three-dimensional view of the clathrin coat has opened a new vista (Marsh and McMahon, 1999).

Symbiosis and Coevolution

Modern evolutionary biologists are deeply entrenched in conceptual frameworks of group selection, genetic kinship, inclusive fitness, and gradual versus punctuated equilibrium. What has been missing from their inquiries is the role of symbiosis in the evolutionary explanations. Lynn Margulis (1992a) expressed her frustration by wondering how one can talk about the evolution of the cow without mentioning its cellulose-digesting microbes. Most biologists have accepted the concept of a serial endosymbiosis origin of eukaryotes, but other coevolutionary themes have yet to receive the attention they deserve. The evolutionary impact of biological interactions is only now beginning to be appreciated in terms of its influence on speciation and biodiversity, microbial pathogenesis, and the evolutionary arms race between a host and its symbiont (Maynard-Smith, 1991).

Generating Novelty by Symbiosis

Plants and animals have acquired new metabolic capabilities through symbioses with bacteria and fungi (Olff and Ritchie, 1998). Mammalian herbivores and termites digest cellulose with the help of microbial symbionts. Marine bioluminescence in some fishes and squids is produced by luminescent bacteria contained in specialized light organs. Diverse animal life around the deep-sea vents is based on symbiosis with bacteria that oxidize hydrogen sulfide and chemosynthetically fix carbon dioxide into carbohydrates. Associations between fungi and algae have resulted in unique morphological structures called *lichen thalli* (Margulis and Fester, 1991).

Early Life and Origin of the Eukaryotic Cell

A major event in which symbiosis played a crucial role is the evolution of cell organelles such as mitochondria and chloroplasts through associations involving prokaryotic organisms. In this manner, eukaryotes acquired the metabolic machinery of cell respiration and photosynthesis from endosymbionts. Another ancient symbiotic union may have produced the microtubule-based motility of the modern cell (Margulis and McMenamin, 1990).

Colonization of Land by Plants

All living things need phosphorous to make nucleic acids (DNA and RNA) and ATP. On land, early plants were faced with the diffi-

culty of absorbing phosphorous. They formed associations with mycorrhizal fungi, which greatly facilitated their phosphorous uptake. This crucial event occurred some 400 million years ago and may have played a significant role in the plant's ability to colonize terrestrial habitats (Atsatt, 1991; Newsham et al., 1995).

Horizontal Gene Transfer: Symbioses that Affect Gene Pools

Evolutionary changes in organisms and their gene pools are not restricted to nuclear events and sexual mechanisms. Horizontal gene transfer between species has been documented in all forms of life. Bacterial cells possess plasmids and prophages that transfer new genetic properties from one cell to another (Dubnau, 1999). Many virulence factors in pathogenic bacteria are expressed through plasmid-borne genes. Similarly, bacteria become resistant to antibiotics when they take up plasmids with genes for antibiotic resistance (fig. 1.1). Horizontal gene transfer has been suggested in the evolution of flowers, fruits, and storage structures from gall-forming insects and fungi. The role of viruses as genetic engineers is gaining importance in evolutionary biology (Amabile-Cuevas and Chicurel, 1993).

Bacterial Symbionts that Behave as Cell Organelles

The *Rhizobium*–legume symbiosis is an example of how host cells and bacterial symbionts within root nodules undergo transformations which allow the bacterial cells to fix nitrogen, which then is transferred to the host plant. Newly formed root nodules continuously replace old nodules throughout the host plant's life span. Rhizobia as bacteroids within the host cortical cells behave as temporary cell organelles that fix nitrogen.

In the rice weevil–bacteria endosymbiosis, bacteria are permanent "organelles" of female germ line cells. The host insect benefits by receiving vitamins, and through rapid development and increased fertility. The bacterial symbionts cannot live independently and occur temporarily in larvae from where the future egg cells become infected.

The Evolution of Sex

Intragenomic conflict as an evolutionary force is providing new insights into the history of

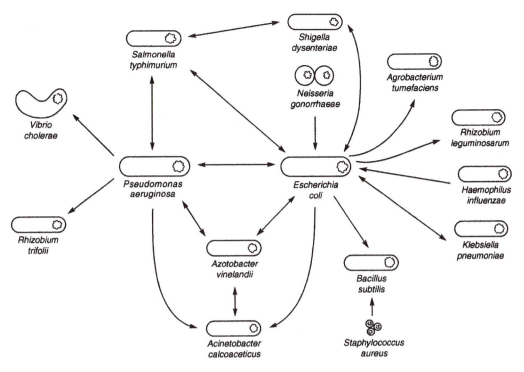

Fig. 1.1   Plasmid-mediated horizontal gene transfer of antibiotic resistance among gram-negative bacteria. In a similar way, genes encoding for virulence factors in bacterial pathogenesis can be transferred among bacterial species. Adapted from Atlas (1997).

life on earth. The evolution of sex is a form of genomic conflict management (Lively, 1996). Uniparental inheritance of cytoplasmic genes, mating types, and many features of sexual behavior may have evolved as a result of evolutionary conflict (Partridge and Hurst, 1998). The two-sex model which is widespread throughout eukaryotic organisms may have been the result of ancient intracellular symbioses (Hutson and Law, 1993; Vollrath, 1998; Moller et al., 1999).

The *Red Queen hypothesis* suggests that harmful parasites and virulent pathogens exert selection pressure on their hosts so that sexual reproduction is maintained, while asexual reproduction becomes an unstable strategy. So how do we account for asexual reproduction among organisms that form symbioses? According to this hypothesis, ancient asexual reproduction was a viable strategy only when asexual hosts did not have a long generation time compared to their parasitic symbionts (Ladle et al., 1993).

In 1930 Ronald Fisher proposed that the development of male characteristics and female preference for certain characteristics

speeds up the evolutionary process. In birds, female choice is responsible for bigger tails, brighter colors, and exaggerated displays; matings with males with these traits will produce sons with the preferred characteristics. During the 1980s, Maynard Smith and Hamilton and Zuk examined this issue in light of parasitic load, host fitness, and the degree of host sexual dimorphism. They argued that parasites and pathogens and their hosts are involved in a microevolutionary arms race and that in time the symbiont's offense and host defense produce cycles of coadaptation. An accurate measure of female choice is for healthy males that are free of harmful symbionts, and displaying that character in males communicates viability (Hamilton, 1982; Hamilton and Zuk, 1982; Maynard-Smith, 1989; McLennan and Brooks, 1991; Zuk, 1996).

Symbiotic interactions occur even at the molecular level, with the concept of "selfish DNA" being the ultimate parasite, spreading around as transposons without causing too much harm (Kunze et al., 1997). Computer models have been designed to study the influence of symbiotic interactions on ecosystems

(O'Callaghan and Conrad, 1992). Looking into the future, some have envisioned a type of "cybersymbiosis" or the interactions between human and manufactured body parts in new life forms (Margulis and Sagan, 1986b).

1.2 SUBDIVISIONS OF SYMBIOSIS

In this book, we use the term "symbiosis" in a broad sense, as originally intended by Anton de Bary in 1879 (appendix 1) to refer to different organisms living together. Proposals to change this definition and redefine symbiosis, such as equating it with mutualism, have led to confusion (Saffo, 1992a; Lewin, 1995). Various types of symbioses, whether beneficial or harmful, are described by the terms commensalism, mutualism, and parasitism. An association in which one symbiont benefits and the other is neither harmed nor benefited is a commensalistic symbiosis. An association in which both symbionts benefit is a mutualistic symbiosis. A relationship in which a symbiont receives nutrients at the expense of a host organism is a parasitic symbiosis.

Commensalism

The term *commensal* was first used by P.J. van Beneden in 1876 for associations in which one animal shared food caught by another animal. The two animals were considered to be "messmates" that ate from the same table. An example of a commensalistic relationship is that between silverfish and army ants. The silverfish live with the army ants, participate in their raids, and share their prey. They neither harm nor benefit the ants. We use the term commensalism in its broadest sense, where the benefit to one of the symbionts may be nutritional or protective.

Mutualism

In a *mutualistic symbiosis*, both partners benefit from the relationship. The extent to which each symbiont benefits, however, may vary and generally is difficult to assess (Cushman and Beattie, 1991; Bronstein, 1994a). The benefit a symbiont receives from an association should be considered in terms of its costs. As de Bary (1879) suggested, there probably is not an example of mutualism in which both partners benefit equally. In most symbioses, despite numerous experimental studies, we

do not fully understand the complex interactions that take place between the symbionts (Bronstein, 1994b; Knowleton, 1998). In many associations there is a reciprocal exchange of nutrients. For example, in the symbioses of algae and invertebrates, the algae provide the animals with organic compounds, which are products of photosynthesis, while the animals provide the algae with waste products such as nitrogenous compounds and carbon dioxide, which the algae use in photosynthesis. Such a close complementarity between partners increases the success and evolution of the mutualistic association. Marine animals such as corals grow well in nutrient-poor tropical oceans because their symbiotic algae provide them with food and oxygen. Another benefit that may be obtained in a mutualistic association is protection. Some bacteria and unicellular algae are protected from a hostile environment by residing within a host cell. Such protection has resulted in adaptations of the symbiont such as a greatly reduced or absent cell wall and loss of sexual reproduction.

Mutualism, symbiosis, and cooperation

Charles Darwin's theory of natural selection, with its emphasis on competition, struggle for existence, and survival of the fittest, left out the role of cooperation as a factor in biological evolution. Darwin was keenly aware of biological interactions in nature, as is evident from his publications: *The Different Forms of Flowers on Plants of the Same Species* (1877), *The Effects of Cross and Self Fertilization in Vegetable Kingdom* (1876), *The Various Contrivances by which Orchids are Fertilized by Insects* (1862), *Insectivorous Plants* (1875), and *Earthworms* (1881).

Unfortunately, in many academic circles, the terms symbiosis, mutualism, and cooperation have similar meanings and are often used interchangeably. Mutualism also has been used widely to describe intraspecies cooperative behavior in various animal species. The study of cooperation has enjoyed a resurgence during the past three decades (table 1.1). *Cooperation* is defined as a beneficial outcome relative to cost that an individual or a group requires for a collective action (Gadagkar, 1997). Cooperation in an evolutionary perspective can be divided into four categories: (1) cooperation via kin selection, (2) cooperation via group selection, (3) cooperation via reciprocity, and (4) cooperation

Table 1.1 Significant publications on the evolutionary study of cooperation

Author	Publication	Year
A. V. Espinas	*Des Societes Animales*	1878
Peter Kropotkin	*Mutual Aid. A Factor in Evolution*	1902
W. M. Wheeler	"The ant-colony as an organism"	1911
H. Reinheimer	*Evolution by Cooperation*	1913
W.C. Allee	*Animal Aggregations*	1931
W.C. Allee	*The Social Life of Animals*	1938
A.E. Emerson	*Social Coordination and the Superorganism*	1939
A.E. Emerson	*The Biological Basis of Social Cooperation*	1946
A. Montagu	*Darwin: Competition and Cooperation*	1952
W.D. Hamilton	"The evolution of altruistic behavior"	1963
R. Trivers	"The evolution of reciprocal altruism"	1971
E. O. Wilson	*The Insect Societies*	1971
E. O. Wilson	*Sociobiology: The New Synthesis*	1975
R. Axelrod	*The Evolution of Cooperation*	1984
D. S. Wilson	*Reviving the Superorganism*	1989
E. O. Wilson	*Consilience: The Unity of Knowledge*	1998

via byproduct mutualism (Dugatkin, 1997). Although the first three categories have received sufficient attention from modern evolutionary biologists, the case for byproduct mutualism has yet to be established. Byproduct mutualism is generally found in the context of interspecific associations and involves pseudo-reciprocity (J. Brown, 1983; Connor, 1986, 1995).

Parasitism

Parasitism is a symbiosis in which one of the symbionts benefits at the expense of the other. As in mutualism, the primary factor in parasitism is nutrition: the parasite obtains its food from the host. Parasitic symbioses affect the host in different ways. Some parasites are so pathogenic that they produce disease in the host shortly after the parasitism begins. In other associations the symbionts have coevolved into a controlled parasitism where death of the host cells is highly regulated (Ebert and Herre, 1996).

Parasitologists often describe a host in relation to the role it plays in the life cycle of a parasitic symbiont (Toft and Karter, 1990). A definitive or final host is one in which a symbiont reaches sexual maturity. An intermediate host may also be necessary for the completion of a symbiont's life cycle. For example, in the life cycle of the sheep liver fluke, *Fasciola hepatica*, sheep are the definitive hosts and snails are the intermediate hosts; the fluke's larval stages develop in the snail.

Some parasites need more than one intermediate host. Reservoir hosts are species that harbor potential pathogenic symbionts often without showing signs of disease. Wild animals are frequently the source of infection for humans and domesticated animals. Vectors are carriers of transmittable pathogenic parasites. Many blood-sucking arthropods, such as horseflies, tsetse flies, mosquitoes, and fleas, are well-known vectors of viral, bacterial, and protozoan pathogens.

Mutualism, parasitism, and commensalism are the only categories of symbiosis considered in this book. Some scientists recognize other categories, such as phoresis and inquilinism, which are based on the transport or shelter of one of the symbionts. However, we consider such associations to be commensalistic. In *phoretic relationships*, for example, some insects deposit their eggs on the bodies of other animals, which transport them to new habitats. Some marine hydroids and sea anemones attach themselves to the shells of molluscs and crabs and are carried about by their hosts. In *inquilinism*, two or more animals of different species share a dwelling place. The predation of one animal by another has been considered to be a type of symbiosis by some investigators.

Terms such as mutualism, parasitism, and commensalism are used to conveniently categorize associations. But many relationships are not static, and there may be frequent transitions from one type to another. Symbiotic associations may change because of environmental factors or internal influences caused by the development of the symbionts. A parasitic association could evolve into one of mutualism or commensalism. Indeed, it is difficult to conceive of two organisms starting out in a mutualistic association. Most mutualistic symbioses probably began as parasitic ones, with one organism attempting to exploit another one. If we consider parasitism as an antagonistic relationship, then mutualism can be regarded as a standoff or a draw between two antagonists. For example, during the course of a parasitic relationship between two organisms, the host's defenses may be strong enough to slow or stop the growth of the parasite. Two extreme examples of such a mutualistic evolution are mitochondria and chloroplasts in eukaryotic cells. These organelles are transformed bacteria that may have begun as parasitic symbionts in ancient prokaryotic cells. Conversely, a mutualistic association may degenerate into a pathogenic one if the defenses of the host weaken. For example, the common intestinal symbiont of humans, *Escherichia coli*, can become a pathogen when it acquires plasmid-borne virulence factors from other bacteria.

Pathogens and Disease

Because early workers equated the term symbiosis with mutualism, parasitism is generally considered to be separate from symbiosis. Terms such as "parasitology," "parasites," and "parasitism" continue to be used without considering their position in the wider picture of symbiosis. Parasites have a poor public image because scientists and laymen tend to link disease and parasitism (Windsor, 1997). Many parasites, however, do not cause disease; they do not disrupt or seriously diminish the performance of their host even though they take nutrients from the host. Parasites and parasitism should be viewed in the broader context of symbiosis and coevolution (Ewald, 1994). *Pathogens* are defined as entities that produce disease conditions in their host (Ewald, 1993; Read, 1994).

Common themes in microbial pathogenesis

Recent advances have shown that many bacterial pathogens follow similar strategies to infect their host and cause disease (Nee and Smith, 1990). We are ignorant of how diverse microorganisms are and the concepts of microbial disease causation need to be reassessed (Relman, 1999). Common virulence factors include toxins, adhesion mechanisms, how pathogens enter the host cells, and how they survive once inside the host cell (e.g., remaining in the vacuole or escaping into the cytoplasm). Avoiding the host immune response is the key to the success of pathogens, and methods of doing this include antigenic variation, camouflage by host, molecular mimicry, and enzymatic disabling of the host immune system. Virulence factors are maintained on plasmids and pathogenicity islands that are easily acquired. Therefore, new strains of pathogens are constantly evolving (Finlay and Falkow, 1997) (fig. 1.1).

1.3 SYMBIOSIS IN ALL FORMS OF LIFE

One of the earliest systems of classification, developed by Linnaeus, consisted of two kingdoms, plants and animals. Although this classification lasted a long time, it was not satisfactory because some organisms such as *Euglena*, bacteria, and slime molds did not fit well in either kingdom. These odd groups were arbitrarily placed into either the plant or animal kingdom depending on the personal biases of scientists. As techniques in cell biology improved and details of the cell were better understood, it became clear that the two-kingdom system of classification was too restrictive and would have to be expanded. In 1969, R.H. Whittaker proposed a five-kingdom classification, which met with initial resistance but later gained general acceptance. The five kingdoms proposed by Whittaker were the Monera, Protista, Fungi, Plantae, and Animalia. In recent times, phylogenetic lineages based on molecular homologies of more than 60 ancient proteins and rRNA sequences have provided a new perspective on the evolutionary tree of life (fig. 1.2) (Corliss, 1994; Brown and Doolittle, 1997).

Viral Symbioses

Symbiotic associations are common in all biological kingdoms, as well as among viruses. We consider *viruses* as noncellular entities that have some basic characteristics of living organisms. Viruses contain genetic material that is transmittable. Their level of symbiotic interaction is molecular rather than organismal, and they generally are more integrated than other symbioses. Viruses can form long-lasting associations with their hosts; for example, prophage viruses insert part of their DNA into a chromosome of the host cell and replicate along with the host. The genetic colonization of host cells by viruses is a subtle type of symbiosis that has far-reaching implications for people. This natural form of genetic engineering is mirrored by human attempts to create artificial symbioses by manipulating the genetic material of different organisms. Recombinant DNA technology strives to produce symbiotic hybrids that will be of use in medicine and agriculture.

Based on molecular information, the modern domains of life include: (1) Archaea (archaebacteria), (2) Bacteria (eubacteria), and (3) Eukarya (eukaryotes) (Doolittle, 1999) (fig. 1.2). Genetic incongruity between Archaea and Eukarya suggests a sisterhood between these domains. One horizontal gene exchange might have involved the gram-positive bacteria and Archaea, while another might have

taken place between proteobacteria and eukaryotes through endosymbiosis (Brown and Doolittle, 1997).

Archaea

Archaebacteria are a major part of the global biomass (Forterre, 1997) and are useful for studying basic questions in biology (Jarrell et al., 1999). All *archaebacteria* possess ether linkages in the lipids of their cell membranes. These linkages are absent in eukaryotes and most bacteria. Most archaebacteria that have been cultured came from extreme environments, hence their popular name of extremophiles (Horikoshi and Grant, 1998). Some archaebacteria can grow at temperatures > 100°C. Methanogens form symbioses with other microorganisms in sediments of lakes, rivers, and bogs, and they metabolize hydrogen produced by anaerobic bacteria. Methanogens are also endosymbionts of

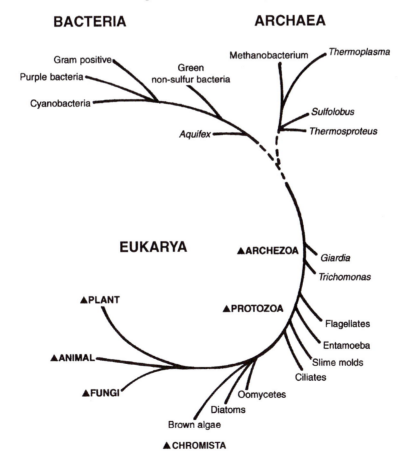

Fig. 1.2 Phylogenetic tree of life based on universal rRNA sequences. Six kingdoms of eukarya are recognized. (Compiled from various sources cited in the text.)

anaerobic ciliates that live in ruminants. *Thermoplasma*, a wall-less archaebacterium, is often considered to be a progenitor host cell of eukaryotes that may have acquired ancient bacteria through endosymbiosis in the evolution of mitochondria and chloroplasts (Aravalli et al., 1998).

Bacteria as Multicellular Organisms

Bacteria in nature exist as films, mats, colonies, aggregates, and chains and rarely as isolated cells. The concept of bacteria as unicellular organisms, a very successful strategy that began with the research successes of Robert Koch, is about to undergo a major paradigm shift in microbiology. Koch invented the pure (axenic) culture method in which monomorphic traits remain constant, but studies on mixed cultures proved difficult and frustrating because they often produced pleomorphism; that is, any microbe could become any other. Pure culture methodology was soon adopted by mycologists to study fungi for essentially the same reasons. The legacy of Koch's method was that it eliminated as many variables as possible from the tangled web that normally exists in nature and opened the way to a reductionist approach to studying the causes of infectious diseases. After more than 100 years, an alternative view of microbiology is gathering momentum. Now it has been shown that microorganisms rarely exist alone under natural conditions but rather interact and form a range of associations with other organisms. Some microbial interactions involve competition, and successful competitors excrete inhibitors. Microbe–microbe interactions form biofilms and are very complex, and tools of molecular biology are now being applied to understanding the dynamics of microbial communities (Cirillo, 1999). Every day we learn how versatile the bacteria are, and appreciate their impact on global ecology and geochemical processes. Bacteria make efficient use of other organisms for their benefit. Bacteria are sensitive, they communicate and integrate information from their neighbors and their environment, and they reproduce and survive as multicellular populations. This view takes us far from the old notion that bacteria are simple, small, and primitive and respond automatically to their changing environment (Margulis et al., 1986; Sonea, 1988; Shapiro and Dworkin, 1997).

The *Aquifex* and hydrogenobacter lineage is believed to be the oldest branch of the bacterial domain. *Aquifex pyrophilus* is an extreme thermophile that occurs in hydrothermal vents and grows well at 85°C. *Proteobacteria* (purple bacteria) are a diverse group of bacteria, and several members of alpha proteobacteria form significant symbioses with plants and animals. They include species of *Agrobacterium*, *Phyllobacterium*, *Rhizobium*, *Rickettsia*, and *Wolbachia*. The *cyanobacteria* are another phylogenetically related group that form symbioses with host organisms. *Cyanelles* are cyanobacteria that are endosymbionts. In the case of one cyanelle, that within *Cyanophora paradoxa*, because some functional genes have been transferred to the host nucleus, it has lost the ability to live independently and thus has become equivalent to a cell organelle like the chloroplast (cyanoplast).

Kingdoms of Eukarya:
Archezoa, Protozoa, Chromista,
Fungi, Animalia, and Plantae

Eukarya is a vast assemblage of organisms that includes protozoans, fungi, animals, and plants grouped into six kingdoms. Although the kingdom Protista is no longer recognized, the term protist is still used for all protozoa, eukaryotic algae, and lower fungi (Corliss, 1994). Most of the primitive lineages of eukarya have yet to be discovered, but some present-day primitive eukaryotes live anaerobically in thermophilic environments.

Archezoa

Archezoa are the earliest eukaryotes that evolved before the acquisition of mitochondria. They include *Enterocytozoon*, *Giardia*, *Hexamita*, and *Trichomonas*, which are traditionally considered to be protozoans. These amitochondrial organisms have evolved nuclei, endoplasmic reticulum, primitive cytoskeleton, and a 9+2 arrangement of microtubules in the flagella. *Giardia lamblia*, a human pathogen, has 70s ribosomes like those of bacteria and also lacks the Golgi apparatus. Sexual reproduction in *Giardia* has not been observed.

Protozoa

Protozoa represent lineages of eukarya that evolved after the acquisition of mitochondria and chloroplasts through endosymbiosis.

Most protozoans have a phagotrophic mode of nutrition, and many are involved in symbiotic associations. The kinetoplastid protozoans such as *Euglena*, *Leishmania*, and *Trypanosoma* represent early lineages among protozoans. This group produced two branches, one leading to the cellular slime molds, such as *Dictyostelium discoideum*, and the other to apicomplexans such as the malarial parasite, *Plasmodium falciparum*. Most of these protists have mitochondria with tubular cristae. In an evolutionary sense, dinoflagellates such as *Symbiodinium* are latecomers as protists. They are closely related to ciliate protozoans. By the time ciliates emerged as an evolutionary group, meiosis and fertilization were regular eukaryotic features. Phylogenetic lineage based on rRNA homology indicates that at about the time ciliates appeared, animals, fungi, plants, chlorophytic algae, and chromophytic algae began to evolve.

Chromista

Phylogenetic analysis indicates that there were several evolutionary lineages of photosynthetic eukaryotes. *Chromista* includes diatoms and brown algae whose chloroplasts are surrounded by a unique periplasmic membrane. Diatoms and brown algae evolved in a two-step symbiosis. It began when a phagocytic protozoan engulfed a photosynthetic protist, which entered the host endoplasmic reticulum by fusion of phagosome and host nuclear membranes. Most chromists possess another unifying feature, the retronemes, which are rigid structures that may have facilitated the endosymbiotic process.

Fungi

The kingdom *Fungi* includes organisms that are heterotrophic and obtain nutrients by absorption from living or nonliving organic sources. Phylogenetically, fungi are more closely related to animals than to plants. They evolved from protozoa some 400 million years ago by acquiring rigid chitinous cell walls. Fungi excrete enzymes that break down organic matter into simpler, usable compounds and, like bacteria, are important decomposers in nature. Most fungi grow by means of filamentous hyphae, although some, such as yeasts, are unicellular. Fungi have diverse reproductive strategies. Many fungi reproduce sexually, but the resulting diploid zygote undergoes meiosis immediately or after a resting period to form haploid spores or offspring. Fungi produce many spores and grow practically everywhere. Fungi possess mechanisms to resist desiccation. Fungi form numerous wide-ranging associations with cyanobacteria, algae, plants, and animals, and some species are important pathogens (fig. 7.1). Fungi even parasitize other fungi. These mycoparasites are common and include different groups of fungi. Marine fungi form close associations, called *mycophycobioses*, with marine algae. Such associations probably have enabled ancient marine organisms to colonize the terrestrial environment. Presumably, the fungi protected the more sensitive algae from drying and helped them obtain water and minerals from the ground. Other highly successful lines of evolution for many fungi have been associations with algae and cyanobacteria to form lichens.

Animalia

The kingdom *Animalia* consists of multicellular organisms that ingest their food. Reproduction consists of fertilization of an egg by a sperm, and the resulting zygote forms a blastula from which the animal develops. Animal cells lack walls and most are diploid, the haploid condition occurring only in the sperm and eggs. Animals, like plants, are hosts to many symbionts and have developed sophisticated defense mechanisms. The host immune response to symbionts in some cases is such that mutualistic relationships have evolved (Clay and Kover, 1996). Insects and flowering plants have evolved many close relationships during a long period of coevolution. We consider pollination to be a type of symbiosis, as did de Bary. The contacts between insects and plants during pollination are transitory, but the repetitive nature of the contacts qualifies these associations as symbiotic. Insects also use plants as a place to deposit eggs or to hide from prey. Seed plants form symbioses with birds to achieve seed dispersal, and some birds such as hummingbirds are efficient pollinators. Numerous symbioses between animals exist. Some birds commonly associate with large land mammals and not only remove insects from the animal's skin but also provide early warning of an impending threat from a predator. In the oceans, small fish and shrimp maintain cleaning stations and remove surface parasites from larger fish that periodically visit the stations.

Plantae

The kingdom *Plantae* consists mostly of photosynthetic, multicellular organisms, such as green and red algae, mosses, ferns, and seed plants. The green and red algae and higher plants evolved from a protozoan ancestor which acquired chloroplasts through endosymbiosis of photosynthetic cyanobacteria. The red algae have two or more membranes around their plastids and store starch in the cytoplasm.

Some plants parasitize other plants. Most parasitic plants lack chlorophyll and obtain food from their hosts, but some, such as mistletoe, can manufacture food and also obtain nutrients from their hosts. Plant cells have walls that contain cellulose and undergo sexual reproduction. They develop from embryos which are surrounded by a jacket of sterile tissues. Plants show alternation of generations in which a haploid generation (gametophyte) alternates with a diploid generation (sporophyte). In mosses and liverworts the haploid generation is dominant; in more advanced plants, such as gymnosperms and angiosperms, the diploid generation is dominant. Most terrestrial plants form symbioses with mycorrhizal fungi. These root–fungus associations are vital for the well-being of plants and are far more complex than originally thought. Many of the symbiotic associations that involve plants are parasitic, as plants serve as hosts to fungi, bacteria, protozoa, and animals.

1.4 CLASSIFICATION OF SYMBIOSES

Mortimer P. Starr (1975) developed a classification of symbiotic associations to standardize the many conflicting terms that have been used to describe different symbioses. Following is a modification of Starr's system.

Location of the Symbionts

A symbiont may be inside or outside another symbiont. An *endosymbiont* is one that resides inside, whereas an *ectosymbiont* resides outside a host organism. The distinction between inside and outside may not be clear. We restrict the term "endosymbiont" to those organisms that reside within the cells of other organisms. Thus, the extracellular microbes of the digestive tracts of animals are ectosymbionts.

Technically, few symbionts lie within the host cytoplasm. Most endosymbionts are surrounded by the host plasma membrane, which separates them from the cytoplasm. Therefore, in a cellular sense, the symbionts are not within the host cell. The structure with the symbiont enclosed has been termed a *symbiosome* and is considered to be an organelle (Roth et al., 1988).

Persistence of the Symbiosis

Most symbioses are persistent; that is, the symbionts remain together for a long time or their contacts are frequent. Associations in which one partner is an endosymbiont are usually the most persistent type. The symbionts remain together through all stages of their life cycles. Many endosymbionts have an arrested life cycle and generally remain in the vegetative stage. In some symbioses, the contacts between the partners are intermittent, as in the cleaning symbiosis of marine fishes or the pollination of flowers by insects and birds. In cases of parasitism, the symbiosis persists throughout the host's life, the duration of which depends on the virulence of the parasite.

Dependence on the Symbiosis

Obligate symbionts are so highly adapted to a symbiotic existence that they cannot live outside it. *Facultative symbionts* can also live in the free-living condition. Sometimes, it is hard to tell whether a symbiont is obligate or facultative. For example, a symbiont may exist free-living but in specialized niches or in small populations that are difficult to identify (Douglas, 1996). Algae such as *Trebouxia* are highly adapted to the lichen symbiosis, but scattered colonies of these algae have been found living separately. Are such colonies truly nonsymbiotic, or are they established by individuals that escape from the symbiosis? Obligate symbionts depend on the symbiosis for nutrients. A symbiosis that involves two organisms, when at least one obtains nutrients from the other, is called a *biotrophic symbiosis*. If one of the symbionts dies and the other uses it as a source of nutrients, the association is called a *necrotrophic symbiosis*. Organisms may obtain physical protection from a symbiotic association or receive some other benefit, such as the light produced by symbiotic bacteria in marine fishes.

Specificity of the Symbionts

Symbionts may be highly specific to one organism, such as the bacterial symbionts of *Paramecium aurelia*, or they may associate with different organisms. How specific an organism is may relate to the evolutionary stage of the symbiosis. Presumably, the more highly evolved a symbiosis is, the longer the symbionts have had to adapt to each other, and the more specific is the association (Douglas, 1995). An extreme example of specificity is seen in mitochondria, which are semi-independent, transformed organisms that cannot exist outside a eukaryotic cell.

Symbiotic Products

In some associations interaction of the symbionts results in the formation of new structures or chemical compounds. For example, a lichen thallus and many of its chemical compounds develop only as a result of the symbiosis. Similarly, legume root nodules and the red pigment leghemoglobin are products that develop from symbiotic associations of *Rhizobium* bacteria and legumes. Cooperation as a byproduct of mutualism occurs in the context of interspecific associations.

1.5 SUMMARY

Symbioses are interspecific associations that have played a significant role in the evolution of plants and animals and in shaping the earth's physical features. There are three main types of symbiosis: mutualism, parasitism, and commensalism. The boundary lines between these categories are not clear, and there are frequent transitions between them. The most common association among living organisms is that of parasitism, and it appears, from an evolutionary point of view, that mutualism and commensalism arose from parasitism.

Parasites that cause disease are pathogens, and they may be localized or spread throughout a host. There are several types of host: intermediate host, final host, and reservoir host.

The most widely accepted classification of living organisms recognizes three domains; Archaea, Bacteria, and Eukarya, and six kingdoms of eukaryotes, Archezoa, Protozoa, Chromista, Fungi, Animalia, and Plantae. All forms of life contain symbiotic associations. Viruses are subcellular symbionts because they have basic properties of living organisms and are commonly found in many life forms.

Because of the wide diversity of symbioses and the growing number of terms used to describe them, a classification system for symbiotic associations has been developed. This classification is based on several features: location of the symbionts, whether ectosymbionts or endosymbionts; persistence of the symbiosis; dependence on the symbiosis, whether obligate or facultative symbionts; specificity of the symbionts and symbiotic products. Biotrophic and necrotrophic symbionts are distinguished on the basis of whether nutrients are obtained from a living or dead partner.

Research on symbiosis is occurring in many disciplines of biology. There is a growing awareness of the fundamental importance of symbiosis as a unifying theme in biology, an awareness that organisms function only in relation to other organisms.

2

VIRAL SYMBIOTIC ASSOCIATIONS

Viruses fascinate us not only because they cause disease, but also because they challenge the way we define life (Radetsky, 1994; Dimmock and Primrose, 1994). Viruses are intracellular, obligate parasites which have evolved numerous methods to survive and reproduce. Ability to reproduce is a central feature of life, and in the living world the only systems that can reproduce are the ones that contain nucleic acids (Lederberg, 1993). *Viruses* consist of nucleic acids that replicate inside living cells and cause the synthesis of virions that can transfer the genome to other cells (Luria et al., 1978).

Viruses lack metabolic activity and therefore depend on host cells for many of their functions. Like most living organisms, viruses replicate, mutate, and adapt to their host cells. Many viruses are pathogenic, but, in some cases, viruses and hosts have evolved together. Today, we know of about 3000 different viruses, but only a few of these have been studied intensively. New viral host relationships are constantly being discovered. Viruses have been found in all major categories of life forms. Despite their diversity, viruses have many features in common (Voyles, 1993).

2.1 VIRUS STRUCTURE

Viral Genome

Viral genes may be clustered together on a single DNA or RNA molecule, or a virus may have a segmented genome consisting of several nucleic acid molecules. Influenza virus has a segmented genome of eight unique strands of RNA, whereas measles and rabies viruses each have a single RNA molecule (Webster et al., 1992).

The simplest virus has a single strand of RNA made up of several thousand nucleotides. If this RNA is a plus strand, its genetic message can be translated directly by the host cell's ribosomes. Examples of plus-stranded RNA viruses are the bacteriophage QB, which infects *Escherichia coli*, and the polio virus in humans. Minus-strand RNA must first be transcribed into a complementary plus strand before viral replication can begin. Influenza A virus in humans is an example of this type.

Retroviruses contain single-stranded RNA which, after entering a host cell, changes into a double-stranded DNA with the help of a viral enzyme, reverse transcriptase. The newly synthesized DNA can then integrate into the host's own DNA. Examples of retroviruses include human immunodeficiency virus (HIV-1), human T-cell leukemia virus (HTLV), and Rous sarcoma virus (RSV) (Varmus, 1988).

Viral capsid

A *capsid* is the protein coat that surrounds the viral genome. It may be composed of only one type of protein or of several different types. Capsids are made up of capsomers, which are arranged in either isometric or helical symmetry. Tobacco mosaic virus (TMV) is a helical virus that forms a spiral around its nucleic acid. The capsid of an isometric virus, such as polio virus, forms an icosahedron of 20 facets (Guo, 1994) (fig. 2.1).

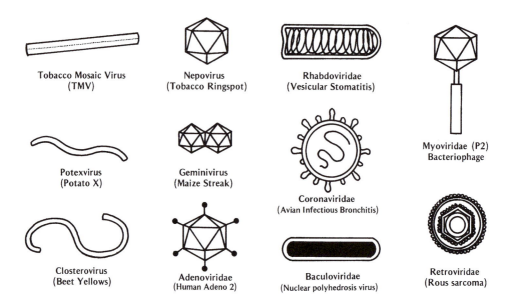

Fig. 2.1   Types of viral particles and their structural features. Adapted from Matthews (1991a).

Viral Envelope

Some animal viruses have a host-derived membrane or *envelope* around the capsid coat. The envelope is made up of phospholipids and proteins. The envelope proteins are specified by viral genes and may have free ends consisting of carbohydrates. Together, they form spikes on the outer surface of the virus (Fields et al., 1996). Some envelope proteins help a virus attach itself to the host cell; others cause the host cell to lyse. The envelope is acquired during the maturation process after the virions have been assembled. An envelope is a common feature of animal viruses but is absent in plant viruses. In herpes virus, the envelope is derived from the nuclear membrane, whereas in vaccinia virus, it comes from the Golgi apparatus. Nonenveloped or naked viruses lack lipids.

Structure of the Human Immunodeficiency Virus

Human immunodeficiency virus is a complex virus which contains two identical copies of a positive-sense RNA (similar to mRNA) of about 9500 nucleotides. The two RNA strands are linked to form a genomic RNA dimer, which associates with a nucleocapsid (NC) protein called p9/6. The ribonucleoprotein, in turn, is enclosed by the capsid protein

(CA), p24. The capsid thus formed also contains other viral proteins such as integrase and reverse transcriptase along with macromolecules derived from the host cell cytoplasm. The capsid has an icosahedral structure and is enclosed by a layer of matrix protein (MA), p17, that is associated with an envelope. The HIV-1 envelope is acquired when the virus exits through the host cell membrane. A viral envelope may carry viral proteins that form spikes. In HIV-1, the most significant of these spike-forming proteins is gp120/41, which attaches the virus to a host cell. Gp41 is a transmembrane protein; gp120 is a surface protein that is attached to gp41. Both proteins are manufactured in the host cell's endoplasmic reticulum and are transported to the cell surface by means of the Golgi complex (Nye and Parkin, 1994) (fig. 2.2).

2.2 VIRUS TAXONOMY

The classification of viruses appears confusing because the system is a hybrid of several traditional approaches. Animal virologists have used a system modeled after Linnaeus's zoological classification, with families, genera, and species, while plant virologists have classified viruses according to their hosts.

Today all viruses are ranked into species, genera, and families. The important taxonomic criteria include host organism(s), virus morphology, and genome type. In 1966, the International Committee on Taxonomy of Viruses developed a system of virus classification and nomenclature, and an outline of its latest edition (1995) is presented in an appendix to this chapter.

Viruses may be investigated by examining infected host cells, observing the effect of viral growth on a cell culture, or measuring the antibody response that a virus evokes in a host animal. Much has been learned about viruses from their cytopathic effects on hosts and from electron microscopic observations of viral particles in host cells. In many naturally occurring virus–host symbioses, there are no recognizable cytotopathic effects, and such infections are called *persistent* (Murphy et al., 1995).

According to the "Baltimore Classification" there are seven different groups of viruses:

1. Double-stranded DNA viruses (dsDNA) (adeno-, herpes, and pox viruses) have DNA which may be linear or circular. Adenoviruses replicate in the nucleus using cellular proteins, whereas others such as poxviruses replicate in the cytoplasm and make their own enzymes for viral DNA synthesis.

2. Circular, single-stranded DNA viruses (ssDNA) (parvoviruses) replicate in the nucleus by forming a minus-sense strand, which serves as a template for plus-strand RNA and DNA synthesis.

3. dsRNA viruses (reoviruses, birnaviruses) have segmented genomes. Each segment produces a distinctive monocistronic mRNA.

4. ssRNA viruses with a plus-sense genome (picornaviruses, togaviruses). Picornaviruses (hepatitis A) have polycistronic mRNA. The naked RNA is infectious and when translated forms a polypeptide chain from which mature protein is cleaved. Togaviruses require two or more rounds of translation to produce viral RNA.

5. ssRNA viruses with minus-sense genome (orthomyxoviruses, rhabdoviruses) have viral RNA-dependent RNA polymerase.

6. "Diploid" ssRNA viruses (retroviruses) have a plus-sense genome that does not serve as mRNA but as a template for reverse transcription.

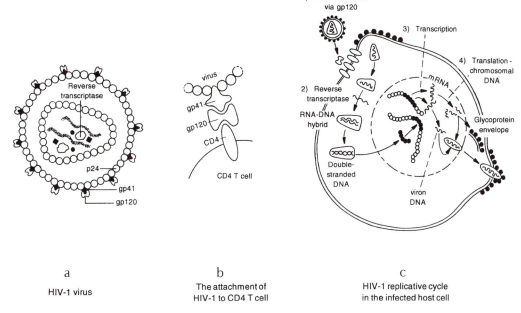

a

HIV-1 virus

b

The attachment of
HIV-1 to CD4 T cell

c

HIV-1 replicative cycle
in the infected host cell

Fig. 2.2   Human immunodeficiency virus, HIV-1: (a) principal structural features, (b) interactions between viral proteins gp 41 and gp 120 with host cell surface receptor, (c) steps involved in HIV-1 entry into and release from host cell. (Compiled from various sources cited in the text.)

7. dsDNA viruses (hepadnaviruses) also need reverse transcription, which occurs inside the mature virus particle.

2.3 VIRAL REPLICATION: VIRUSES AND HOST CELL INTERACTIONS

All viruses penetrate the host cell, replicate, and then exit the infected cell. Six generalized stages are recognized in a typical viral replication (fig. 2.3).

1. *Adsorption* involves the attachment of the virus to a host cell. Specific regions on the host cell surface serve as virus receptors. Viral adsorption involves weak forces such as ionic or hydrogen bonds. In addition, there may be a complex interplay at the receptor sites.
2. *Penetration and uncoating* introduce the viral genome into the host cell. Viral structural identity is lost during this stage. Some viruses inject their nucleic acid into a host cell and leave their protein coats behind as empty shells. Other viruses enter a cell with their protein coats intact. After a whole virus has entered a cell, the viral nucleic acid must be freed from its protein coat before the virus can replicate and initiate viral protein synthesis.
3. *Synthesis of viral proteins and viral genome* depends on the activity of a specific enzyme replicase. Viral mRNA directs the synthesis of enzymatic proteins by mobilizing the host cell's resources. The normal host metabolism is altered after the infection process so that it produces new viral particles.
4. *Synthesis of viral structural proteins* may occur at the same time as the viral genome is being replicated, but it usually lags behind. New viral genome and structural proteins accumulate within the host cell.
5. *Maturation* is the phase during which newly synthesized nucleic acid molecules and protein subunits assemble into virions (virus particles). Symmetry is important in viral assembly. Some viruses such as tobacco mosaic virus can self-assemble, while others such as polio virus require the participation of

cellular chaperones in the viral assembly (Fields et al., 1996). Herpes viruses assemble in the host cell nucleus and acquire an envelope as they pass through the inner nuclear membrane during the budding process. Virions accumulate as inclusions in either the cytoplasm or nucleus.

6. *Release* is the exit of mature virions from the infected host cell and can occur suddenly or gradually. The host cell often disintegrates because viral replication damages the normal cellular functions. Some viruses remain in a resting stage until the cell dies.

2.4 VIRAL PATHOGENESIS

Viruses produce disease in the host in different ways. Capsid proteins are often toxic to host cells and cause cell death and the release of new viral particles. Viruses show a high degree of host tissue specificity because of surface receptor molecules that are used for virus attachment. Adenoviruses, orthomyxoviruses, and paramyxoviruses have spikes on their outer surfaces that attach to the host cell. Viruses such as polio, herpes simplex, togavirus, and pox inhibit protein synthesis of the host cells. Herpes simplex virus can also inhibit and degrade cellular DNA (Cann, 1997).

Virus-infected cells often show distinctive structural and physiological abnormalities which are collectively termed cytopathic effects. The most common of these effects are the *inclusion bodies*, which are usually the sites of viral assembly or cellular damage. Virus-infected cells may detach from the substrate, lyse, or form a *syncytium*, which is a multinuclear giant cell that results from a mass of fused cells. Syncytia are characteristic of cytomegalovirus, measles, herpes simplex, paramyxoviruses, and HIV-1 (T.E. Shenk, 1993).

Some viruses depress host cell activities in general but stimulate biochemical processes that involve viral replication. The viral genomes of herpes viruses, hepatitis B virus, and HIV-1 contain enhancer regions that increase the efficiency of viral transcription in specific host tissue cells. For example, the papilloma virus has an enhancer region in its genome that becomes activated only in skin cells (Bernard, 1994).

Virulence of the influenza virus is controlled by hemaglutinin and neuraminidase, proteins that form HA and N spikes in the envelope of the virus particles. Hemaglutinin needs to be cleaved at a specific amino acid site in order to activate viral virulence. A mutation with a different amino acid at the cleavage site renders the virus avirulent (Webster et al., 1992; Scholtissek, 1996).

After a virus replicates in one type of tissue, it can spread to other host tissues that have the appropriate cell surface receptors. Polio virus replicates first in intestinal cells and then spreads to cells of the central nervous system, where it causes paralysis of the limbs by destroying motor neurons in the spinal cord (White and Fenner, 1994). Respiratory viruses such as influenza replicate first in the epithelial cells of the lungs, where they cause lesions and block air passages. The virus then spreads to cells lining the trachea (Atlas, 1997).

Viral infections that produce diarrhea often interfere with absorption of fluids and loss of lactose-degrading enzymes in the intestine. As lactose accumulates in the digestive system, it lowers the sodium–potassium ATPase activities and leads to acidosis, which changes the rate of potassium ion exchange and causes a buildup of water in the intestine (Atlas, 1997).

In some human viral infections, the disease symptoms are the result of the body's own immune response, such as the skin rashes associated with measles (G. L. Smith, 1994). Cytomegalovirus and HIV-1 cause host cells to fuse and thereby allow the viruses to infect new cells without encountering host antibodies. Glycoproteins in the envelope of HIV-1 inhibit T-cell proliferation (Nye and Parkin, 1994). Hepatitis B virus produces large amounts of antigens that are not associated with the fully assembled virions. These antigens counteract antibodies in circulation so that when the virus particles are released, there are few antibodies present to interact with them (Tiollais and Buenidia, 1991; Bradley, 1993).

Some viruses, such as HIV-1, herpes simplex virus 2, hepatitis B virus, human papilloma virus, and Epstein-Barr virus, have oncogenes. When the oncogenes are transferred to new host cells, the cells become malignant. Cancer results from either activation of oncogenes or repression of tumor-suppressor genes. Oncogenic viruses have v-*onc* genes that are similar to the host's normal developmental genes (proto-oncogenes). An *oncogene* may encode for a growth factor, a signal transducer, or a transcription factor (Hausen, 1991).

Virokines (also called viral cytokines) are viral proteins which are not needed for replication but are necessary for host cell infection. Virokines suppress the host's immune system and often mimic host cell molecules that are active in the immune response. Virokines are produced by retroviruses, pox viruses, herpes viruses, and adenoviruses. Pox viruses produce virokines that mimic receptors for tumor necrosis factor; HIV-1 produces a virokine that inhibits protein kinase C, thereby blocking signal transduction (Atlas, 1997). The herpes simplex virus produces a protein (UL6) similar to one in the cornea, causing a human autoimmune disease, herpes stromal keratitis, which leads to blindness. T-cells with UL6 protein from virus-infected animals produced a severe corneal autoimmune disease in healthy animals. This study suggests that molecular mimicry by viral proteins may be a significant component of virus-induced autoimmune disease (Zhao et al., 1998).

Case Study: HIV-1 Pathogenesis

The success of HIV-1 infection is largely due to (1) a "Trojan horse" mechanism which allows the virus to escape detection by replicating in lymphocytes; (2) latency, which allows the virus to control its own expression by means of virus-encoded regulatory proteins (tat and rev); and (3) antigenic variation, which allows new strains to emerge.

The depletion of CD4+ cells of the immune system is a characteristic feature of AIDS (Levy, 1993). HIV-1 can replicate in blood, brain, intestinal, and skin tissues. With the help of the envelope proteins gp120 and gp41, an HIV-infected cell can fuse with other infected CD4+ cells. This leads to multinucleate syncytia, which die from altered permeability of their cell membranes. Cell death also occurs when viral DNA accumulates in the host cell cytoplasm, as has been observed with brain cells. Gp41 envelope protein may be cytotoxic to the infected host cell. HIV-infected cells produce large amounts of gp120, some of which is stablized in the membrane by gp41, while the rest is released from the infected cell. The freed gp120 binds to the CD4+ receptors of healthy T-cells and brings about their programmed death or apoptosis (Johnson et al., 1992). Gp120 also can bind with immature T lymphocytes in the thymus and cause their early death.

Decline of CD4[+] T-cell populations in an HIV-infected person alters the immune system. HIV-1–infected T-cells produce viral cytokines, which may cause Kaposi's sarcoma, B-cell lymphoma, and anal carcinoma. As HIV-infected individuals survive longer, other cancers such as cervical carcinoma and Hodgkin's lymphoma may appear, and other viruses may become involved (Wei, 1995). Cytomegalovirus, human papillomavirus, and hepatitis B virus are all associated with Kaposi's sarcoma. Similarly, the Epstein-Barr virus is correlated with the development of B-cell lymphomas (Levy, 1993; Weiss, 1993; Masucci and Ernberg, 1994).

2.5 PERSISTENT VIRAL INFECTIONS

Not all viruses kill their host cells or trigger an immune response. Some viruses form long-lasting symbioses with their host by altering a specialized function of the infected cells such as hormone production. Viruses may be responsible for human illnesses such as growth retardation, diabetes, neuropsychiatric diseases, autoimmune diseases, and heart disease (Oldstone, 1989).

Viruses are genetic parasites that have evolved different strategies for survival and reproduction. They must infect a host cell and exploit its cellular machinery in order to replicate. Both polio and common cold viruses cause acute infection and rapid cell death, whereas herpes simplex virus can remain dormant in a cell for long periods before being activated. Oldstone (1989) has investigated the lymphocytic choriomeningitis virus (LCMV), which is endemic in some wild mice populations. Virus-infected neuroblastoma cells in tissue culture cannot synthesize and degrade acetylcholine. The infected cells continue to grow and metabolize normally. Microscopically, virus-infected cells are indistinguishable from noninfected cells. In laboratory studies, the pituitary glands of LCMV-infected mice produced 50% less growth hormone than normal glands (Oldstone, 1989).

Most persistent viral infections are asymptomatic, but a few cause cancers or severe organ dysfunction. When human cytomegala virus infects early embryonic cells, a few viral genes are expressed in the host cells, and the infection becomes persistent. In an immunocompromised host, however, the virus is reactivated and causes a life-threatening disease. Herpes simplex virus normally kills the epithelial cells it infects, but in neurons the virus becomes latent (White and Fenner, 1994).

The introduction of myxoma virus to control the expanding rabbit population in Australia is an excellent example of virus–host coevolution. European rabbits were introduced into Australia in the mid-1850s as a source of meat, and they multiplied uncontrollably in the absence of natural enemies. Myxoma virus is endemic in its natural host, the South American rabbit, in which the virus produces wartlike, benign skin tumors. The virus is spread in nature by mosquitoes or rabbit fleas, but it does not multiply in its vector. In European rabbits the virus causes the disease myxomatosis, which is characterized by fatal skin lesions. When the myxoma virus was first introduced into Australia in 1950, it killed more than 99% of the infected rabbits, but it did this so rapidly that the most virulent strains of the virus died with the rabbits. Virus-resistant rabbits multiplied, and 7 years after the virus was introduced, only 25% of infected rabbits died. The virus is now established within the Australian rabbit population as a persistent, nonlethal infection (Fenner, 1983).

2.6 INTERFERONS

Interferons are protein molecules that are secreted by animal cells in response to infection with viruses. Although all types of viruses can stimulate interferon production, the most significant response is obtained with double-stranded RNA viruses. Interferon (INF) does not affect viral metabolism in the host cell, but when neighboring healthy cells are treated with interferon, they become resistant to viral infection. There are three types of interferons, a, b, and v, of which INF V is the most effective cell regulator. The interferon molecules bind to the cell membrane and inhibit synthesis of viral nucleic acid and protein by inducing the host cell to produce two enzymes, 2,5-oligoA synthetase and an RNA-activated protein kinase, to create an "antiviral state." Interferons are powerful drugs against virus infection but produce severe side effects; therefore, their use as antiviral therapy agents has been limited to life-threatening infections. Interferon produced by an infected cell may constitute its primary defensive response until the host can produce antibodies. Several biotechnology companies are attempting to clone human interferon genes in order to mass-produce interferons. Large quantities of specific interferons will allow studies to be made on their effectiveness against human cancers (Cann, 1997).

2.7 RETROELEMENTS: RETROVIRUSES AND SELFISH DNA

Retroelements are genetic sequences that occur as DNA and RNA and can move through an RNA molecule and make DNA by reverse transcription. The discovery of the enzyme reverse transcriptase in the 1970s revolutionized ideas about genomic evolution. Retroelements are grouped into three major categories:

1. Viral retroelements (e.g., retroviruses, hepadnaviruses, and caulimovirus).
2. Eukaryotic, nonviral retroelements: transposable retroelements, (e.g., copia and gypsy elements in *Drosophila*; copialike elements in plants); retroposons (e.g., long interspersed nucleic acid elements [LINEs] in mammals; mitochondrial introns and plasmids); and retrosequences (e.g., short interspersed repetitive elements [SINEs]).
3. Bacterial retroelements (*retrons*). All retroelements are made up of a *gag-pol* replication core, to which are added other genes that adapt the retroelements to its host.

   Retroelements, in many copies, are widely dispersed throughout the genomes of animals, plants, fungi, protozoa, and bacteria.

The discovery of retrons in the archaebacterium *Myxococcus xanthus* and the myxobacterium *Stigmatella aurantiaca* in 1981 established the reverse transcriptase genes as the oldest known genetic elements. Retron genes may have served as a genetic link between prokaryotes and eukaryotes and thus have been the possible ancestor of all retroelements (Patience et al., 1997).

Retroviruses are widespread throughout vertebrate species. Retroviruses copy their own RNA genome into DNA with the help of the virally encoded enzyme reverse transcrip-

tase. The newly synthesized DNA (provirus) then becomes inserted into the host chromosomal DNA with the help of another viral enzyme, integrase. The integration site of the provirus is random and often triggers disruptions of adjacent genes. Oncoviruses, for example, involve insertion activation of cellular oncogenes that closely resemble the host cell's regulatory genes. In 1989, Howard Temin proposed that all viral DNA or RNA, which share similarities, were once transposons or retrotransposons that arose from the host cellular genome by acquiring the ability to replicate autonomously. Accordingly, some ancient transposable elements, by possessing additional genes that allowed them to be transmitted horizontally, evolved into infectious particles. The infectious particles may infect the same cell type or a new type of host cell, making it possible for them to spread across the species barrier (Varmus, 1988; Temin, 1989).

Reverse transcriptase associated with eukaryotic nonviral retroelements generates extra gene copies by using mature mRNAs as a template. In addition, reverse transcriptase produces DNA from all types of nonmessenger RNAs, which upon insertion in the host genome become long repeated sequences of LINEs/SINEs. Novel genes may emerge when such retrogenes move into new genomic neighborhoods and thus become a powerful driving force of evolutionary change (Brosius and Tiedge, 1996).

Retroelements can be regarded as molecular parasites with selfish nucleic acid, which may enhance their own survival and also that of the host (Leib-Mosch and Seifarth, 1996). Thus, there has been a kind of mutualism between the selfish genes and the host genome that gains genetic variability as a consequence of retroelement translocation. The coevolution of retroelements and their host DNA has been going on, perhaps, throughout the history of life on earth (Hull and Covey, 1996; Patience et al., 1997).

2.8 VIRUSES IN BACTERIA

Bacterial viruses (*bacteriophages* or *phages*) are distributed widely throughout the world of bacteria and cyanobacteria. They are excellent models of virus–host interplay and have played a significant role in bacterial evolution. Bacteriophages serve as vectors in the genetic engineering of bacteria, being used to transfer genes of interest to areas near particular promotor sites.

There are about 12 groups of bacteriophages, including the most widely studied phage system, the T-series (T1 to T7) phages, which infect *Escherichia coli*. T-even phages (T2, T4, and T6) have elaborate capsid proteins and contractile tails, whereas T-odd phages (T3, T7) lack contractile tails. T-even phages have terminal redundancies; that is, some base pairs are repeated at both ends of a linear DNA molecule. After adsorption to a specific receptor in the bacterial cell wall, the phage DNA enters the cell, while capsid proteins remain outside the infected cell. When an infected bacterial cell undergoes lysis, it releases 70–300 new phage particles. The entire infection process from initial contact to the release of new bacteriophages (latent period) takes less than 30 min (fig. 2.3).

Virulent phages such as T-even phages do not integrate their DNA into the host cell chromosome and usually cause the host cell to lyse. Temperate phages can integrate into the host DNA, causing lysogeny. This condition of the bacteriophage is the prophage state. As a prophage, the viral DNA replicates along with the bacterial DNA and, through such an integration, transforms the genetic properties of the bacterium that determines virulence. This phenomenon is called *viral conversion*. Some environmental stresses such as UV radiation, X-rays, or chemicals may activate the temperate phage and cause it to replicate. Lysis of the host cell results, and a new generation of viral particles is ready to infect other host cells. Bacterial cells containing prophages are lysogenic and immune to superinfection by the same phage, due to repression of transcription by the resident prophage. Some temperate phages, such as Mu, act as transposons and move from site to site in the bacterial chromosome.

The λ phage establishes a stable lysogenic relationship as a prophage. This phage can be made to alternate between the lysogenic and lytic growth cycles and is one of the best understood genetic control systems. When phage λ injects its DNA into the host cell, the bacterial cell has neither repressor nor Cro proteins. The balance between the λ repressor and Cro protein determines whether a lysogenic or lytic pathway is followed. The *C1* repressor gene of the λ phage encodes for a well-characterized repressor protein that binds to two operator regions called OL and OR. When repressor protein is bound to these two operators, RNA polymerase is unable to bind and initiate transcription of λ lytic genes. Thus, the lytic genes are repressed and the dormant prophage is passed from one generation to another. The lysogenic state is maintained as long as the lytic pathway genes are tightly repressed. In addition, the presence of lambda

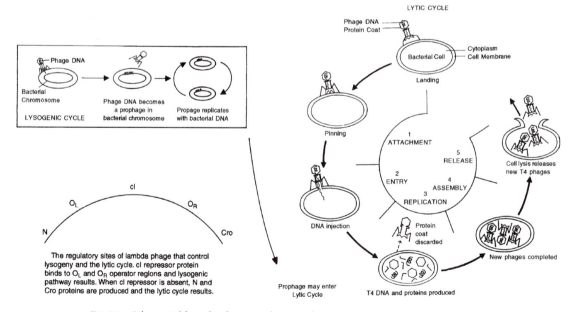

Fig. 2.3   Schematic life cycle of a T-even bacteriophage on its host *Escherichia coli*. The bacteriophage enters the lytic or lysogenic cycle depending on the synthesis of repressor cl protein.

repressor stimulates the synthesis of more repressor in a self-regulation process. Exposure of lysogenic cells to UV light activates a protease which cleaves the λ repressor protein and renders it nonfunctional.

In the absence of λ repressor protein, two new proteins, N and Cro are synthesized. Mutations in the operator regions, OL and OR, will prevent the repressor protein to bind and thus end any possibility for lysogeny. Similarly, mutation of Cro or N will abolish any potential for host cell lysis.

Lysogenic conversions, which result in new genetic properties for the host, have been observed in *Clostridium botulinum*, *Corynebacterium diphtheriae*, *Salmonella* sp., and *Vibrio cholerae* (Atlas, 1997; Voyles, 1993). In *Salmonella*, ε phages were the first viruses known to alter the antigenic properties of host bacteria. The cholera pathogen, *Vibrio cholerae*, though rare in Western countries, threatens the rest of the world with sporadic outbreaks. These sudden outbreaks of the disease have been linked to the conversions of surface antigens induced by phages. Significant discoveries have been made on the nature of proteinaceous toxins of *C. botulinum* and *C. diphtheriae*. It has been confirmed that these toxins are the products of prophages residing on the host chromosome. Currently, scientists think that several toxins previously believed to be of bacterial origin may in fact be the products of phage genes.

Some intron-containing bacteriophages have been described as parasites of bacteriophages (Belfort, 1989). The bacteriophage introns contain a gene for endonuclease that allows the introns to splice themselves and thus behave as mobile genetic elements within the phage DNA. These self-splicing introns confer no particular selective advantage to a T4 phage. In a strict sense, the introns act like genetic commensals. Endonuclease genes may be viewed as symbionts of introns, as introns are genes within genes, with their own transcriptional and translational signals. The bacteriophage introns containing genes for endonuclease can thus be viewed as mutualists because introns get propagated, and a gene for endonuclease is provided with a safe haven by the intron. Together, they form a highly efficient mobile element (Derbyshire et al., 1995).

Cyanobacteria contain viruses that are similar to T-phages in morphology and life cycle but are not genetically related. *Cyanophages* inhibit growth of the infected cyanobacteria.

In this respect, cyanophages may have a useful role, as "blooms," or population explosions of some cyanobacteria, can kill aquatic organisms and diminish water quality in ponds, lakes, and reservoirs.

Viruses have been reported in several actinobacteria such as *Actinomyces bovis* and *Nocardia farcinia* which are abundant in soil. These viruses resemble a T-phage but differ in the structure of their tail pieces and in the absence of base plates and contractile fibers.

2.9 VIRUSES IN INSECTS

Under natural conditions, most viral infections of insects are subclinical or latent and are often difficult to detect. Insects may have evolved an efficient mechanism that prevents viral infections from reaching pathogenic levels. Under ecological stresses, however, these viral infections may be expressed, a situation that often results in host death. Below are some examples of well-established virus–insect symbioses.

Wasp Polydnavirus Symbiosis

One of the most complex symbiotic associations in nature involves a wasp, a caterpillar, and a virus (Beckage, 1997b). The wasp is an endoparasite of the caterpillar, and when it deposits its eggs it transmits a virus that disables the caterpillar's immune system. The association is so close that if the caterpillar dies before the wasp larvae have matured, the wasp also dies. One of the polydnavirus-encoded proteins called early protein 1 (EP1) is produced within minutes by the virus-infected hemocytes. As long as EP1 protein is produced, the wasp eggs will continue to develop within the caterpillar. Additional studies have indicated that immediate, short-term protection against the host's immune system is conferred by wasp ovarian protein molecules which are injected during oviposition (Lavine and Beckage, 1995).

The virus also delays caterpillar development by stimulating its production of juvenile hormone, which suppresses the caterpillar's pupation. Each polydnavirus particle is composed of up to 28 separate, circular DNAs, which are scattered throughout the chromosomes of both male and female wasps. Virus-free wasps have not been found. The integration of polydnavirus into wasp chromosomes raises some interesting questions about the origin of this virus. Researchers have noted a similarity between the viral protein EP1 and wasp venom. Antibodies against wasp venom can also recognize the viral protein (Flemming, 1992). In an evolutionary sense, polydnavirus may have picked up the wasp venom genes, or the wasp may have found an efficient way to use its venom protein by copying these genes and packaging them in the caterpillar's hemocytes. Both wasp and virus seem to share a genetic evolutionary history (fig. 2.4).

Baculoviruses as Viral Pesticides

Baculoviruses are one of the largest groups of insect pathogenic viruses. They can be isolated from vegetation and soil, where they remain viable for several years. Field studies have demonstrated the usefulness of baculoviruses as pesticides. Two subgroups of

baculoviruses are the nuclear polyhedrosis viruses and granulosis viruses. These viruses have a narrow host range, and their replication is limited to insects. Therefore, these viruses are considered safe and ecologically acceptable as pesticidal agents.

Nuclear polyhedrosis viruses (NPVs) form polyhedral bodies in cells of insects such as the caterpillars of butterflies and moths and larvae of bees and flies. The rod-shaped virus, which contains DNA, multiplies within the host cell nucleus, which enlarges, and this ultimately causes the cell to burst. Infected larvae often stop feeding and move, en masse, to different areas of the plant. For example, infected nun moth larvae gather on tree tops. There they change color, become fragile, and die when their epidermis ruptures and releases virus inclusion bodies. The polyhedral inclusion bodies contain virus particles, which are enclosed either singly or in bundles

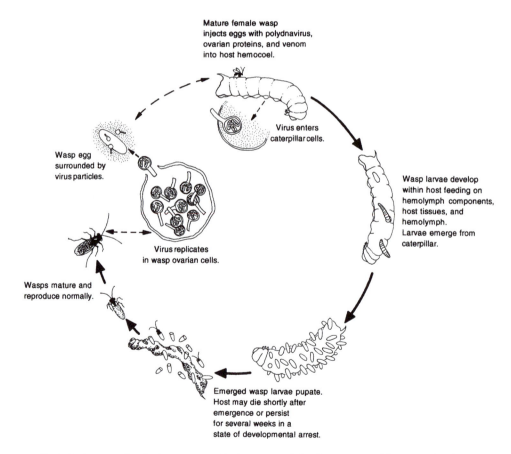

Fig. 2.4   An example of a three-way symbiosis among polydnavirus, a parasitoid wasp, and its host the caterpillar. Adapted from Beckage (1997b).

within a crystalline protein. Because of their size (up to 15 μm in diameter) and their ability to refract light, inclusion bodies are easily seen with a light microscope. Upon liberation, the virus is spread to neighboring plants by rain and wind. When ingested by larvae, the virus migrates from the intestine to susceptible tissues in the hemocoel. Nuclear polyhedrosis viruses may also be transmitted by eggs of adult insects (Wood and Granados, 1991).

The United States and Canada have registered seven baculoviruses for use against forest and agricultural pests. Gypchek, for the control of Gypsy moth, and Tm-BioControl-1, for the control of Douglas-fir tussock moth, are two successful viral pesticides used by the U.S. Forest Service (Federici and Maddox, 1996). Baculoviruses are also being used as viral pesiticides in Europe, Russia, and China. The most successful commercial use of a baculovirus has been to control the velvet bean caterpillar (*Anticarsia gemmatalis*), which is a serious pest of soybeans in Brazil (Wood and Granados, 1991).

Iridescent Viruses

One of the best known viral infections of insects involves the larvae of the crane fly. These larvae have a blue-green iridescence from the crystalline arrays of viral particles in their tissues. The virus also occurs in beetles, mosquitoes, and midges. Most insects acquire the virus during feeding. The virus, in addition to DNA, contains lipids and proteins, resides in the host cytoplasm, and in the advanced stage of infection may account for one-quarter of the insect's weight. This is the highest weight proportion known for an animal viral infection. Iridescent viruses have been used as a biological control for mosquitoes.

2.10 MYCOVIRUSES: VIRAL INHABITANTS OF FUNGI

Mycoviruses share the following features: (1) they are endogenous to the host cell not only during replication but also during transmission; (2) their genomes consist of dsRNA; and (3) many cause asymptomatic infections and can be described as latent and persistent. Mycoviruses occur in all of the major groups of filamentous fungi. Some examples of mycoviruses include:

1. *Agaricus bisporus* viruses that cause degenerative diseases of cultivated mushrooms. These viruses are transmitted through hyphal fusions and are also carried by spores of the fungus.
2. Viruses of *Penicillium*, including species that produce the antibiotic penicillin, of *Gaeumannomyces graminis*, a causal agent of take-all disease in cereals, of *Helminthosporium maydis*, a causal agent of corn blight, and of *Helminthosporium victoriae*, which causes blight of oats.
3. Killer system viruses in the yeast *Saccharomyces cerevisiae* and in the corn smut fungus *Ustilago maydis*.

Agaricus bisporus viruses have been studied for their virulence and pathogenicity (Lemke, 1979). Disease severity is correlated with the concentration of viral particles. The virus accumulates in the vegetative mycelium as well as in the fruiting body and spores. In spores, the virus particles occur in the cytoplasm; in the vegetative mycelium they accumulate in the vacuoles in crystalline arrays. Disease symptoms include deformities of the mushrooms because of degeneration of infected mycelium.

Saccharomyces cerevisiae consists of several strains. The killer strain produces a toxin that kills the cells of sensitive strains. The killer system in this yeast was earlier thought to have been a genetic phenomenon with three phenotypic expressions: killer, sensitive, and neutral. It is now believed that a virus is closely integrated with the yeast cells. The infected host cells contain segments of dsRNA, which represent the viral genome and contain the determinants for toxins and immunity in the killer system. In the yeast, more than 20 nuclear genes may be required to regulate the dsRNA segment, which is maintained as a wild-type allele. The viral genome is frequently lost through mutation, and the daughter cells then become sensitive to the killer strains.

A killer system also exists in *Ustilago maydis*, where extra dsRNA genomes have been noted in killer and immune strains of the fungus. The killer strain secretes a protein toxin that is lethal to sensitive strains. The killer toxins are encoded by satellite dsRNAs, which are dependent on a helper virus.

Many fungal viruses are associated with their hosts as latent infections. Several viruses that infect plants are transmitted by fungi. It has been suggested that fungi may be alternate hosts for plant viruses. Tobacco mosaic virus-like particles have been found in fungi such as rusts and powdery mildews. Virulence of plant pathogenic fungi may be altered because of their mycoviral symbionts. The chestnut blight fungus, *Cryphonectaria* (*Endothia*) *parasitica*, causes a devastating disease of the American chestnut. The tree now survives as stump sprouts, whose shoots become infected with the fungus and die. The strain of *C. parasitica* commonly found in Europe shows a reduced level of virulence along with a few fungal spores that are depigmented. This hypovirulent strain contains a dsRNA virus. In 1992, virus-infected strains of *C. parasitica* were isolated from recovering chestnut trees in New Jersey (Ghabrial, 1994). Another mycoviral infection reduces the pathogenicity of *Gaeumannomyces graminis*, which causes take-all disease in cereals. Mycoviruses have been considered as a possible means of biological control of pathogenic fungi.

Viruslike structures called *concentric bodies* have been found in more than 100 species of lichenized fungi as well as in some common plant pathogenic fungi. Concentric bodies occur in all parts of a lichen thallus and do not appear to harm the mycobiont (Ahmadjian, 1993).

Mycoviruses offer an excellent model for understanding the molecular basis of the evolution of virus–host relationships. Mycoviruses produce stable associations with their host. Viral latency benefits host survival, whereas viral persistence benefits the virus. In this mutualistic symbiosis, the virus gains a more efficient means of transmission when the integrity of a host system is maintained. There is a close interaction between host genes and fungal viruses (Ghabrial, 1994). Viral replication in fungi is controlled in such a way that its rate does not exceed the rate of host cell division. Host-mediated transmission of mycoviruses may sometimes lead to viral degeneration. For example, many viral particles in mushrooms lack nucleic acids.

2.11 VIRUSES IN ALGAE

Viruses or *viruslike particles* (VLP) have been reported in a number of algae (Van Etten 1999). A virus capable of degrading the nuclei of a green alga, *Uronema gigas*, was first described in 1972 (Mattox et al., 1972). The virus consists of a large polyhedron capsid bounded by an outer membrane and dsDNA; some viral particles have a prominent tail. Algae in the two- to four-cell stage of division are the most susceptible to viral infection.

Virus or VLPs have been observed in vegetative cells of the red alga *Sirodotia tenuissima*, in germlings of the green alga *Oedogonium* sp., and in spores of the brown alga, *Chorda tomentosa*. Algal viruses or VLPs are difficult to study because (1) of low concentrations of viral particles, (2) usually few algal cells contain particles, (3) particles are found only in certain stages of the algal life cycle, (4) algal cells containing particles seldom lyse, and (5) in most cases, the particles are not infectious. The virus-infected filamentous brown alga *Feldmannia* showed deformed sporangia and significantly reduced photosynthesis (Robledo et al., 1994). Viruslike particles have been observed in *Chlorella* symbionts of *Hydra viridis* and *Paramecium bursaria*, suggesting that they may play some role in their symbiosis. *Chlorella* viruses are similar to iridescent viruses that infect insects, and their infection process resembles that of bacteriophages. *Chlorella* viruses are excellent subjects for studying gene regulation in photosynthetic eukaryotes. The viral genomes are thought to be excellent sources of control elements such as origins of replication or promoters for plant genetic engineering (Van Etten et al., 1991).

Algal viruses are a major factor in marine ecology. They have been linked to rapid changes in algal populations and are thought to have a significant role in genetic exchange among algal species (Suttle et al., 1990). Algae may serve as reservoirs or vectors for pathogenic viruses that infect animals (Van Etten et al., 1991).

2.12 VIRUSES IN PLANTS

The "breaking of colors" in tulips, a condition that is produced by a virus, was beautifully illustrated in the still-life oil paintings by Dutch artists in the early seventeenth century. The streaking effect produced on infected flowers was much prized in Holland. In the heyday of "tulip mania," farmers traded cattle, grain, and cheese for the much desired bulbs. Infected tulips have been known for more than 400 years. It was not until 1892, however, that a Russian scientist, Dimitri Iwanovski, showed in the tobacco plant the existence of a disease agent that could pass through a filter capable of holding back the smallest known bacteria (Matthews, 1991). In 1898, a Dutch biologist, Martinus Beijerinick, confirmed Iwanovski's observation and described the infectious agent as *contagium vivum fluidum*, meaning "contagious living fluid." Since 1900, scientists have described more than 600 plant disorders in which the causal agent is suspected to be a virus.

Viral infections in plants vary from clearing of leaf veins and chlorotic mottling to severe necrosis. Symptoms of a viral disease may be generalized throughout the plant or localized. A common sign of a virus infection is retarded growth, which results in stunted plants, lowered productivity, and leaf deformation. Tumors are characteristic of several viral infections, notably, the wound tumor of clover and swollen shoot of cacao. The mechanism of tumor formation resulting from wound tumor virus infection is unknown. Viral plant infections vary widely. At one end of the spectrum are virulent infections that result in necrotic lesions—the cells die so rapidly that they are unable to pass on viral particles to neighboring cells. At the other end are asymptomatic or latent infections, in which the cells are not seriously damaged.

Most plant viruses are spread by fungi, nematodes, and arthropods such as leaf hoppers and aphids. Leaf hoppers, because they feed on phloem tissue, are vectors of viruses that become systemic throughout the plant. This type of viral infection is usually called yellows. Aphids feed on superficial parenchyma cells and are vectors of mosaic viruses. Most plant viruses produce inclusion bodies, which may be amorphous or crystalline and reside either in the cytoplasm or the nucleus. These inclusions, in most instances, consist of aggregates of viral particles. *Plant inclusion bodies*, which are similar to those of insect viruses, are large structures that are visible under a light microscope. Plant cells infected with tobacco mosaic virus contain two kinds of inclusions, hexagonal crystals and irregular masses called X bodies.

More than 95% of all plant viruses possess ssRNA. Tobacco mosaic virus (TMV) is one of the best studied of the plant viruses. In 1946 Wendell Stanley was awarded a Nobel Prize for crystallizing TMV particles, and in 1957 Heinz Fraenkel-Conrat showed that the RNA of TMV was infectious (Atlas, 1997; Fraenkel-Conrat and Williams, 1955). When inoculated into plants, TMV RNA initiates synthesis of more viral RNA and also of viral protein subunits, which, together with the RNA, assemble into typical rods of TMV (fig. 2.5). An example of a dsRNA plant virus is the wound tumor virus. Studies have confirmed the existence of DNA plant viruses such as the cauliflower mosaic virus (dsDNA) and the bean mosaic virus (ssDNA).

Many plant viruses carry more than one

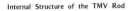

Internal Structure of the TMV Rod

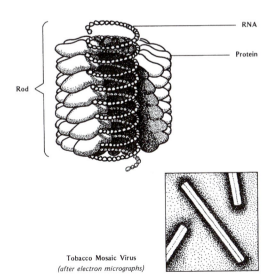

RNA

Protein

Rod

Tobacco Mosaic Virus
(after electron micrographs)

Fig. 2.5 Morphology and detailed organization of the protein subunits and nucleic acid of tobacco mosaic virus. Adapted from Fraenkel-Conrat and Kimball (1982).

RNA strand in each viral particle. There are four strands of RNA in alfalfa mosaic and cucumber mosaic viruses. Viral infection and the production of new viruses require the genetic information of three of the four RNA strands. Cowpea mosaic, tobacco ringspot, and tobacco mosaic viruses have two RNA strands, and both are necessary for viral reproduction.

Gemini viruses and cauliflower mosaic viruses are the only plant viruses that contain DNA genomes. Gemini viruses, such as bean mosaic virus and corn streak virus, occur chiefly in the tropics and form viral aggregates in the nuclei of plant cells. The use of DNA viruses as vectors of beneficial genes into plants is being studied (Matthews, 1991).

Associated with the tobacco necrosis virus (TNV) is a smaller, unrelated virus, the satellite virus. The *satellite virus* is an obligate symbiont and requires the presence of TNV, the host virus, for its replication. The nature of the dependency is not known. The satellite virus is not needed by the TNV, but when present it modifies the infectious nature of the virus. Some virologists predict that similar satellite viruses will be discovered among bacteriophages and animal viruses.

Although virus-infected plants often recover and new growth may be free of symptoms, the virus still remains in the plant. Such a phenomenon is called *acquired immunity*. Some viruses can develop intracellular symbiotic associations in both plant and animal cells. These viruses have become adapted to a "double life" of molecular biosynthesis in two distant life forms. Many plant viruses multiply in the tissues of their insect vectors. Leaf hoppers, for example, are well-known hosts of many systemic plant viruses. Development of the wound tumor virus in a leaf hopper's intestinal filter chamber and in its hemolymph has been well documented. The ability of insects to transmit a plant virus is genetically controlled.

2.13 VIROIDS AND SATELLITE RNA VIRUSES

Viroids are the smallest and simplest naked RNA plant pathogens. They are noncoding genomes that require host enzymes for their replication and transmission. About 20–25 different viroids have been identified, including potato spindle tuber viroid, coconut cadang-cadang viroid, and chrysanthemum stunt viroid. The viroid RNA genome usually consists of a single strand of 246–375 nu-

cleotides, whereas RNA viruses are 10–100 times larger. Viruses often have methylated ribonucleotides in their genomes, whereas viroid RNA lacks modified bases. All virus RNAs are subject to a host cell's enzymatic degradation; viroid genomes are resistant (Voyles, 1993). Viroids produce disease in their host plants either by viroid RNAs or by other RNAs generated during the infection process. The coconut cadang-cadang disease, caused by the smallest of viroids, is a virulent pathogen that has killed thousands of coconut palms in the Philippines. Other viroids have little (e.g., hop latent viroid) to mild (e.g., apple scar skin viroid) pathogenic effects. Viroids cause diseases mostly in agricultural crops and are believed to have originated from the ancestral plants as a result of past plant breeding practices (Diener, 1996).

Even viruses have their own subviral symbionts. *Satellite RNAs* consist of long, circular, single strands of several hundred nucleotides and require a helper virus for their replication and transmission.

Hepatitis δ virus (HDV) has some common features with satellite RNA and viroids. Hepatitis δ virus was first described as a nuclear antigen (δ antigen) in the 1970s and later was found to cause a virulent form of hepatitis (type D hepatitis) among indigenous inhabitants of South America. Delta antigen association is necessary for the expression of the hepatitis B virus (HBV). Viral RNA of HBV/HDV are packaged together in hepatitis B virion particles. Unlike viroids, the RNA of HDV encodes for a nuclear phosphoprotein that is essential for viral replication as well as virion particle formation. Viral hepatitis is a worldwide disease with increased mortality when HDV and HBV co-infections are involved. The transmission of viral hepatitis is linked with drug abuse and transfusion of blood products. Because HDV depends on HBV for its replication, vaccination against HBV protects against the disease (Tiollais and Buenidia, 1991).

2.14 PRIONS

Prions are infectious proteins that are not associated with nucleic acids (Prusiner, 1994). Some scientists believe that prions may be the cause of several diseases such as scrapie, a disease of sheep and goats; transmissible mink encephalopathy (TME) and chronic wasting disease (CWD) of mule deer and elk; Gerstmann Straussler and Scheinker syndrome (GSS) (the new name of Creutzfeldt-Jacob disease [CJD]); kuru, a disease of the New Guinea Fore tribe; and mad cow disease.

During the 1960s, similarities between kuru and scrapie became evident, and later on kuru, CJD, and GSS were shown to be similar in their infectiousness and familial transmission. All these diseases are associated with neuronal cell death and spongiform changes in brain tissue, along with proliferation of glial and astrocyte cells. Many regions of the brain were affected, and amyloid plaques of insoluble protein aggregates, called PrP protein, were always observed. The function of normal cellular PrP protein is not known, but mutant forms of PrP change shape and attach to other prions and thus damage brain tissue. Stanley Prusiner of the University of California Medical School at San Francisco received a Nobel Prize in 1997 for his research on prions, which he began in 1972 to resolve cases of degenerative brain disease. The prion hypothesis remains controversial, with critics contending that a protein could not cause a disease without DNA or RNA to direct its replication and synthesis. Some also believe that a virus co-factor may be involved. No one has ever isolated pure prions and injected them into an animal to cause disease. Disease in experimental animals occurs only after they are injected with ground brain tissue from sick animals. PrP is a brain glycoprotein, and the disease is associated with the accumulation of an abnormal form of PrP protein. Recent studies indicate that PrP protein can be produced in several forms (Prusiner and Scott, 1997). The normal cellular PrP protein, called PrP^C, can be converted into an abnormal form, termed $PPrP^{Sc}$, and its accumulation in the brain leads to the GSS disease. In mice, a PrP protein called PrP^{ctm} produces neurodegenerative symptoms that are similar to the prion diseases. This significant research offers the first experimental model for the study of prions (Hedge et al., 1998).

2.15 THEORIES ON THE ORIGIN OF VIRUSES

Regressive theory postulates that viruses are degenerate forms of intracellular parasites, similar to mitochondria and chloroplasts. Progressive theory suggests that viruses evolved from normal cellular DNA and RNA that gained the ability to replicate autonomously. DNA viruses may have evolved from plasmids or transposons, which later evolved protein coats and became infectious. Retroviruses may have evolved from retrotransposons, while RNA viruses may have had their origin as mRNA (Becker, 1996a).

Coevolutionary theory argues that viruses and living organisms coevolved through ancient biological interactions. Small DNA viruses (papilloma viruses and parvoviruses) are species specific and genetically stable. They show a close link between virus evolution and host speciation. Experimental observations strongly suggest that viral replicators are ancient systems with their own origins of replication and replication proteins that have co-evolved with their host (Shadan and Villarreal, 1996).

2.16 SUMMARY

Viruses are the smallest of all symbionts, and they occur intracellularly in all forms of life. Some viruses kill their host cells, whereas others may live for years in concert with their hosts, often conferring new genetic properties. There are DNA viruses and RNA viruses, retroviruses, satellite viruses, viroids, and viruslike particles such as prions. Bacterial viruses, or bacteriophages, carry DNA from one host to another. Insect viruses have been studied more closely than viruses of other animals.

Some scientists believe that invertebrates may be natural reservoirs of viruses that infect humans and livestock. Viruses cause a variety of cytopathic effects in vertebrate hosts, including cancers. Vertebrate host cells respond to viral infections by producing defensive compounds such as interferons and antibodies. Mycoviruses usually form stable associations with their host, whereas many plant viruses cause virulent infections of host plants. Concentric bodies are viruslike structures that are common in lichen fungi. Retroelements are widely dispersed in the genomes of living organisms.

Viruses are significant disease agents of plants and animals and are among the most difficult parasites to control. Viruses may have a useful role in human affairs. Some viruses are used as biological control agents for various types of pests; others such as λ phages are commonly used as vectors in recombinant DNA research. Viruses in hosts that belong to different kingdoms have common patterns of behavior in their life cycle. Viruses represent the ultimate stage in the evolutionary development of host–parasite relationships. Their extreme reduction in structure and their ability to integrate themselves into the genome of the host cell have parallels with mitochondria and chloroplasts as well as bacterial symbionts, such as κ, which also have developed close relationships with the host genome. Satellite viruses modify the infectious ability of the TNV virus, showing that even viruses are not immune from parasitic agents. In a similar fashion, it is becoming increasingly recognized that the traits of some organisms are the result of viruses that infect these organisms. For example, pathogenic virulence of certain bacteria and protozoans has been correlated with the presence of viral genomes.

Advances in structural analysis of capsid protein and sequencing of the animal viral genome have provided new insights into the basic nature of viral symbioses, enabling scientists to discover how a virus enters, replicates, and assembles inside a host cell and to identify the molecular mechanisms of viral pathogenesis. The presence of viruses in many different forms of life suggests that viruses may have a much greater role in symbiotic associations than previously realized. There are several theories on the origin of viruses, including the regressive theory which states that viruses are degenerate forms of intracellular parasites.

APPENDIX: AN OUTLINE OF VIRUS
CLASSIFICATION

In 1966, the International Committee on Nomenclature of Viruses produced the first unified system for virus classification. In 1973, this committee expanded its objectives and renamed itself the International Committee on Taxonomy of Viruses (ICTV). The ICTV meets every 4 years. The sixth report, published in 1995, recognized 3465 virus species belonging to a single order (Mononegvirales), 50 families, 9 subfamilies, and 164 genera (Murphy, F.A. et al., eds. *Virus Taxonomy: Classification and Nomenclature of Viruses: Sixth Report of the International Committee on Taxonomy of Viruses. Archives of Virology*, supplement 10, Springer-Verlag, 1995).

Some of the significant rules for virus taxonomy include:

1. A virus name should be short and meaningful. Latin binomial names for viruses have been abandoned.
2. A virus species is represented by a cluster of strains, all of which have some stable properties that separate the cluster from clusters of other strains.
3. A genus is a group of virus species sharing common features.
4. A family is a group of genera with common characteristics. Groups of related viruses are placed in families whose names are capitalized, italicized and end in the suffix -viridae.

Following are some examples of viruses discussed in this text which are arranged according to the Baltimore Classification System.

Family	Genus	Species
	Group 1. dsDNA Viruses	
Adenoviridae	*Mastadenovirus*	Human adenovirus 2
Baculoviridae	*Nucleopolyhedrovirus*	Nucelopolyhedrovirus
	Granulovirus	Granulovirus
Herpesviridae	*Simplexvirus*	Human herpesvirus 1
	Cytomegalovirus	Human *Herpesvirus 5*
Iridoviridae	*Iridovirus*	Chilo iridescent virus
Myoviridae	T4-like phages	Coliphage T4
Papovaviridae	*Papillomavirus*	Cottontail rabbit papillomavirus
Phycodnaviridae	*Phycodnavirus*	*Paramecium bursaria* Chlorella virus 1
Polydnaviridae	*Ichnovirus*	*Campoletis sonorensis* virus
Poxviridae	*Orthopoxvirus*	Vaccinia virus
	Leporipoxvirus	Myxoma virus
	Entomopoxvirus A	Entomopoxvirus

(*continued*)

Family	Genus	Species

Group 2. ssDNA Viruses

Family	Genus	Species
Geminiviridae	*Geminivirus*	Maize streak virus

Group 3. dsRNA Viruses

Family	Genus	Species
Partitiviridae	*Partitivirus*	*Gaeumarmomyces graminis* virus
	Chrysovirus	*Penicillium chrysogenum* virus
Reoviridae	*Rotavirus*	Simian rotavirus SA 11
	Phytoreovirus	Plant wound tumor virus
Totiviridae	*Totivirus*	*Saccharomyces cerevisiae* virus
	Giardiavirus	*Giardia lamblia* virus
	Leishmaniavirus	*Leishmania* RNA virus

Group 4. Plus-sense RNA Viruses

Family	Genus	Species
Barnaviridae	*Barnavirus*	Mushroom bacilliform virus
Bromoviridae	*Alfamovirus*	Alfalfa mosaic virus
	Ilarvirus	Tobacco streak virus
	Cucumovirus	Cucumber mosaic virus
Flaviviridae	*Flavivirus*	Yellow fever virus
	Pestivirus	Bovine diarrhea virus
	Hepatitis C-like virus	Hepatitis C virus
Leviviridae	*Luteovirus*	Barley yellow dwarf virus
	Necrovirus	Tobacco necrosis virus
Picornaviridae	*Enterovirus*	Poliovirus 1
	Rhinovirus	Human *rhinovirus*
	Hepatovirus	Hepatitis A virus
	Potexvirus	Potato virus X
Tetraviridae	*Tobavirus*	Tobacco mosaic virus
	Tobravirus	Tobacco rattle virus
Tombusviridae	*Tombusvirus*	Tomato bushy stunt virus
Togaviridae	*Alphavirus*	Sindbis virus
	Rubivirus	Rubella virus

Group 5. Minus-sense RNA Viruses

Family	Genus	Species
Paramyxoviridae	*Paramyxovirus*	Human parainfluenza virus
	Morbillivirus	Measles virus
	Rubulavirus	Mumps virus
Rhabdoviridae	*Lyssavirus*	Rabies virus
	Nucleorhabdovirus	Potato yellow dwarf virus
Arenaviridae	*Arenavirus*	Lymphocytic choriomeningitis virus
Bunyaviridae	*Hantavirus*	Hantaan virus
Orthomyxoviridae	*Influenzavirus* A, B	Influenza A virus
	Influenzavirus C	Influenza C virus

Group 6. RNA Reverse-Transcribing Viruses

Family	Genus	Species
Retroviridae	Mammalian type B retrovirus	Mouse mammary tumor virus
	Avian type C retrovirus	Avian leukosis virus
	Lentivirus	Human immunodeficiency virus 1

Group 7. DNA Reverse-Transcribing Viruses

Family	Genus	Species
	Caulimovirus	Cauliflower mosaic virus
Hepadnaviridae	*Orthohepadnavirus*	Hepatitis B virus

3

BACTERIAL ASSOCIATIONS OF BACTERIA, PROTOZOA, AND ANIMALS

Prokaryotes are simple, primitive organisms that dominate life on earth in terms of numbers of individuals. They also support life by recycling carbon, nitrogen, and sulfur in the biosphere. Prokaryotes, in particular the bacteria, have adapted to almost every ecological niche, ranging from plant surfaces to the skin and digestive system of animals. Bacteria are key players in the Gaia hypothesis ("life regulates life"), and their controversial presence in a Martian meteorite has spurred interest in the possibility of life on other planets (Gibson et al., 1997). Bacteria are important in human affairs. Some cause disease, but others are used to make vaccines and antibiotics which cure disease. Bacteria are an integral part of the current biotechnology revolution.

The origin and evolution of prokaryotes has long been a mystery (Woese, 1994). Today, such mysteries are being addressed by scientists such as Carl Woese of the University of Illinois, whose findings have raised new and perplexing questions of phylogeny (Morell, 1997). Scientists classify life into three domains, archaea, bacteria, and eukarya. The first two domains are prokaryotes; the third one includes all the eukaryotes. Archaeans are represented by specialized microbes such as methanogens, thermophiles, and species that live in high salt and highly acidic environments.

Prokaryotes provide significant insight into our rich and complex biochemical heritage that began 3.5 billion years ago (Mathieu and Sonea, 1995). Some bacteria obtain energy from photosynthesis, others by means of chemosynthesis or by fermentation. Many prokaryotes and some eukaryotes grow without oxygen and may even be killed by its presence. Core samples collected from a depth of 2.75 km in the earth's crust contained *Bacillus infernus* which was living without oxygen at a temperature of 75°C. Anaerobic deep life accounts for about 0.1% of the earth's total biosphere (Kerr, 1997). Examples of anaerobic communities include those found in ancient microbial mats, subsurface soils, freshwater and marine sediments, and the digestive systems of animals. Anaerobic communities have been important participants in symbioses with plants and animals (Fenchel and Finlay, 1995). Interactions between bacteria and their hosts involve a specialized protein secretion system that exports and translocates the bacterial protein to the host cell and induces or interferes with host cell signal transduction pathways (Galan and Bliska, 1996; Yang He, 1998).

Prokaryotes differ from eukaryotes in several ways (Watt, 1999). The prokaryotic cell divides by binary fission or by budding, but never by mitosis. Each cell has only one chromosome that lacks histone proteins. Sexuality occurs in some prokaryotes, but it is distinctive in that the cytoplasm of the cells never fuse; only the chromosome, or part of it, passes from one cell to another. Eukaryotic cells divide by mitosis and meiosis, the latter being part of a sexual process that involves cell fusion and is well developed and regulated. The prokaryotic cell does not contain membrane-bound organelles such as nuclei, mitochondria, and chloroplasts and on the average is smaller than eukaryotic cells, usually $< 10\mu m$ long. The ribosomes of prokaryotic cells have a sedimentation coefficient of 70S, compared to 80S for the ribosomes of eukaryotic cells. Motile prokaryotes have flagella that lack microtubules and consist of the protein flagellin. The cell walls of prokaryotes are made of peptidoglycans and never cellulose or chitin. Motile structures of eukaryotic cells consist of microtubules, which contain the protein tubulin, in a 9+2 paired arrangement. Prokaryotic cells have a tubulinlike protein called FitsZ (Erickson, 1997). In this chapter and in chapters 4 and 5 we focus on prokaryotes as symbionts.

3.1 BACTERIAL SYMBIONTS OF BACTERIA

Bdellovibrio: Predatory Bacteria

Most of the known bacterial symbionts of bacteria have been placed in three genera: *Bdellovibrio*, *Microvibrio*, and *Vampirovibrio*. In 1962, Heinz Stolp, while searching for bacterial viruses from soil, observed plaques on cultures of the bacterium *Pseudomonas phaseolicola* (Stolp, 1979). The plaques continued to grow in size, a feature not found in plaques caused by bacteriophage infection. Stolp examined the colonies with a phase-contrast microscope and observed small bacteria attacking the pseudomonads. These unusual bacteria infect and kill other Gram-negative bacteria.

Bdellovibrios are bacteria (1–2 μm long) whose natural habitat is the periplasmic space of other bacteria. *Bdellovibrios* attack sensitive bacteria, often 10–20 times larger than themselves, by penetrating their cell walls and growing and multiplying in the periplasmic space. The symbiont breaks down the host's cellular constituents, and when the host cell dies, progeny of the symbiont are liberated. When bdellovibrios are first introduced into a bacterial colony, they move quickly, about 100 cell lengths per sec, and collide with a host bacterium. Just before colliding, the symbiont spins rapidly, making contact with the surface of the host cell within seconds. The end of the cell opposite the flagellum attaches to the host cell wall by means of tiny filaments. The symbiont penetrates the cell wall by making a hole, but exactly how it does this is not fully understood. Penetration is completed in 5–20 min. Young host cells are penetrated in less time than older or metabolically inactive cells.

After infection, the bdellovibrio grows in size, elongates into a large, helical-shaped structure, and divides to produce vibrio-shaped, motile individuals. The life cycle of a bdellovibrio is completed in 1–3 h. The host cell, or part of it, swells into a globular body, the bdelloplast, because the cell wall is weakened by enzymes produced by the symbiont. Infected bacteria stop moving within seconds after a bdellovibrio attaches to their cells; their synthesis of nucleic acids and proteins stops after penetration. Permeability of the host cell membrane changes, allowing degraded products to leak into the periplasmic space. In the final stage of the infection, the host membrane breaks down, and bdellovibrio progeny are liberated (fig. 3.1). The number of progeny per cell depends on the host species: for example, 5 for *Escherichia coli*, 10 for *Pseudomonas fluorescens*, and 20 for *Spirillum serpens*. The differences in the number of symbiont offspring per cell reflect the size and metabolic rate of the host and its ability to provide nutrients to the symbionts. Bdellovibrios are not true endosymbionts because they are outside the host cytoplasm. Some strains attack a wide range of bacterial species, whereas others have a limited host range. Three species of *Bdellovibrio* are recognized on the basis of nucleic acid homology and DNA base composition analysis: *B. bacteriovorus*, *B. starrii*, and *B. stolpii*. The predator–prey relationship between *Bdellovibrio* and its host cells is unique among the bacteria (Guerrero, 1991).

Bdellovibrios possess exoenzymes that degrade nucleic acids, proteins, lipids, and peptidoglycan, and they are noted for their high rate of respiration, which is seven times that of *E. coli*. *Bdellovibrio's* life cycle includes a parasitic phase, during which it infects the host bacterium, and a reproductive phase in which it grows and multiplies within the periplasmic space. *Bdellovibrio* can be cultured axenically. *Microvibrio* is an ectosymbiont of *Xanthomonas maltophilia*, which kills its host without penetrating it (Atlas, 1997).

Bdellophages, or viruses that attack bdellovibrios, were first noted in 1970 (Varon, 1974). Bdellophages develop only in a sensitive bacterial host or in the presence of its extract. This is a unique compound symbiotic association that involves a bacterial virus, a small prokaryote (the bdellovibrio), and a Gram-negative prokaryote (the host bacterium).

Pseudomonads and Cyanobacteria

Studies have suggested that a mutualistic relationship exists between nitrogen-fixing,

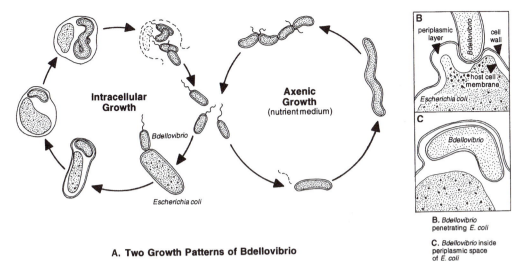

B. *Bdellovibrio* penetrating *E. coli*

C. *Bdellovibrio* inside periplasmic space of *E. coli*

A. Two Growth Patterns of Bdellovibrio

Fig. 3.1 *Bdellovibrio—Escherichia coli* symbiosis. (a) Schematic diagram of *Bdellovibrio* life cycle stages as symbiont of *E. coli* or growth under axenic conditions. (b) penetration of *Bdellovibrio* into *E. coli* showing the invagination of host cell membrane. (c) *Bdellovibrio* symbiont inside periplasmic space of host cell. (Drawn from several sources cited in the text.)

blue-green bacteria and pseudomonads attached to the host filaments. Nitrogen compounds excreted by the host are used by the attached bacteria, which in turn consume oxygen and liberate carbon dioxide when they respire organic compounds. Growth and nitrogen fixation of the host are stimulated by the reduced oxygen and increased CO_2 concentrations caused by the pseudomonads.

3.2 BACTERIAL SYMBIONTS OF PROTOZOA

Paramecium and Other Ciliates

In 1856, Johannes Muller observed rod-shaped particles (bacteria) inside *Paramecium caudatum*, and since then the literature on endosymbionts of ciliates and protozoa has grown considerably (Lee et al., 1985). Bacteria have been observed in the micro- and macronuclei as well as in the cytoplasm of protozoans. The bacterial symbionts of *Paramecium aurelia*, or κ *particles*, have been studied the most thoroughly. In 1943, Tracy Sonneborn described these particles as cytoplasmic factors that killed *P. aurelia*. The particles were considered to be an example of cytoplasmic inheritance (Sapp, 1994b).

All symbionts of *Paramecium* are rod-shaped, gram-negative bacteria that cannot live outside their host (fig. 3.2). Thousands of symbiont particles may exist in one host cell. Previously, the bacteria were designated by Greek letters, but now they are assigned to the genera *Caedibacter*, *Holospora*, *Lyticum*, *Pseudocaedibacter*, and *Tectibacter* (Heckmann and Gortz, 1992). Some strains of the bacteria, called *killers*, are toxic to sensitive paramecia. Killer strains develop only when the bacterial symbionts contain phage or plasmids. Below is a list of some well-known symbionts of *P. aurelia*. All symbionts, except α, occur in the cytoplasm of the host.

α (*Holospora caryophila*). These appear in the macronucleus as short rods, crescents, or spirals, the shape depending on the growth rate of the host; a nonkiller. Infection of *Paramecium bursaria* with *Holospora acuminata* is highly specific and complex (Rautian et al., 1993).

δ (*Tectibacter vulgaris*). Generally non-motile rods, which are often associated with other symbionts such as κ and μ; a nonkiller.

γ (*Pseudocaedibacter minuta*). Frequently appear as small doublets; a strong killer; death of sensitive paramecia results from vacuolization of the cell.

κ (*Caedibacter taeniospiralis*). Kappa symbionts contain unusual inclusions called *R bodies* which form tightly

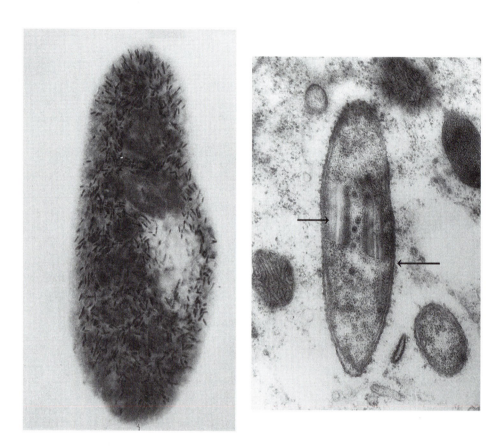

Fig. 3.2 Bacterial symbionts of *Paramecium*. (Left) *Lyticum flagellatum* (λ), endosymbiont of *Paramecium tetraurela*; λ appears as dark-stained rods in the cytoplasm. (Right) *Caedibacter varicadens* (κ), endosymbiont of *Paramecium biaurelia*. Electron micrograph of a longitudinal section of a bright κ particle (long arrow) with R body (short arrow) and spherical phages inside the coiled body. (Courtesy Louise B. Preer, Indiana University.)

coiled proteinaceous ribbons that are refractile and shine brightly. Kappas with R bodies are toxic; those without R bodies are nontoxic. Sensitive paramecia, which includes most strains, are killed by the R protein. Killing is highly specific, each killer being resistant to its own κ symbiont but sensitive to other symbiont strains. R bodies are also found in other bacteria such as *Pseudomonas avenae* (a plant pathogen), *P. taeniospiralis* (a hydrogen-oxidizing soil bacterium), and *Rhodospirillum centenum* (a photosynthetic bacterium). R bodies are classified into five groups based on their size, morphology, and unrolling behavior of the ribbons. Genes coding for R bodies have been cloned in *Escherichia coli*. The genes are encoded in either bacteriophages or DNA plasmids that are present in nonbright κ particles. Induction of the genes may occur spontaneously or result from stimuli such as UV light. R bodies are unusually resistant to drying and are similar to ejectosomes of *Cryptomonas* and *Pyramimonas* (Pond et al., 1989).

λ (*Lyticum flagellatum*). Motile, large rods that are covered with flagella and cause rapid lysis of sensitive paramecia.

μ (*Pseudocaedibacter conjugatus*). Mate killer; elongated rods whose killing action depends on the cell-to-cell contact that occurs during mating. Particles of μ are transferred from killer to sensitive mate.

σ (*Lyticum sinuosum*). The largest symbiont and a close relation of λ; a curved rod and a rapid killer.

o. The ciliate *Euplotes aediculatus* contains particles of the symbiont o, which is essential for survival of the host.

Amoeba: Evolution of a New Symbiosis

In 1966, Kwang W. Jeon noted a bacterial infection in a laboratory culture of *Amoeba proteus* (Jeon, 1991). The bacteria initially killed most of the host amoebae, but some survived. By 1971, the association had become so integrated that the amoebae could not survive without their symbionts. New amoebae could be induced to establish symbioses after about 200 cell generations, or 18 months. The bacteria were Gram-negative rods which resembled *Escherichia coli*. In the amoebae, large numbers of symbionts became enclosed in membrane-bound vesicles of different sizes (symbiosomes). The bacterial symbiont was named x-bacteria (fig. 3.3). It infected only amoebae and in doing so released large amounts of a protein into the host cytoplasm. This protein appeared to prevent lysosomes from fusing with the symbiosomes, thus protecting the bacteria (Jeon, 1992; Pak and Jeon, 1997). Attempts to grow x-bacteria apart from the host amoebae have failed. The x-bacteria contain two types of plasmids. Newly infected amoebae grow faster and are more sensitive to starvation than symbiont-free amoebae. The nuclei of amoebae with x-bacteria are not only incompatible with the original amoebal strain but are now lethal to them. Jeon has suggested that the symbiont-bearing amoebae represent either a new strain or species of *Amoeba* (Jeon, 1991). The bacterial endosymbionts produce high levels of stress proteins, suggesting that the symbiotic relationship may be less benign than supposed (Choi et al., 1991). The amoeba–bacteria association is an excellent model for studying the evolution of new symbiotic associations (Jeon, 1995).

Amoeba–Bacterium Symbiosis in *Entamoeba histolytica*

Pathogenicity of the intestinal protozoan *Entamoeba histolytica* (amoebic dysentery) is determined in part by its association with specific bacteria. Trophozoites usually live as harmless commensals in humans but can become virulent pathogens because of changes in the intestinal bacteria which trigger development of the disease. Isozymes of cloned cultures of the bacteria can change from one state to another (Mirelman, 1987).

3.3 BACTERIAL SYMBIONTS OF ANIMALS

Bacterial Luminescence

Photobacterium, *Photorhabdus*, *Vibrio*, and *Xenorhabdus* contain luminous species that form symbioses with animals. *Photorhabdus* and *Xenorhabdus* are pathogenic bacteria that associate with soil nematodes and insects. *Photobacterium* and *Vibrio* occur in marine organisms such as cephalopod molluscs, teleost fish, and tunicates (Meighen and Dunlap, 1993; Forst and Nealson, 1996; Ruby, 1996).

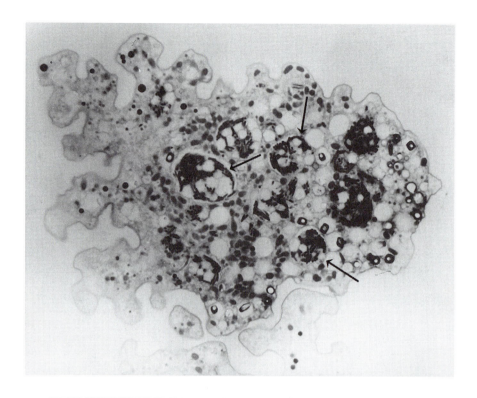

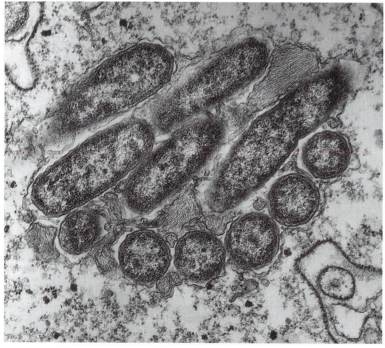

Fig. 3.3 Bacteria–*Amoeba proteus* symbiosis. (Top) an amoeba with symbiont-containing vesicles (arrows). (Bottom) Electron micrograph of a vesicle with bacteria. (Courtesy Kwang W. Jeon, University of Tennessee.)

Marine bioluminescence

More than 1500 marine animal species are bioluminescent, and they form two groups: those that produce their own light and those that use bacteria to produce light (Morin, 1999). Light that is emitted by symbiotic bacteria, although apparently of no direct benefit to the bacteria, is important for the host. *Luminescent bacteria* occur free-living, as plankton, and they also grow on the surface of dead marine animals. The light produced by these bacteria attracts feeding organisms. Luminescent bacteria that are ingested by other organisms resist digestion and are expelled into the ocean and in this way are dispersed. Bioluminescence may have an evolutionary selection value because it enhances a symbiont's survival and propagation (T. Wilson and Hastings, 1998).

Photobacterium and *Vibrio* live as ectosymbionts on teleost fish and as endosymbionts in squids and tunicates. They are heterotrophic on dead and decaying organisms, parasites of marine crustaceans, and commensals in the digestive tract or surface of marine fishes and crustaceans. *Vibrio* is the more cosmopolitan and nutritionally versatile genus. Luminescent bacteria are important symbionts in the alimentary canals of marine animals. The bacteria produce extracellular chitinase, which helps invertebrates such as mussels, scallops, and crabs digest chitin.

Some fish have light-emitting organs that contain large numbers of luminescent bacteria (Haygood, 1993). The bacteria receive nutrients from the host and are in a protected environment. Light produced by the bacteria frightens or diverts predators of the fish and also helps the fish communicate with mates. Some teleost fish, such as the flashlight fish, harbor dense colonies of bacteria in specialized light organs. The bacteria produce light continuously, but the fish controls light emissions by means of shutterlike structures on the organs. The angler fish has a modified dorsal-fin ray that contains a light organ at its tip. The organ contains luminescent bacteria and is dangled in front of the fish, like bait, to lure prey (Haygood, 1993) (fig. 3.4). The gas bladder of leiognathid fishes serves as a complex light organ system for camouflage, communication, schooling behavior, and sexual displays. Purine crystals covering the gas bladder reflect light; other structural modifications control the direction and timing of the emitted light (Haygood, 1993). The squid *Loligo* harbors bacteria in an ink sac near its anus and when alarmed discharges lumines-

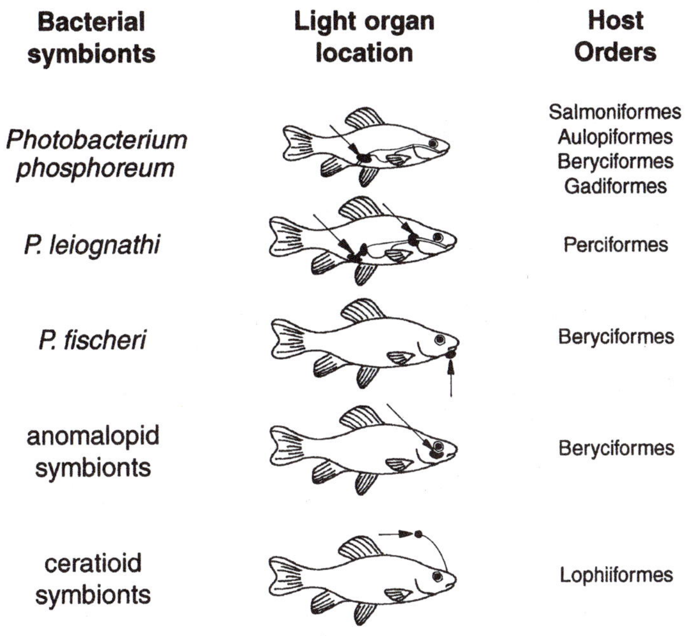

Bacterial symbionts	Light organ location	Host Orders
Photobacterium phosphoreum		Salmoniformes Aulopiformes Beryciformes Gadiformes
P. leiognathi		Perciformes
P. fischeri		Beryciformes
anomalopid symbionts		Beryciformes
ceratioid symbionts		Lophiiformes

Fig. 3.4 Location of light organs (arrows) in selected marine fish. Adapted from Haygood (1993).

cent secretions from this organ. The secretions distract or frighten other organisms and are effective substitutes, under dark conditions, for the black ink that the squid usually secretes (Montgomery and McFall-Ngai, 1995).

Biochemical pathways for bioluminescence are similar in most luminescent bacteria. One common feature is regulation of the enzyme *luciferase*. This enzyme mediates a reaction in which reduced flavin mononucleotide (FMNH) and a long-chain aldehyde are oxidized, and blue light (490 nm) is emitted in the oxidation process (Nealson and Hastings, 1992). Bacterial symbionts also produce small amounts of an autoinducer molecule (*N*-acyl homoerine lacotone), which must be present in high concentrations for the light-emitting system to function. Luminescence is controlled by a number of environmental signals such as salinity, oxygen level, nutrients, and transcription of the *lux* operon. Free-living bacteria produce low levels of autoinducers and therefore do not synthesize luciferase. When bacteria are concen-

trated in a host, however, autoinducers accumulate in sufficient quantity to generate the light-producing compounds. *LuxA* and *LuxB* encode for α and β subunits of luciferase. Other genes located here are *LuxC*, *LuxD*, and *LuxE* which regulate synthesis of the aldehyde. When a high cell density is detected (also called quorum sensing), the *lux* operon is activated (Meighen, 1991; Ruby, 1996) (fig. 3.5).

Squid–Vibrio fischeri *light organ symbiosis*

The small Hawaiian squid *Euprymna scolopes* uses the light emitted by its bacterial symbiont, *Vibrio fischeri*, to camouflage itself at night in a strategy called counter-illumination (Ruby and McFall-Ngai, 1992; McFall-Ngai, 1999). Light from the ventral organ simulates moonlight and starlight and makes it difficult for predators to see the squid. In this association, both the squid and the bacterial symbiont can be grown separately and re-

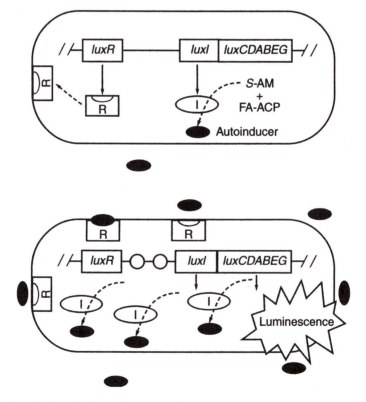

Fig. 3.5   The role of autoinducer in *lux* gene regulation and biochemical production of bioluminescence in *Vibrio fischeri*. Autoinducer is synthesized from S-adenosyl-methionine (S-AM) and fatty acyl-acyl carrier protein (FA-ACP). Adapted from Gray (1997).

combined to form the symbiosis (McFall-Ngai and Ruby, 1998). The light organ of the adult squid is a bilobed structure with epithelial tissue within the mantle cavity that contains luminous bacterial symbionts. Other accessory tissues include the ink sac, which controls the light intensity, a reflector, and a lens that directs and diffuses light over the ventral surface of the squid. The host can thus control and modify the light emitted by the bacteria. The squid embryo develops specific tissues that become infected with *V. fischeri* cells, which are carried with the flow of water through the mantle cavity. The surface of the light organ is covered with cilia which direct the sea water toward a central opening. The porelike opening leads to a duct lined with cilia. The juvenile light organ thus creates dead-end crypts that become populated with bacteria. Within 10 h after colonization, the autoinducer is activated and luminescence begins. The bacteria cause death of the ciliated epithelium and at the same time change the shape and size of the crypts. Each morning the squid expels more than 90% of its symbionts from the light organ crypts, and by nightfall the remaining bacteria have multiplied to maximum numbers again (fig. 3.6). Genetic and biochemical analyses of host and

symbiont have provided insights into how a complex light-emitting organ is formed through molecular mechanisms that involve recognition, remodeling of host tissue, and a sustained symbiotic dialogue. Mutations that lead to loss of the symbiosis offer exciting opportunities to explore how the host species and the bacteria have coevolved (McFall-Nagi and Ruby, 1998).

Xenorhabdus and Photorhabdus: symbionts of nematodes

Species of *Xenorhabdus* and *Photorhabdus* form symbioses with soil nematodes that are virulent pathogens of insects (Nealson et al., 1988; Forst et al., 1997). The Gram-negative bacteria live harmlessly in the gut of the nematodes and together with insects they form a three-way symbiosis. The bacteria produce pigments, antibiotics, and intracellular protein crystals that are toxic when released into the insect host. When nematodes in the infective stage are ingested by an insect, they migrate into the hemocoel and release their bacterial symbionts to rapidly kill the host. While the nematodes feed and reproduce in the dead host, the bacteria stop multiplying. Bacteria then infect the new generation of nematodes.

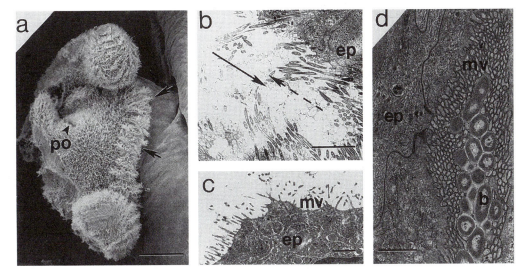

Fig. 3.6 Squid—*Vibrio fischeri* symbiosis. Bacteria move in a specific manner as they pass from the mantle cavity to the brush border of the crypt. (a) Scanning electron micrograph showing the ciliated surface of a juvenile light organ. The arrows point in the direction the symbionts move toward the pore (po) on the light organ surface. (b) The duct connecting the mantle cavity with the crypt epithelium (ep). Bacterial symbionts enter the crypt spaces. (c) Bacterial association with the microvilli (mv) of the crypt epithelium. (d) An electron micrograph of bacteria (b) associated with the adult light organ. From McFall-Nagi and Ruby (1998), with permission.

Bioluminescence and red pigments produced by the bacterial symbionts in the dead insect attract new insects in search of food. In addition to limiting the growth of other bacteria, antibiotics produced by the bacterial symbiont retard decay of the host. Toxin genes from *P. luminescens* are being considered for use in making insect-resistant (transgenic) plants (Bowen et al., 1998).

Bioluminescence in this case is due to bacterial luciferase that resembles the luciferases of other marine luminous bacteria. The *lux* operons from *P. luminsecens* have been cloned, sequenced, and expressed in *Escherichia coli*. Nematodes may protect bacteria from the insect's immune system. The bacteria are pathogenic to the insect host only in the presence of the nematodes. Similarly, the nematode cannot complete its life cycle without the bacteria (Nealson et al., 1988).

Bacterial Symbionts of Insects

There are more species of insects than any other group of animals; thus it is not surprising that there are numerous examples of symbiotic associations between insects and bacteria (Schwemmler, 1989, 1991, 1993; Moran and Telang, 1998). Insects that depend on a restricted diet such as blood or plant sap contain bacterial symbionts that provide them with nutrient supplements such as amino acids, vitamins, and nucleic acids. The bacteria may be extracellular or intracellular within the insect tissues.

About 10% of the known species of insects, including cockroaches, leaf hoppers, and aphids, have well-defined organs called *mycetomes* (bacteriomes). A mycetome consists of mycetocytes (bacteriocytes), or highly specialized cells that contain symbionts. Mycetomes are usually in the body cavity mixed with fat tissues, although in some insects they occur in the intestines and Malpighian tubules. Several generalizations can be made about insect symbioses: (1) The primary intracellular symbionts belong to the eubacterial class *Proteobacteria* and are round or oval. (2) In most cases, the bacteria are enclosed within host-derived vesicles. (3) Some species have secondary Gram-negative, rod-shaped symbionts located in cells that surround the mycetome. Secondary endosymbionts coexist in the same host with primary endosymbionts. How the secondary endosymbionts affect their hosts is not clear, but they are believed to be of more recent origin than primary symbionts. (4) All symbionts reproduce by binary fission. (5) The symbiont's ribosomes are about the same size as the ribosomes of free-living bacteria. (6) The bacterial symbionts are cytoplasmically inherited when eggs become infected with bacteria, thus ensuring continued host association from one generation to another. (7) The symbionts excrete vitamins and nutrients that are used by the host (Gutnick, 1992).

Paul Buchner (1965), an authority on insect symbioses, believed that the mycetocyte, or host cell, controlled the rate of division of the primary symbionts. The number of dividing symbionts increases significantly just before embryogenesis and transovarial infection. Mycetocytes destroy some symbionts by lysosomal enzymes and in this way control the size of the bacterial population. Lysosomal breakdown of secondary symbionts is more pronounced and is a way of removing nonviable individuals, a means by which the host acquires nutrients from the symbionts, or both. The insect hemocoel is a hostile environment for primary mycetocytal symbionts, and they are rarely found there except during a brief period of transovarial infection. Secondary symbionts, however, are often found in hemocytes, which may indicate their recent symbiotic association with the host.

Buchnera aphidicola–aphid mutualism

Some 250 million years ago, a bacterium infected an ancestor of modern-day aphids, and it has since been a constant companion of these insects. Like eukaryotic mitochondria and chloroplasts, all aphids today carry *Buchnera aphidicola* in their cell cytoplasm. A mature aphid contains about 5.6 million bacterial symbionts, and the symbiosis is obligate for both partners (Douglas, 1998). Aphids die when treated with antibiotics. The amino acid tryptophan, which is rare in plant sap, is provided by *Buchnera* to its host (Douglas and Prosser, 1992). The tryptophan gene has been isolated from a *Buchnera* chromosome and amplified on a plasmid. Similarly, the gene for leucine production is plasmid amplified (Bracho et al., 1995). Genetic analysis of *Buchnera* shows that it has additional copies of genes that are involved in vitamin synthesis. *Buchnera* endosymbionts possess high concentrations of GroEL proteins, a chaperonin that functions in the refolding of polypeptides (Baumann et al., 1995, 1997). *Buchnera* reproduction and host growth rates are tightly coupled. For example, although most living bacteria have several copies of rRNA genes, *Buchnera* contains only one gene, suggesting that its growth rate is at a minimum (Baumann et al., 1995). During embryogenesis, aphid mycetocytes release their bacteria, which then enter the developing eggs. The infection process during embryogensis is complex.

Parthenogenesis is common among aphids. Some aphids face extinction because the *Buchnera* they carry has a heavy mutation load and dies out. In sexual reproduction, genes for harmful mutations are reshuffled during meiosis and mating, but in parthenogenetic reproduction there is no gene recombination, and lethal mutations are expressed in each generation. Similarly, mutations in mitochondria and chloroplasts over a long period of time can also cause extinction for their host organisms. In modern eukaryotes, many mitochondrial and chloroplast genes have become incorporated with the host nuclear genome (Lynch, 1996).

Wolbachia–*insect symbiosis and the diversity of life*

Wolbachia is an intracellular bacterium that infects a wide range of invertebrates, including insects, crustaceans, and nematodes (Leong, 1999; Rigaud, 1999). About 16% of all insect species carry this bacterium, and, like mitochondria, it is inherited through the maternal cytoplasm (Werren, 1997; Wilkinson, 1998). The *Wolbachia*–insect symbiosis is more recent than the *Buchnera*–aphid association. It contributes little to its host, and the bacteria do not reside in specialized bacteriocytes. *Wolbachia* infects several types of host cells and can live under different conditions (Stevens and Fialho, 1999). It was first described in 1936 (O'Neill, et al., 1997) as an unculturable bacterial symbiont which infected ovaries of the mosquito *Culex pipens*. Its role as a source of cytoplasmic incomptability in mosquitoes has been studied extensively (Yen and Barr, 1973). Recent DNA sequence analysis shows that *Wolbachia* is closely related to *Rickettsia*, and its two major groups, A and B, diverged some 50 million years ago (Werren et al., 1995). *Wolbachia* can be transferred among different arthropod species by hemolymph injection.

Wolbachia affects the insect host in some interesting ways. In woodlice, *Wolbachia* infects the salivary glands and converts embryonic males into females. In wasps, which can lay eggs with or without mating, *Wolbachia*-infected females produce only daughters, while uninfected eggs develop into males (Hurst et al., 1993). In a most complex symbiosis, a *Wolbachia* strain in a male insect modifies the sperm DNA so that when it fertilizes eggs, the sperm DNA is cut to pieces and the egg dies. Insects carrying different strains of *Wolbachia* are mutually incompatible and cannot interbreed. It has been suggested that *Wolbachia* may have been responsible for speciation among insects (Werren, 1997). *Wolbachia*-infected insects that reproduce parthenogenetically can be "cured" with antibiotic treatment and produce sexually reproducing offspring (Stouthamer et al., 1990).

Wolbachia has attracted attention as an insect pest control because of its ability to cause cytoplasmic incompatibility, thereby introducing sterility into wild populations of mosquitoes (Bourtzis and O'Neill, 1998). Some of the most devastating diseases are transmitted by arthropod vectors. Alternative strategies for vector control include introducing foreign genes into insects to make them unable to transmit disease causing agents, genetic transformation of bacterial symbionts, and foreign gene expression in symbiotic bacteria (Beard et al., 1993; Federici and Maddox, 1996).

Microbial Fermentation in Animal Digestive Systems

Cellulose is a major constituent of plant cells and accounts for approximately half of the global carbon dioxide that is fixed by photosynthesis. Cellullose is degraded by microbial symbionts under the anaerobic conditions that occur in digestive systems of animals where dense populations of bacteria, protozoans, and fungi carry out a variety of metabolic pathways. The digestive tracts have anatomical, biochemical, and behavioral modifications that allow for fermentation either in a foregut or hindgut (Troyer, 1984).

Most herbivorous mammals have postgastric fermentation in the colon or cecum in which ingested fibers are degraded by microbial symbionts to release nutrients. Cecal fermentation occurs in small mammals such as rodents and rabbits, where plant substrates are fermented to fatty acids that are absorbed in the hindgut. The colon is the fermentation site in primates, horses, elephants, and manatees.

Ruminants are specialized herbivores that

retain ingested plant food in their pregastric fermentation chambers. True ruminants such as sheep, cattle, goats, and deer have stomachs that consist of four pregastric chambers (rumen, reticulum, omasum, and abomasum), whereas camels and llamas have three-chambered stomachs. Anaerobic fermentation occurs in the rumen and reticulum, which are usually considered as a single organ. The rumen is lined with fingerlike projections called papillae that increase its absorptive surface. In sheep and cattle, the rumen occupies more than 65% of the volume of the digestive system. Pregastric fermentation allows the host animal to use volatile fatty acids and vitamins as well as microbial proteins. Secondary metabolites that are ingested are detoxified by microoganisms in the pregastric chambers. Nitrotoxins that occur in certain legumes are converted by rumen bacteria into harmless amines. *Acidaminococcus fermentans* converts *trans*-aconitate, which is present in grasses, into harmless acetate (Van Soest, 1994; Douglas, 1994a).

There are more than 3.5 billion domesticated ruminants worldwide, and they are an important resource to agriculture. Human domestication of cattle during the dawn of the agricultural revolution some 10,000 years ago represents another example of behaviorial symbiosis between humans and ruminants (Rindos, 1984). Biotechnology firms are using transgenic sheep to make pharmacologically active proteins. Even though obligate anaerobic organisms have been difficult to study in culture, analysis of 16S rRNA sequences and polymerase chain reaction-based detection and profiling techniques have revealed the complex nature of microbial ecosystems and new species from the specialized niches have been described (Amman et al., 1995; Flint, 1997).

Fermentation, whether in the rumen or the lower digestive tract, is a continuous system. According to the feast-famine hypothesis, when an animal eats, pulses of potential substrates are sent down the digestive tract and the microbial population dynamics shift because of death by lysis of those who exhaust their substrate (McBurney et al., 1987). Ruminants use lysozymes to digest bacteria in the gastric chamber. The normal expression of the gene that codes for this enzyme is restricted to cells in the tear glands. In ruminants, however, duplicate copies of this gene are expressed in digestive functions.

Adhesion is a microbial strategy where bacteria attach to their substrates and avoid being washed away by the digestive flow. Mechanisms for bacterial attachment include fimbria, lectins, and glycocalyx. Many rumen microbes attach to surfaces of food particles. Competition for surface attachment sites may result in old cells being replaced with new ones.

Cellulose digestion, although generally rare among insects, has been demonstrated in silverfish, roaches, beetles, and wood wasps. Termites are the most efficient cellulose digesters. Recent evidence indicates that in some cockroaches and in higher termites cellulose is not digested by microbial symbionts. However, anaerobic microbial symbionts in the hindgut are fully exploited in the lower termites *Kalotermes flavicollis* and *Retculitermes flavipes* and in the wood roach, *Cryptocerus punctulatus* (Breznak, 1999). Symbiont-mediated cellulose fermentation occurs commonly among insect species that are scavengers and detritivores. It is not yet understood how a symbiont-free cellulolytic system evolved in some insect species (M. M. Martin, 1991). The curious South American bird hoatzin has a diet of leaves which it ferments in a foregut similar to that of ruminants (Grajal and Strahl, 1991).

Rumen microbial ecosystem

Rumen microbiology began in 1947 when Robert Hungate developed the roll-tube technique to study anaerobic organisms (Hungate, 1950, 1966). Microorganisms can live in most parts of an animal gut, but certain regions, such as the rumen, are more suitable. Bicarbonate ions from saliva are continuously being added to the rumen to maintain a pH of about 6.5. Peristalsis occurs at 1- to 2-min intervals and mixes and moves the rumen contents. The rumen is almost free of oxygen and has a relatively constant temperature of about 39°C. Carbon dioxide and methane are the two principal gases of the rumen. Methane is a direct product of fermentation; carbon dioxide is produced by fermentation and also by neutralization of bicarbonate ions (Leschine, 1995).

Cellulolytic bacteria in the rumen degrade the cellulose of plant material into its constituent subunits, cellobiose and glucose. The sugar molecules are then fermented into volatile fatty acids, such as acetic acid, butyric acid, and propionic acid, which are then metabolized by the host. The ruminant converts propionic acid into carbohydrates such as lactose and glycogen. *Syntrophy* is a beneficial association where the metabolites of one organism serve as substrates for another organism. The interdependence of several species in rumen fermentation results in beneficial associations (fig. 3.7a). For example, carbohydrate-fermenting bacteria produce hydrogen, carbon dioxide, and formate as metabolic end products, which methanogens use to produce methane. Similarly, plant proteins that enter the rumen are digested and fermented by bacteria, and ammonia is liberated. Nearly all of the bacterial species in the rumen can use ammonia as a major source of nitrogen. Small amounts of amino acids are also produced and are used by the bacteria in protein metabolism. Interactions between cellulose-degrading bacteria and the nitrogen-fixing bacterium *Clostridium* and a noncellulolytic anaerobe *Klebsiella* illustrate the phenomenon of microbial interdependence. In this association, *Clostridium* fixes nitrogen and provides *Klebsiella* with amino acids, while *Klebsiella* provides two vitamins, biotin and *p*-aminobenzoate, which are needed by *Clostridium*. Both bacteria use the energy molecules obtained from cellulolytic degradation by *Clostridium* (Cavedon and Canale-Parola, 1992) (fig. 3.7b).

Bacteria and protozoa in the rumen benefit each other. Digestion of starch and proteins by the protozoa releases sugars and amino acids, which the bacteria use to grow and reproduce. The protozoa, in turn, ingest bacteria and use their protein as a source of amino acids. The protozoa in the rumen stimulate bacterial growth, which in turn increases bacterial ingestion by the protozoa. As a result, there is a greater turnover of bacterial carbon and nitrogen. The ruminant participates in this symbiosis by providing nutrients for the rumen symbionts and by regulating the physical and chemical conditions of its fermentation chamber. The ruminant meets its protein needs and also obtains many vitamins by digesting some of its rumen residents.

Interdependent obligate anaerobic bacterial species make up about half of the total biomass of a normal rumen. Species of *Butryovibrio*, *Fibrobacter*, and *Ruminococcus* are the prinicipal cellulose-degrading bacterial symbionts. In addition, a chytrid, *Neocallimastix frontalis*, secretes cellulases that are more ac-

tive than those of other cellulolytic microorganisms (Bauchop, 1989). Anaerobic chytrids are new to rumen microbiology because they were first mistakenly identified as flagellates. These anaerobic fungi have hydrogenosomes instead of mitochondria and occur only in the gut of herbivorous animals. The cellulolytic bacteria and protozoans also compete with cellulose-degrading ciliates. Anaerobic ciliated protozoa are uniquely adapted to the rumen environment and make up about 40% of the rumen biomass (Williams and Coleman, 1991; Tannock, 1995) (fig. 3.7a).

The role ciliates play in rumen fermentation is controversial. Their predation on rumen bacteria is thought to cause rumen inefficiency. Ciliates have slow turnovers. They feed on bacteria and can ingest small plant fragments that enter the rumen. Digestion occurs in the food vacuoles and produces volatile fatty acids, ammonia, and hydrogen, which in turn serve as a substrate for the methanogens which are ectosymbionts of the ciliates. There are two major groups of ciliates in the rumen, the holotrichs and the entodiniomorphs. The *holotrichs* have cilia all over their cell surface and include the genera *Dasytricha* and *Isotricha*. Holotrichs convert sugar into a type of starch that is chemically similar to plant starch. *Entodiniomorphs* have a firm pellicle that is often drawn out posteriorly into spines and cilia and are represented by the genera *Diplodinium*, and *Entodinium* (see fig. 9.1). Rumen protozoa do not survive in culture.

Clostridium and *Peptostreptococcus* produce ammonia, which is required by many rumen microbes. Rumen fermentation of cellulose degradation products results in hydrogen, formate, succinate, and lactate.

Methanogens are obligate anaerobic archaebacteria. They use hydrogen and carbon dioxide as substrates for their energy-yielding metabolism and release methane into the environment. They were first observed as ectosymbionts of rumen ciliates (Krumholz et al., 1983). Methanogens as endosymbionts of *Dasytricha ruminantium* and *Entodinium* sp. occur in membrane-bound vacuoles; they escape digestion and grow and divide synchronously with their host cells (Finlay and Fenchel, 1993). Methane is the most abundant hydrocarbon in the trophosphere, and its emission from ruminants is second only to emission from rice cultivation. Termites are

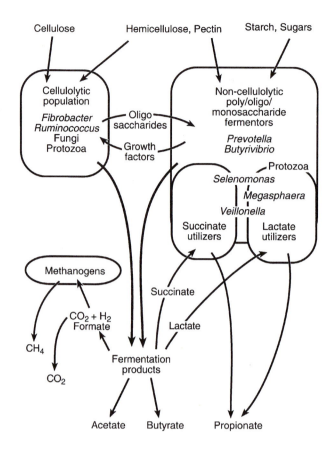

Fig. 3.7 Rumen microbial fermentation of plant material. (a) Metabolic interactions among major rumen microorganisms. (b) Nutritional symbiosis between *Clostridium* and *Klebsiella*. (c) Scanning electron micrographs of (left) cells of *Ruminococcus flavefaciens*, (right) the anaerobic protozoan *Polyplastron multivesiculatum* ingesting rumen fungal material. (Parts a and c from Flint (1997), with permission; b adapted from Fenchel and Finlay (1995).

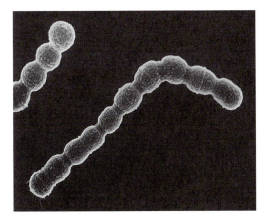 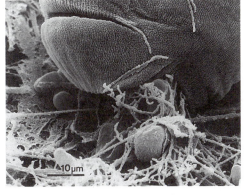

Fig. 3.7   (continued)   c (left and right)

also major contributors of global methane. The role of methane in global warming has attracted the attention of climatologists.

The rumen system is an excellent model for studying symbiotic associations under stable environmental conditions (Duncan and Edberg, 1995). On a practical basis, rumen fermentation is of special significance to humans. Cattle and other domesticated ruminants are efficient converters of the bioenergy of green plants into animal protein, woolen fiber, and leather products.

Chemolithotrophy: Life without Oxygen

The finding of luxuriant biological communities around deep-sea hydrothermal vents and in other sulfide-rich environments such as mudflats and subtidal sediments were important discoveries in biology. An obvious question about these strange, new animals was the nature of their food source (table 3.1). It was soon discovered that endosymbiotic sulfur-oxidizing chemoautotrophic bacteria lived within tissues of the organisms (Cavanaugh 1994). Hydrogen sulfide in the bacterial symbionts is oxidized in the process of oxidative phosphorylation to ATP (Doeller, 1995). A more common pattern occurs when the hydrogen sulfide that enters the host body is oxidized to thiosulfate, which is then supplied to all the symbionts.

Methanotrophic bacteria also form endosymbioses with members of the vent animal community as well as with protozoans that live in the anaerobic marine sediments. Such associations are found in marine bivalves, which use methane as a source of carbon and energy (Distel and Cavanaugh, 1994; Distel, 1998).

Anaerobic marine ciliates and their symbionts

Free-living ciliates are common inhabitants of marine sandy sediments and often form intimate associations with bacteria. Sulfate-reducing bacteria are ectosymbionts of anaerobic ciliates such as *Caenomorpha levanderi*, *Metopus contortus*, and *Parablepharisma pellitum*. The bacteria occupy pits on the outer wall of *P. pellitum*, whereas in the other ciliates they form layers that attach to host cell membranes. Ectosymbionts may add more than 10% to the biovolume of *P. pellitum* (Fenchel and Finlay, 1995). Purple nonsulfur photosynthetic bacteria and methanogens live as endosymbionts in their protozoan hosts (fig. 3.7). In general, all anaerobic protozoans enhance their energy metabolism by acquiring hydrogen-consuming symbionts. Methanogens live as endosymbionts of anaerobic ciliates because they have a steady source of hydrogen that is generated in the hydrogenosomes.

Strombidium purpureum is reddish brown because of the photosynthetic nonsulfur bacteria in its cytoplasm. The ciliate can be cultured anaerobically with frequent exposure to light. Other closely related species are aerobic, suggesting that *S. purpureum* evolved from an aerobic ancestor. *Strombidium purpureum* produces large amounts of hydrogen, which then is consumed by the bacterial symbionts. It is not known whether *S. purpureum* receives any products of the symbiont's photosynthesis. The bacterial symbiont is closely related to *Rhodopseudomonas*, which in the dark and at low oxygen concentration can carry out oxidative phosphorylation by using hydrogen and fatty acids as substrates. In the dark, the symbiont benefits from oxygen, and the host ciliate reduces its threat of oxygen toxicity. *Strombidium purpureum* and its cytoplasmic symbionts provide an evolutionary window into how an anaerobic host can convert into an aerobic ciliate (Fenchel and Finlay, 1995) (fig. 3.8). Modern-day mitochondria may be descendents of ancient ancestors that are similar to present-day nonsulfur photosynthetic bacteria (Woese, 1977).

Pelomyxa palustris is an interesting ciliate because, although lacking hydrogenosomes, it contains three different rod-shaped bacterial endosymbionts, two of which are methanogens, a third type which surrounds the nucleus. The endosymbiont associated with the nucleus is thought to perform the hydrogenosome function. If true, the cytoplasm of *Pelomyxa* would demonstrate a three-step food chain consisting of eukaryotic fermentation to volatile fatty acids, prokaryotic hydrogen production, and methanogensis (Fenchel and Finlay, 1995).

Psalteriomonas lanterna, an anaerobic amoeboflagellate, also contains hydrogenosome-like organelles that form tight complexes with methanotrophic symbionts. Inactivation of methanogenic symbionts in *Metopus contortus* and *Plagiopyla frontata* retarded the host's growth rate by 25–30% (Fenchel and Finlay, 1995). Methanogen symbiosis in *Plagiopyla frontata* is the most complex. The methanogenic symbionts and host hydrogenosomes form alternating layers. The host regulates the number of methanogens so that when it divides both methanogens and hydrogensomes increase in number, an example of co-evolution between this specific host and its symbionts.

A typical *Metopus contortus* has about 7000 methanogens as endosymbionts and about 4000 sulfate-reducing bacteria on its outer surfaces (fig. 3.9). The ciliate breaks down its food into pyruvate and NADH, which in turn are oxidized in hydrogenosomes to produce acetate and hydrogen. Methanogens use hydrogen and carbon dioxide to liberate methane and water. The ectosymbionts metabolize acetate, lactate, or alcohol that diffuse out of the host cell.

Bacterial symbionts of tubeworms and other animals

The deep-sea vents of the Pacific Ocean are unusual habitats of symbiotic associations (Childress et al., 1987; Fisher, 1990; Childress

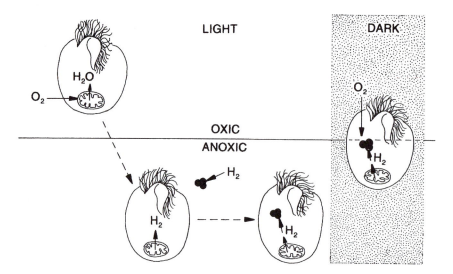

Fig. 3.8 *Strombidium purpureum*, a marine ciliate, shows early stages of mitochondrial evolution. Ancestors of this ciliate possessed mitochondria and lived aerobically. This ciliate evolved into an anaerobe that produced hydrogen in its fermentation. The ciliate, by acquiring photosynthetic non-sulfur purple bacteria as endosymbionts, found a sink for its hydrogen and thus transformed into an aerobic organism. Adapted from Fenchel and Finlay (1995).

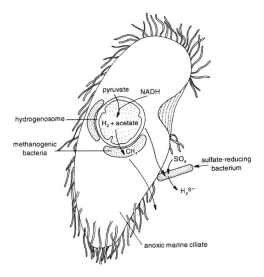

Fig. 3.9 Marine anaerobic ciliates such as *Plagiopyle frontata* and *Metopus contortus* possess hydrogenosomes and release hydrogen as metabolic waste, which in turn is used by methanogens that live as endosymbionts. Redrawn from Fenchel and Finlay (1994).

and Fisher, 1992; Van Dover, 1996). Hot water rises from these vents as a result of water seeping through cracks in the ocean floor and being boiled by the underlying molten rock. The water has a high concentration of hydrogen sulfide. These vents are identical to the hot springs found on land, the only difference being that the marine vents are about 2 miles below the surface of the ocean. Surprisingly, these remote and seemingly inhospitable habitats contain a rich and diverse group of organisms (Tunnicliffe, 1991, 1992; Childress, 1995). One of the strangest members of this group is the red tube worm, *Riftia pachyptila* (phylum Pogonophora). Large colonies of this giant worm are attached to rocks along the path of the water that flows from the hydrothermal vents (Hessler and Kaharl, 1995).

Pogonophorans live inside stiff, chitinous shells, which they secrete around themselves. Their body consists of a long trunk, a short anterior piece, and a short posterior piece, which anchors the animal to the tube and to the ground. Numerous long, cilia-bearing tentacles project from the anterior end of the animal. Pogonophorans that live in regions such as the Galapagos rift may be up to 3 m long. They glide partially in and out of their tubes and move with the currents of water that flow from the vents. The animals lack a mouth and a digestive tract.

Part of the body cavity of a tube worm, the trophosome, is highly vascularized and contains dense colonies of endosymbionts belonging to the γ group of Proteobacteria. The bacteria are chemoautotrophs, which means that they obtain energy from the oxidation of sulfides and fix carbon dioxide through the Calvin-Benson cycle. The symbionts are most likely acquired from the environment, but they have not been detected outside of their host. The well-developed circulatory system in the trophosome and the high oxygen-binding capacity of the blood hemoglobin are adaptations that ensure anaerobic conditions for the bacterial symbiont. In the bivalves *Bathymodiolus thermophilus* and *Calyptogena magnifica* and the gastropod *Alvinconcha hessleri*, endosymbionts are located within the gill tissues. These vent molluscs have a digestive system which allows them to feed on free-living bacterial mats. Filaments of *Beggiatoa* and *Thiotrix* form thick mats through which the hydrothermal water percolates. The free-living sulfur-oxiding species *Thiobacillus hydrothermalis* and *Thiomicrospira crunogena*, as well as heterotrophic bacterial species, have been described.

Shrimps and annelids have only ectosymbionts, whose role in the symbiosis is not clear. In the annelids *Alvinella caudata* and *A. pompejana*, the ectosymbionts consist of complex bacterial filaments of the ε group of *Proteobacteria*, which are inserted into the host epidermis at specific regions (Nelson and Hagen, 1995).

Hydrothermal vent communities continue to be a source of new research challenges. The mechanisms of symbiont transmission, genetic exchange between host and symbionts, and the upper temperature limits to life are some of the unanswered questions.

Table 3.1   Animals with chemoautotrophic bacterial symbionts

Pogonophora
   Riftia pachyptila, vestimentiferan tube worm

Molluscs (bivalves)
   Bathymodiolus thermophilus, hydrothermal vent mussel
   Calyptogena magnifica, hydrothermal vent vescomyid mussel
   Lucina pectinata, lucinid mussel in mangrove sediments
   Lucina floridana, warm sulfide-rich shallow sediment
   Lucinoma aequizonata, Santa Barbara basin
   Solemya togata, mudflats
   Solemya borealis
   Solemya velum
   Thysaria, subtidal sediments

Annelids (alvinellid polychaetes)
   Alvinella sp.
   Paralvinella sulfinocola, sulfide worm in subtidal sediments

Nematodes
   Stilbonematid, free-living marine nematode in subtidal sediments

3.4 SUMMARY

Prokaryotes form two domains, the archaea and the bacteria. Bacteria are prokaryotes whose cells are fundamentally different in organization and chemistry from those of eukaryotes. Bacteria are symbionts in diverse organisms, from one-celled paramecia to humans. Bdellovibrios are parasitic bacteria that forceably enter other bacteria. Their mode of entry is being used as a model of how ancestral prokaryotes may have become colonized by symbiotic organisms.

Paramecium aurelia is host to different types of rod-shaped, Gram-negative bacteria such as κ, strains of which are killers. Why organisms such as *P. aurelia* are more predisposed to form symbiotic unions than others is not clear.

Laboratory studies have shown that *Amoeba proteus* can establish artificial symbioses with bacteria within a few years. The artificially induced symbiosis between the bacteria and *A. proteus* illustrates how symbiotic associations may have evolved and suggests that even today symbiotic unions are being formed and dissolved continually in natural situations. Within 5 years, bacteria that began as pathogens of *A. proteus* assumed a role of organelles and became indispensable to their hosts. Pathogenicity of *Entamoeba histolytica* is influenced by its bacterial symbiont.

Luminescent bacteria occur in different marine organisms and terrestrial insects, and the light they produce may lure prey, attract feeding organisms, or frighten predators. The diversity of organisms that house luminescent bacteria suggests that bioluminesence has a selective value in terms of the evolution of organisms.

Bacteria are common in insects and are housed in specialized cells called mycetocytes, which form unique organs called mycetomes. The life cycles of bacterial symbionts inside insects is complex and not clearly understood. Aposymbiotic insects have been freed of mycetocytal symbionts, usually by means of antibiotics fed to the insects.

The *Buchnera*–aphid mutualism is 250 million years old and very complex. An adult aphid may contain more than 5 million bacterial symbionts. *Wolbachia* is an intracellular parasite that determines the sex of its insect host. Its ability to cause cytoplasmic incompatibility in mosquitoes has gained it attention as an insect pest control.

Ruminant animals contain large populations of cellulolytic bacteria in their stomachs along with various types of protozoa, which also digest cellulose. Rumen bacteria, in addition to their relationship with their host, also interact mutually with rumen protozoa. The hoatzin is a bird with foregut fermentation that eats leaves. Anaerobic marine ciliates often form associations with bacteria.

Sulfur-oxidizing bacteria are symbionts of tube worms, which live near deep-sea vents in the Pacific Ocean. The bacteria synthesize carbon compounds, which are used by the host. Methanotrophic bacteria occur in marine bivalves.

4

BACTERIAL PATHOGENESIS

Molecular Mechanisms

De Bary's (1879) broad definition of symbiosis includes parasitism and disease, areas in which significant discoveries are being made. This has been most evident in bacterial pathogenesis. In this chapter we introduce some of the recent findings of molecular mechanisms of bacterial pathogens. During the past decade, scientists have introduced innovative approaches and concepts from disciplines such as bacteriology, cell biology, and immunology in an attempt to understand the molecular basis of infectious diseases. Stanley Falkow calls such attempts the "Zen" of bacterial pathogenesis, believing, in the Zen Buddhist tradition, that true enlightenment is achieved through meditation and insight (Falkow, 1988, 1990).

"Virulence" and "pathogenicity" refer to the ability of bacteria to cause disease. The traditional criteria for establishing that a bacterium is responsible for a disease have been Koch's postulates, which were developed in 1882. Although serving well for many years, these postulates have limitations: (1) not all bacteria can be cultured, (2) not all members of a species are equally virulent, and (3) adequate animal hosts are not always available. Host susceptibility is an important virulence factor for bacteria.

Falkow has proposed the "molecular Koch's postulates" to examine the potential role of genes and gene products in pathogenesis. The new postulates are:

1. The gene or its product should be found in strains of bacteria that cause the disease and not in bacteria that are avirulent.
2. A mutation of a virulent gene should reduce its virulence, and, conversely, introduction of a virulent gene into an avirulent bacterial strain should make it virulent.
3. The gene should be expressed by the bacterium during the infection process.
4. Antibodies to a gene product should be protective, or the gene product should elicit an immune response.

Molecular biology techniques now make it possible to detect and identify bacteria and their genes even though they cannot be cultured (Finlay and Falkow, 1989; Relman et al., 1992).

4.1 HOW BACTERIA COLONIZE A HOST

The first important step in bacterial pathogenesis is adherence to a host cell. This occurs by means of *pili*, which consist of long rods that extend out from the bacterial surface. The tips of the pili contain proteins that attach to host cell receptors. In some cases, pilin, the protein subunits of the pilus shaft, attach to the host cell's receptors. In addition to pili, bacterial surface proteins called *adhesins* attach firmly to the host cells. Sometimes the host makes antibodies against pili or adhesin proteins, and this induces the bacteria to make different types of adhesins. Only Gram-negative bacteria make adhesins; the mechanism of how Gram-positive bacteria attach to a host cell is not known.

Bacterial surface proteins called *invasins* cause cytoskeletal changes in host cells that result in pseudopodial formation and engulfment of bacteria in phagocytic vesicles. Some bacterial pathogens escape from the vesicles and multiply in the host cell cytoplasm. Bacterial invasion can become systemic after the pathogens cross the mucus membrane by moving through M cells, but they must survive the attack of phagocytes in the lymphatic system (Cundell and Tomanen, 1995). Once established in the host tissue, bacteria obtain nutrients from the host in different ways. For example, iron, which is an essential mineral for living cells, occurs in complexes with iron-binding proteins. Some bacteria concentrate the iron they need by producing a *siderophore*, a compound which has a high affinity for iron. Other bacteria use surface proteins to remove iron from host iron-binding proteins such as transferrin, lactoferrin, ferritin and hemin.

Bacteria survive by avoiding the host's immune system. Some bacteria have a capsule that consists of a loose network of polysaccharides or a protein–carbohydrate mixture. Capsules protect the bacteria from the host's inflammatory response after phagocytosis. Antibodies that bind to a capsule help a host facilitate phagocytosis. Some bacteria evade antibodies by covering themselves with host proteins such as fibronectin, changing their surface antigens, or mimicking host molecules such as hyaluronic acid and sialic acid (Finlay and Cossart, 1997).

4.2 HOW BACTERIA DAMAGE HOST CELLS

The virulence of many bacterial pathogens is due to the toxins they produce, which disrupt normal cell functions and cause cell death. *Exotoxins* are proteins that are excreted by dividing bacteria. Exotoxins that attack a variety of cell types are called *cytotoxins*; those that attack a particular cell type or tissue have specific names, such as neurotoxin, leukotoxin, hepatotoxin, or cardiotoxin. Exotoxins can be associated with a specific bacterial disease. For example, cholera toxin is produced by *Vibrio cholerae*, diphtheria toxin by *Corynebacterium diphtheriae*, and tetanus toxin by *Clostridium tetani*. Microbiologists place exotoxins into three groups:

1. *A-B exotoxins* consist of one enzymatically active portion (A) and another portion (B) that binds to the host cell. The B portion is recognized by a host cell-surface receptor; the A portion enters the cell cytoplasm and disrupts either protein synthesis or cAMP regulation.
2. *Membrane-disrupting exotoxins* make holes in the host's endosomal membrane by hydrolyzing phospholipids, thereby enabling bacterial pathogens to escape from phagocytic vacuoles into the cytoplasm. These toxins are also responsible for apoptosis of the host cell. Two examples from this group of toxins are listeriolysin O produced by *Listeria monocytogenes* and α-toxin produced by *Corynebacterium perfringens*.
3. *Superantigens* bind to major histocompatibility complex (MHC) receptors and to T-cells and cause them to produce protein cytokines such as interleukin-2 (Il-2). High levels of Il-2 in the blood result in toxic shock syndrome, which is characterized by nausea, vomiting, malaise, and fever. Examples of superantigen exotoxins are found in staphylococcal foodborne diseases (Johnson et al., 1992).

An example of an endotoxin is the lipopolysaccharide (LPS) component of the Gram-negative bacterial cell wall. The lipid A portion of this endotoxin is exposed only when the bacteria lyse. LPS triggers the release of cytokines and activates the complement cascade in the host, which leads to fluid leakage from blood vessels. This reaction, along with blood clotting in the peripheral circulation, often leads to septic shock and death.

In addition to toxic proteins, pathogenic bacteria also produce hydrolytic enzymes that degrade host tissues and disseminate bacteria within the host. Heat-shock proteins produced by bacteria stimulate autoimmune responses so that host antibodies and T-cells attack healthy host cells (Salyers and Whitt, 1994).

Role of Adhesins in Pathogenesis

Adhesins are surface proteins involved in protein–carbohydrate or protein–protein interactions between bacteria and host cells. Bacteria can attach only when their adhesins match receptors of the host cell membrane. This association of bacteria with a eukaryotic cell often brings about changes in the host and pathogen. *Bordetella pertussis* (whooping cough), *Hemophilus influenzae*, *Psuedomonas aeruginosa*, and *Streptococcus pneumoniae* (bacterial pneumonia) are pathogens that destroy the ciliated epithelium and alveolar macrophages. Damage to the ciliated cells occurs only when bacterial adhesins attach to the cilia and cause bacteria to release toxins. Two adhesin proteins of *B. pertussis*, filamentous hemagglutinin (FHA) and pertussis toxin (Ptx), have been characterized (Cundell and Tomanen, 1995). Filamentous hemagglutinin is a 220-kDa protein with several epitopes that recognize receptors called *integrins* on host cells. Attachment of pathogenic bacteria to eukaryotic integrins involves three strategies: lectin binding, masking, and mimicry. During *lectin binding*, as seen in *E. coli* and group B *Streptococci*, bacterial proteins recognize the carbohydrate moiety of integrins. *Masking*, as seen in *Mycobacteria* and *Legionella* sp., involves adsorption of serum protein, such as C3bi, onto the bacterial surface, which then binds to a C3bi binding site on the integrin. *Mimicry* occurs when the bacteria produce a protein containing an epitope that binds directly with integrin. Mimicry is the binding mechanism for the three epitopes of FHA of *B. pertussis*. Mutations in the genes for adhesin proteins result in loss of virulence in the pathogen (Isberg, 1991; Cundell and Tomanen, 1995).

Role of Iron in Pathogenesis

Bacteria need iron to grow, reproduce, and make toxins, but iron concentrations in nature are usually low. Acquiring host iron is an important virulence factor. The concentration of free iron in the human body is low because compounds such as ferritin, hemin, lactoferrin, and transferrin strongly bind to most of the available iron. Increased iron levels in individuals have been associated with susceptibility to bacterial infections. Overdose of iron tablets, excess red meat consumption, alcoholism (resulting in accumulation of iron in the liver), and smoking all contribute to iron overload. The ability of macrophages to phagocytize bacteria is compromised because high iron concentrations reduce their adherence and endocytic capabilities. Patients with a high iron overload have been given injections of siderophore desferrioxamine to reduce the levels of iron in circulation.

Pathogenic bacteria have developed several strategies for getting iron from the host: (1) lysis of red blood cells, digestion of hemoglobin, and binding and assimilating the heme, (2) cell surface binding of transferrin, (3) production of siderophores to gather iron from ferrated transferrin, and (4) assimilation of intracellular iron from pools of low-molecular weight host iron-binding compounds.

Pathogenic strains of *Neisseria gonorrhoeae* can obtain iron from human transferrin, whereas commensal strains of this bacterium cannot gather iron from the host environment. A wide variety of bacteria synthesize siderophores to gather iron from transferrin or lactoferrin. Catechols and hydroxamates are two classes of siderophores that form tightly chelated complexes with iron. These then bind to siderophore receptors on the bacterial surfaces and are internalized in the cytoplasm. The siderophore–iron complex is cleaved to free the iron molecule. Some bacteria not only make their own siderophores and surface receptors, but they also have siderophore receptors for other bacteria. Highly virulent strains of *Escherichia coli* produce siderophores called aerobacterin. Pathogens such as *Legionella pneumophila*, *Listeria monocytogenes*, and strains of *Mycobacterium*, *Salmonella*, and *Shigella* obtain iron intracellularly and do not need siderophores (Weinberg, 1997). Lactobacilli prevail over other bacteria in host environments where iron is scarce because they use cobalt or manganese in their metabolism instead of iron.

4.3 BACTERIAL PATHOGENS

The term *parasite-specified endocytosis* describes microbial entry into phagocytic cells such as macrophages (Moulder, 1985). Alternatively, the bacterial entry may be through a *host-specified endocytosis*. Phagocytic cells have many lysosomes with hydrolytic enzymes that can digest microbial pathogens. Once a pathogen is engulfed, it is contained in a vacuole called a *phagosome*. The phagosome fuses with a lysosome to make a phagolysosome, and digestion of the engulfed microbes then occurs by one of the following mechanisms:

1. Oxygen-dependent phagocytosis, which requires increased oxygen consumption by phagocytic cells and conversion of oxygen into toxic

intermediate forms such as the super-oxide anion, hydrogen peroxide (H_2O_2), singlet oxygen, and hydroxyl radicals.

2. Oxygen-independent mechanisms, which involve the activities of lysosomal enzymes such as phospho-lipases, proteases, RNase, and DNase.

3. Nitrogen-dependent mechanisms, which use reactive forms of nitrogen intermediates such as nitric oxide (NO), nitrite (NO_2^-) and nitrate (NO_3^-). Nitric oxide is the most effective killing agent of the three and is produced from arginine when the phagocytes are stimulated by interferons or tumor necrosis factor (Lancaster, 1992).

Lactic acid production, a result of altered metabolism in phagocytes, lowers pH and enhances the activities of many lysosomal enzymes. Examples of intracellular pathogens are given below.

Yersinia Entry into Mammalian Cells

Yersinia pestis, the causal agent of plague, enters the host from the bite of an infected flea. *Yersinia enterocolitica* and *Y. pseudotuberculosis* enter the host's digestive system from contaminated food or water (fig. 4.2b). All yersiniae eventually end up in lymphatic tissue. The invasin protein of *Y. pseudotuberculosis* was originally identified by its giving *E. coli* the ability to enter cultured mammalian cells. Later, it was found that this membrane protein confers invasive properties by binding to β-1 integrin receptors on the host cell membrane. The gene (*inv*) for this protein has been cloned. Another *Yersinia* gene located on the virulence-associated plasmid, YadA, mediates another pathway into the host cell. In animal models mutations for *inv* and YadA have resulted in loss of virulence (Miller et al., 1994).

During the infection process, pathogenic *Yersinia* spp. export proteins known as the *Yersinia outer membrane proteins* (YOPs). The YOPs are encoded by 70–75 kb virulence plasmids (pYV) that are highly conserved among the three pathogenic species of *Yersinia*. *Yersinia* strains that lack pYV can enter host cells but are unable to multiply. *Yersinia* strains with virulent pYV can overcome host inflammatory responses and multiply extra-

cellularly in host tissue (Salyers and Whitt, 1994).

Pathogenicity Islands and How Salmonella Became a Pathogen

Salmonella is usually acquired from contaminated food or water. Because there is about a 90% DNA–DNA homology between *Salmonella* sp. and *Escherichia coli*, some microbiologists want to reclassify them within a single genus. Another proposal is to place all pathogenic *Salmonella* in one species, *Salmonella enterica*. The traditional approach is to include the following three species in the genus *Salmonella*: (1) *Salmonella typhi*, which causes typhoid fever in humans. Bacteria enter the bloodstream following their uptake in M cells of the Peyer's patches and are spread throughout the body. Death results from septic shock. (2) *Salmonella enterditis*, which causes diarrhea in humans and animals and has more than 2000 serotypes, including *Salmonella typhimurium*, whose infection in humans is characterized by diarrhea, fever, and abdominal pain. The disease is usually self-limiting. Young children and immunocompromised adults are most susceptible to it. *Salmonella typhimurium* infections are usually acquired from poultry products; (3) *Salmonella cholerasuis*, which causes systemic infections in humans, but is primarily a pathogen of pigs.

Recent studies have shown that at least 60 genes are involved in virulence of *Salmonella*, perhaps because of the complexity of its life cycle. The bacteria must survive the acidic pH of the stomach before they attach to intestinal epithelial cells, thereby bringing about their own phagocytosis and replication in macrophages (Groisman and Ochman, 1997).

Laboratory studies have shown that *Salmonella* strains adhere to cultured mammalian cells and cause cytoskeletal changes. The cell's outer membrane is modified to produce pseudopodia which engulf the bacteria and internalize them inside endocytic vesicles (fig. 4.1b). Bacteria multiply inside the vesicles which may fuse together. How *Salmonella* triggers cytoskeletal changes of the host cell is not known. One explanation is that a host cell signal-transduction pathway increases intracellular calcium, and this activates actin depolymerization enzymes (Pace et al., 1993).

Many *Salmonella* strains, with the excep-

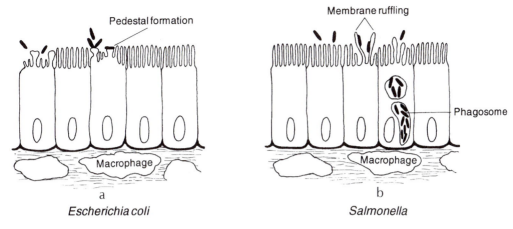

Fig. 4.1 *Escherichia coli* and *Salmonella* cause cytoskeletal changes in the host cell membranes. (a) Infection with enteropathogenic strain of *Escherichia coli* induces the host cell membrane to form dome-shaped pedestals on top of which bacteria attach. (b) *Salmonella* strains cause the host cell membrane to form pseudopodia before the bacteria are engulfed. Adapted from Salyers and Whitt (1994).

tion of *S. typhi*, contain a large virulence plasmid. Loss of this plasmid significantly reduces virulence in mice. Ability to survive inside the phagocytes depends on the expression of many genes. These include genes that help bacteria withstand exposure to reactive forms of oxygen, low pH, and a number of host defenses. Only a few of these genes have been identified, such as the *PhoP/PhoQ* operon, which regulates *prg* and *pag* genes.

Most of *Salmonella*'s virulence genes occur in large clusters in several regions of the bacterial chromosome called *pathogenicity islands*. The best known pathogenicity island of *Salmonella* is SPI-1, a cluster of 25 genes. Experimental evidence has shown that this island was acquired through horizontal gene transfer from *Shigella* (Baumler, 1997). The size, order, and orientation of the *inv* and *spa* genes in the SPI-1 island are similar to the invasion genes on *Shigella*'s virulence plasmid. Homologs of the *inv* and *spa* genes occur in a broad range of bacterial pathogens, such as *Erwinia*, *Pseudomonas*, *Xanthomonas*, and *Yersinia*. Mutations in the SPI-1 gene cluster prevent *Salmonella* from invading epithelial cells in vitro (Groisman and Ochman, 1997).

Shigella: Subversion of Host Cellular Cytoskeleton

Shigella spp. and certain strains of *Escherichia coli* cause dysentery in humans and produce blood and mucus in diarrheal stools.

The disease is self-limiting in healthy adults, but among infants and children in developing countries it often causes death.

Shigella spp. adhere to mammalian cells in a culture medium and disrupt the host's cytoskeleton (Theriot, 1995). The bacteria become engulfed within vacuoles. Once inside the host cell the bacteria escape into the cytoplasm, where they multiply. Bacterial surface proteins, the invasion plasmid antigens (IpaB-D), are essential for adherence and invasion. The release of Ipa proteins from the bacteria is controlled by another set of proteins called the membrane excretion of Ipas (Mxi). Bacteria move within the cell cytoplasm and from one cell to another by polymerizing the host's actin filaments. Shigellae are unable to invade the mature cultured and normal intestinal columnar epithelium of a host animal. The free surfaces of these host cells lack receptors called *integrins*. However, integrins are present on the lateral surfaces of mucosal cells and free surfaces of M cells of the Peyer's patches. The M cells are phagocytic and are positioned along the mucosal lining to sample antigens from the lumen of the intestine. Phagocytic uptake of *Shigella* (bacterial) protein by M cells are then presented to lymphatic tissue of the Peyer's patch. *Shigella* invades other body tissues by escaping the M cell phagocytosis (Jepson and Ann Clark, 1998) (fig. 4.2a).

Shigella invasion of macrophages in tissue culture causes *apoptosis* or programmed cell

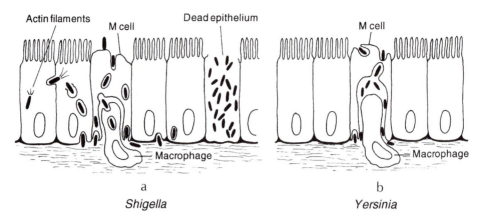

Fig. 4.2 *Shigella* and *Yersinia* bacteria enter the host through M cells of Peyer's patches, which are common throughout the intestinal mucosa. (a) *Shigella* is unable to invade the normal columnar epithelial cells, and their phagocytosis through M cells, followed by escape from the infected macrophages, allow the bacteria to spread to epithelial cells. The bacteria spread through epithelial cells by propelling themselves by using the host cell's cytoskeletal proteins. (b) *Yersinia* invasion of M cells depends on the production of *Yersinia* outer proteins. The bacteria then escape and multiply extracellularly in host tissue. Adapted from Salyers and Whitt (1994).

death. Apoptosis is characterized by DNA breakdown and condensation of chromatin at the nuclear boundary while the cell organelles remain structurally sound. On the other hand, in accidental cell death (cell necrosis), the chromatin becomes granular and the cytoplasmic organelles disintegrate. The ability to trigger apoptosis in macrophages explains how *Shigella* sp. survive a phagocytic attack of macrophages (Sansonetti, 1992).

Shigella dysenteriae produces an A-B exotoxin called shiga toxin. This toxin inactivates ribosomes and stops protein synthesis in the infected host cells. *Shigella* strains possess large plasmids that have virulence genes for invasion and intracellular movement (Zychlinsky et al., 1994).

Listeria monocytogenes

Listeria monocytogenes is a foodborne pathogen that attacks the cells of the intestinal lining and produces mild influenzalike symptoms in healthy adults. The disease is commonly associated with unpasteurized milk or foods such as coleslaw and shrimp. The bacterium can pass through the placental barrier in pregnant women and can cause stillbirth, premature birth, and systemic infection in newborn infants. Up to 5–10% of the human population may be asymptomatic carriers of this pathogen.

Listeria monocytogenes enters a variety of mammalian cells in situ and in culture. A surface protein, *internalin*, mediates entry into intestinal epithelial cells in culture. The host cell surface receptor of internalin is a transmembrane adhesion protein called *E-cadherin*, which is bound to actin filaments of the cellular cytoskeleton through a complex with catenines (Ireton and Cossart, 1997). A mutation in the gene for internalin, *InlA*, prevents the pathogen from invading host cells. Internalin induces phagocytosis of *L. monocytogenes* and creates an endocytic membrane-bound vacuole containing bacteria. The bacteria escape from the vacuole by producing a hemolysin, listeriolysin O (LLO), which is a sulfhydral-activated, pore-forming cytotoxin. The gene for LLO protein is a major virulence factor in *L. monocytogenes*. In addition, two other hemolysins called phospholipases disrupt host cell membranes and contribute to the spread of bacteria to other cells (Farber and Peterkin, 1991; Tilney et al., 1992).

The movement of *Listeria monocytogenes* in the host cell cytoplasm is the result of a bacterial protein called ActA, which causes the polymerization of actin and the formation of long tails. ActA is an essential determinant of pathogenicity in *Listeria* because bacteria use the actin tails to pass from one host cell to another (G.A. Smith and Portnoy, 1997; Dramsi and Cossart, 1998). *Listeria* virulence genes are clustered together on the bacterial chromosome.

Mycobacterium tuberculosis

More than 2 billion people are infected with *Mycobacterium tuberculosis*, and about 8 million new cases occur each year. Tuberculosis is the most common infectious disease of humankind, accounting for more than 26% of the avoidable deaths in the developing world. *Mycobacterium tuberculosis* grows in alveolar macrophages of the lungs. Macrophages that ingest *M. tuberculosis* use the bacterial antigens to evoke T-helper cell response (CD+4) and cytotoxic T-cell response (CD+8). The CD+4 cells stimulate antibody response in B-cells and release interferon, which activates the macrophages. The CD+8 cells kill macrophages that contain bacteria. The cell wall of *Mycobacterium* consists of a thick peptidoglycan layer containing arabinogalactan, lipoarabinomannan (LAM), and mycolic acids that protect it from host cell phagocytosis. LAM helps reduce the receptor-mediated phagocytosis of mycobacteria by suppressing T-cell proliferation. The mylonic acid component of mycobacterial cell walls is toxic to the host cells and stimulates the host's inflammatory response. The ability of *M. tuberculosis* to survive in macrophages is a major virulence factor. Bacteria survive in phagocytic vacuoles by preventing the acidification of the phagosome (MacDonough et al., 1993; Salyers and Whitt, 1994). Examining the genetics of *M. tuberculosis* is difficult because of its slow growth under laboratory conditions.

Legionella pneumophila:
Phagocytosis with a Twist

Legionnaire's disease is a form of pneumonia caused by *Legionella pneumophila* and is usually spread by inhaling aerosols from contaminated water that undergoes rapid cooling, as found in air-conditioning cooling towers. *L. pneumophila* occupies a unique intracellular niche inside a macrophage. The macrophage takes up the bacterial pathogen by means of a novel mechanism called "coiling phagocytosis." A long, thin pseudopod from the macrophage coils around the bacterium and engulfs it. Once internalized, the host cell's mitochondria gather around the phagosome. The phagosome does not acidify and thus avoids fusion with lysosomes. An outer membrane protein of *L. pneumophila*, called MIP (macrophage invasion protein) aids in the bacterial invasion. Bacterial genes designated as DOT (defect in organelle trafficking) encode proteins involved in intracellular survival. Damage to the lungs is caused not only by accumulation of macrophages but also by the production of several proteases by *L. pneumophila*. A zinc metalloprotease, in particular, is very destructive to the host tissue.

Legionella spp. are common in soil and freshwater and parasitize *Acanthamoeba* and *Naeglaria*, which may serve as reservoir hosts of the bacterial pathogen (Barker et al., 1992).

Diphtheria Toxin and Temperate Bacteriophages

Diphtheria is primarily a disease of children and is caused by *Corynebacterium diphtheriae*. It can be fatal if not treated, but an effective vaccine is available. Diphtheria toxin is an example of how a single molecule can produce disease symptoms, and its study helped give rise to the concept of *virulence factor*, which is defined as any product or strategy of a bacterial pathogen that enhances infection of a host. The toxin suppresses the protein synthesis in the host cell. The use of diphtheria toxin to kill tumor cells or HIV-infected cells has attracted attention from the scientific community (Salyers and Whitt, 1994). The toxin gene is carried by a temperate bacteriophage, and only those strains of *C. diphtheriae* that have this phage can cause the disease. The toxin gene is regulated by an iron-dependent repressor protein, DtxR. It is suggested that the role of diphtheria toxin is to provide iron for the bacteria (Roth et al., 1995).

Diphtheria toxin selectively binds to the host cell receptor protein, the heparin-binding epidermal growth factor. Cardiac muscles and neurons have many more toxin receptors, and therefore heart failure and brain damage are common symptoms of this disease (Salyers and Whitt, 1994).

Vibrio cholerae: Cholera as a Paradigm

Vibrio cholerae is a motile, Gram-negative, curved rod with a single flagellum. The bacterium grows in freshwater and seawater and can colonize the human intestine, where it produces an exotoxin. The toxin disrupts ion transport across the epithelial lining and causes diarrhea and water loss. Cholera is most severe among children. Individuals who recover from cholera may become asymptomatic carriers and spread the bacteria during a subsequent epidemic.

The principal virulence factors in *V. cholerae* are pili and cholera toxins. Filamentous bundles of pili on the bacterial surface are used to colonize the intestinal mucosa. The pili are called toxin coregulated pili (TCP) because the regulation of genes for pili formation is similar to the way cholera toxin genes are transcribed. The TCP pili are important cholera adhesins. Mutants deficient in TCP pili are avirulent. *Vibrio cholerae* also makes a hemoagglutinin, which may serve as a cholera adhesin.

Cholera toxin, an A-B type of exotoxin, is the main virulence factor responsible for the disease. The toxin has five B subunits and one A subunit, which in turn is made up of two parts, A1 and A2. Genes *ctxA* and *ctxB* are organized into an operon and are responsible for transcription of the toxin subunits. The excreted toxin binds to GM1 ganglioside on host mucosal cells. An unknown mechanism translocates the A1 subunit into the host cell cytoplasm, where it triggers adenylate cyclase activity, resulting in high levels of cAMP. Sodium ion transport is blocked, creating an ion imbalance that leads to diarrhea (fig. 4.3). Mutants for cholera toxin genes fail to produce the disease. *Vibrio cholerae* also produces Zot toxin, which disrupts the tight junctions binding the mucosal cells together, and Ace toxin, which produces a milder form of diarrhea. The genes for Zot and Ace toxins are located next to each other on the bacterial genome. Together they increase virulence during infection (Atlas, 1997).

Cholera and El Niño

Cholera is an ancient disease with historical records in Greek and Sanskrit tracing back 2000 years. The disease has existed for centuries on the Indian subcontinent. Medical historians have noted that since 1817, seven major pandemics have killed millions of humans worldwide. Six pandemics believed to have started in the coastal waters of the Bay of Bengal were caused by the classic *Vibrio cholerae* 01 serotype. The seventh and most recent outbreak of cholera in the 1990s was caused by *V. cholerae* 0139, and the first outbreak of this new serotype occurred in the port city of Chancay in Peru. Many *Vibrio* sp., including *V. cholerae*, occur in chitinaceous zooplankton (copepods) and shellfish. The as-

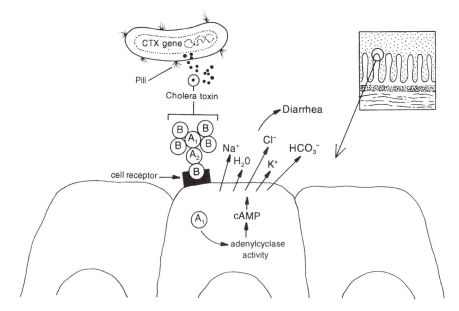

Fig. 4.3   Cholera toxin, an A-B exotoxin which is encoded by the *ctx* gene, binds to the host cell receptors. A portion of the toxin triggers adenylate cyclase activity which in turn results in a high level of AMP. Sodium channels are blocked, causing a loss of electrolytes and the disease symptoms. Redrawn from Atlas (1997).

sociation of *V. cholerae* and copepods is key to understanding the global nature of cholera epidemics. It is now believed that cholera is carried away from its endemic area by means of ocean currents that transport copepods and their bacterial symbionts to new coastal communities.

The Peruvian outbreak of cholera has been correlated with plankton blooms triggered by the climatic phenomenon known as El Niño, which causes rainfall that washes nutrients from the land and brings about warmer sea surface temperatures. A single copepod may contain 10,000 bacterial cells, a dose sufficient to infect humans. The probability of ingesting several copepods with drinking water increases with planktonic blooms (Colwell, 1997).

The birth of Vibrio cholerae 0139

In 1993 a cholera outbreak by *Vibrio cholerae* serogroup 0139 in South Asia was a significant turning point in the history of the disease. Evidence shows that this strain arose as a result of genetic recombination and horizontal gene transfer. With the exception of *V. cholerae* 0139, all known cholera epidemics have been caused by the classic serotype 01 of *V. cholerae*.

The cause of the sixth cholera epidemic, which originated in Indonesia in 1991, was designated as *Vibrio cholerae* 01 E1 Tor. *Vibrio cholerae* non-01 serotypes do not cause epidemics of diarrhea. An outbreak of a choleralike disease occurred in Madras, India, in October 1992 and was identified as *Vibrio cholerae* non-01. As the epidemic spread along the coastal communities on the Bay of Bengal, a new serotype, *Vibrio cholerae* 0139, was identified. The new serotype was identical to *V. cholerae* 01 E1 Tor except that it had a capsule. *Vibrio cholerae* 0139 not only had the virulence factors of *V. cholerae* 01 but could better survive external environmental conditions. The 1991 *V. cholerae* 01 E1 Tor strain has virtually disappeared from many areas. The *V. cholerae* 0139 lipopolysaccharide (LPS) virulence factor has an additional O-antigen. *Somatic antigens* or *O antigens* are those associated with the surface of LPS of the cell wall. Genetic analysis of *V. cholerae* 0139 revealed a deletion of about 22 kb DNA from chromosome 1 and an insertion of a new 35-kb DNA fragment that encodes the 0139 LPS and capsule proteins. Molecular evidence suggests that 01 E1 Tor is transformed into 0139 by acquiring new DNA through a horizontal gene transfer (Mooi and Bik, 1997).

The modern study of cholera has engaged scientists from such disparate disciplines as oceanography, marine biology, microbiology, molecular biology, immunology, medicine, epidemiology, and space satellite imagery to produce a new conceptual understanding of this ancient disease.

4.4 *ESCHERICHIA COLI*: A VERSATILE HUMAN PATHOGEN

Escherichia coli can cause many different types of diseases in humans. Most strains of *E. coli* are avirulent because they lack virulence genes. *E. coli* is classified into the following five "virotypes" on the basis of how the bacterium attaches to host cells, how host cells are affected following attachment, and the kind of toxins produced:

1. Enterotoxigenic *E. coli* (ETEC). Strains of ETEC resemble *Vibrio cholerae* in that they adhere to the mucosa of the small intestine and produce diarrhea in the host by secreting toxins. They are unable to invade the host tissue. Infants and children are most susceptible to this disease, the adult version of which is called traveler's diarrhea. A choleralike toxin, called heat-labile toxin, and a diarrheal toxin, heat-stable toxin, are involved. Genes for these toxins are carried on plasmids.
2. Enteroaggregative *E. coli* (EAggEC). The bacteria adhere to the mucosal cells in clumps and produce a heat-stable–like toxin and a hemolysinlike toxin. The EAggEC strains cause diarrhea in children.
3. Enteropathogenic *E. coli* (EPEC). The EPEC strains produce dramatic changes in the host mucosal cells. The epithelial cells to which the bacteria bind lack the normal microvilli and instead have a cup-shaped pedestal under each bacterium. EPEC strains are the major cause of an often fatal diarrhea in children and infants. The EPEC strain invades cultured host cells in three stages. First, bacteria bind to the host cell surface via bundle-forming pili (BFP). (The BFP gene is carried on a plasmid.) Second, a signal

transduction pathway is triggered that activates host cell tryosine kinases and increases Ca^{2+} levels. Third, the host actin cytoskeleton is rearranged, and a pedestal where the bacterium become firmly attached is formed (Donnenberg et al., 1997) (fig. 4.1a).

4. Enterohemorrhagic *E. coli* (EHEC). The EHEC strain such as O157:H7 causes dysentery and hemolytic uremic syndrome, which can result in death from acute kidney failure. EHEC strains produce a toxin similar to shiga toxin, the gene for which is located on a temperate phage.

5. Enteroinvasive *E. coli* (EIEC). The EIEC strains cause a disease identical to those caused by *Shigella* sp. but do not produce shiga toxin. The genes for virulence are located on a large plasmid and are regulated in a manner similar to those of *Shigella* species.

4.5 *PSEUDOMONAS AERUGINOSA*: QUORUM SENSING AND VIRULENCE GENE REGULATION

Pseudomonas aeruginosa is a metabolically versatile Gram-negative bacterium found in soil, fresh water, plants, and animals. In humans, *P. aeruginosa* is an opportunistic pathogen which causes infections in burn patients or immunocompromised patients and those with cystic fibrosis. The bacteria produce pili and nonpilus adhesins. Pili adhesins bind to ganglioside receptors of the epithelium, whereas the nonpilus adhesins bind to mucin associated with the epithelial cells. *P. aeruginosa* also produces a thick coat of exopolysaccharides or alginate, which covers the bacterial outer surfaces. In strains associated with cystic fibrosis, alginate is a mechanism for adherence and also prevents phagocytosis of the bacterium. Alginate synthesis is energy consuming, and its gene expression is tightly regulated (Deretic et al., 1994).

P. aeruginosa secretes several cellular products into the surrounding environment which supply nutrients and create conditions suitable for bacterial survival. Three extracellular proteases, LasB elastase, LasA elastase, and alkaline protease, have been identified, and the genes responsible for their expression have been cloned and characterized (Passador et al., 1993). These proteases cause lung damage in cystic fibrosis patients. Exotoxin A,

transcribed by the *toxA* gene, is the most toxic substance produced by *P. aeruginosa*. Recent experiments with *P. aeruginosa* show that gene *regA* regulates *toxA* expression. Another gene, *LasR*, though not required for toxA, contributes significantly to enhance toxin production. The amino acid sequence of LasR protein is similar to another bacterial protein, LuxR, which controls the bioluminescence genes of the marine microbe *Vibrio fisheri*. The LuxR protein binds DNA and controls a signal molecule known as the *Vibrio* autoinducer. A *Pseudomonas* autoinducer has been isolated, purified, and identified as *N*-3-oxododecanoyl homoserine lactone, which is similar to other autoinducers (Gray, 1997; Pesci and Iglewski, 1997).

Quorum sensing (a cell-density–dependent phenomenon) is mediated through autoinducer molecules. Many organisms make autoinducer molecules to monitor their population size, which in turn controls the expression of density-dependent genes. Classic examples of cell-to-cell cooperation are fruiting body formation in the slime mold *Dictyostelium discoideum* and sporulation in *Bacillus*. Quorum sensing also ensures the proper timing in producing virulence factors that overwhelm the host defense response. Quorum sensing mechanisms have been identified in other bacteria (Strauss, 1999). In the plant pathogen *Agrobacterium tumefaciens*, the Ti plasmid directs the production of opines, which serve as food for the bacterium. *Erwinia carotovora*, a plant pathogen that causes bacterial soft-rot in vegetables, possesses a Lux-homologous system that regulates the synthesis of carbapenum antibiotics and cell wall degrading enzymes. The autoinducer of *E. carotovora* is similar in structure to that of *Vibrio* (Passador et al., 1993; Pesci and Iglewski, 1997) (fig. 4.4).

4.6 *HELICOBACTER PYLORI*: MOLECULAR MIMICRY BETWEEN PATHOGEN AND HOST

Until 1991, medical microbiologists had been teaching that the stomach environment was too harsh to support microbes. This changed following the discovery of *Helicobacter pylori*, which opened an exciting new chapter in our understanding of bacterial life (Rabeneck and Ranshoff, 1991). *Helicobacter pylori* not only grows and thrives in the stomach, but it may also be responsible for 90% of gastric and

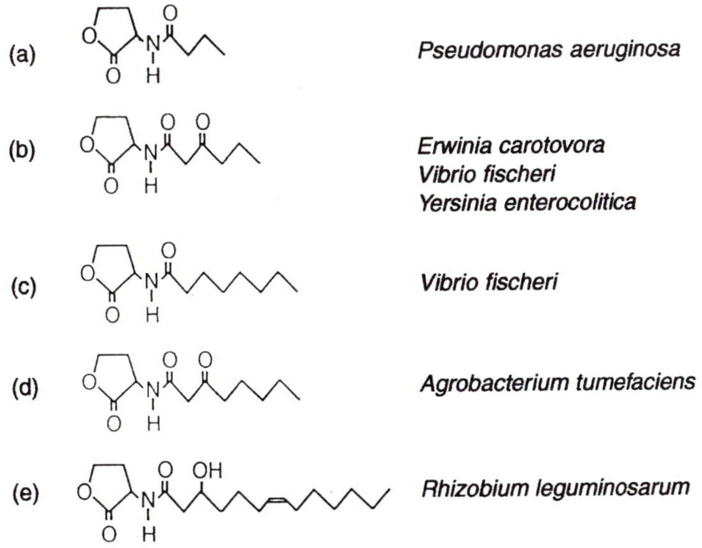

(a) *Pseudomonas aeruginosa*

(b) *Erwinia carotovora*
Vibrio fischeri
Yersinia enterocolitica

(c) *Vibrio fischeri*

(d) *Agrobacterium tumefaciens*

(e) *Rhizobium leguminosarum*

Fig. 4.4   Structural comparison of N-acylhomoserine lactone autoinducer molecules in five species of bacterial symbionts. Autoinducers play an important role in quorum sensing, a cell-density–dependent phenomenon. Adapted from Gray (1997).

duodenal ulcers, though most infections are asymptomatic.

Helicobacter pylori produces large amounts of urease, which converts urea into ammonia and carbon dioxide. Bacteria survive the stomach environment by being surrounded with ammonia molecules, which neutralize stomach acids. Many bacteria invade the mucin layer, but only *H. pylori* has the adhesins that bind to mucosal cells (fig. 4.5). These adhesins include Lewis blood group O antigens, phosophatidyl-ethanolamine, sialic acid, and laminin. *H. pylori* also produces cytotoxins that cause the symptoms of peptic ulcers. The cytotoxins produce vacuoles within the mucosal cells lining the stomach and small intestine. When the injected mucosal cells die, gastric acids and digestive enzymes cause the formation of ulcers. *H. pylori* isolated from patients with peptic ulcer disease and with gastric cancer contained a 38-kb fragment of DNA that was not present in asymptomatic carriers. A cytotoxin-associated gene, the *CagA* gene, occurs in this stretch of DNA called the pathogenicity island. Another virulent gene, the vacuolating toxin gene, the *VacA*, is located 300 kb away from *CagA*. The genes are coexpressed in the most severe forms of gastroduodenal diseases. Strains lacking the *CagA* gene also lack the vacuolating activity of *VacA*. Mutations of

CagA show that virulence in *H. pylori* has evolved through the inheritance of one or more DNA insertions (Covacci et al., 1997).

Molecular mimicry is observed between pathogen and the host when certain protein sequences are compared. The LPS of *H. pylori* is unusual in that it expresses Lewis x and y blood group antigens. Such antigens are also expressed on the host mucosal cell. Sequence similarity exits between the vacuolating toxin of *H. pylori* and the host gastric H^+K^+-ATPase. Gastric H^+K^+-ATPase is the principal target of the autoimmune response in pernicious anemia. Whether *H. pylori* plays a role in triggering antibodies against host gastric proteins remains to be determined (Appelmelk et al., 1997).

The DNA sequence of *Helicobacter pylori* strain 26695 has also provided new insights into its pathogenesis (Zhongming and Taylor, 1999). The *H. pylori* genome contains the two best-known virulence determinants, the vacuolating cytotoxin allele and the 38-kb *cagA* pathogenicity island. Though *H. pylori* is a Gram-negative pathogen, many of its proteins correspond to proteins in eukaryotes, archaea, and Gram-positive bacteria. This suggests that horizontal gene transfer from disparate phylogenetic groups into *H. pylori* lineage occurred during evolution (Berg et al., 1997).

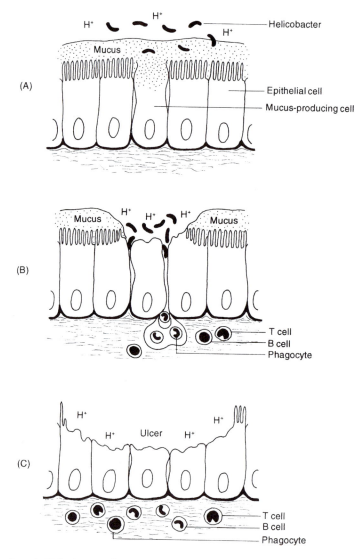

Fig. 4.5   Steps in *Helicobacter pylori* invasions of mucus-producing cells of stomach epithelium and the subsequent development of ulcers. Redrawn from Atlas (1997).

4.7 SUMMARY

Much progress has been made toward understanding the molecular basis of infectious diseases. Koch's postulates, the standard test for determining if a bacterium causes disease, have limitations. A new set of postulates, the molecular Koch's postulates, has been proposed and focuses on the role of genes and gene products in disease. Adherence to a host cell is an important step in bacterial pathogenesis. This is accomplished by means of pili and surface proteins such as adhesions and invasions. Attachment of bacteria to eukaryotic integrins involves three strategies: lectin binding, masking, and mimicry. Bacteria use siderophores to obtain the iron they need from a host. Bacterial pathogens produce exotoxins that disrupt host functions and cause cell death. Digestion of microbes in a phagosome occurs by several methods including nitrogen- and oxygen-dependent phagocytosis.

Intracellular pathogens include *Yersinia pestis* (plague), which uses different ways to enter into host cells. About 60 genes are involved in virulence of *Salmonella*, and many of them occur in clusters called pathogenicity islands. When *Shigella* invades macrophages, it causes apoptosis. Up to 5–10% of the human population may be asymptomatic carriers of *Listeria monocytogenes*. *Mycobacterium tuberculosis* causes one of the most common of human diseases. *Legionella pneumophila* occupies a unique niche inside a macrophage, which engulfs the pathogen by means of a novel mechanism called coiling phagocytosis. The gene which codes for diphtheria toxin is carried by temperate phages, and only strains of *Corynebacteria diptheria* that carry the phage can cause the disease.

The main virulence factors in *Vibrio cholerae* are pili and cholera toxins. Outbreaks of cholera in Peru have been linked to planktonic blooms (copepods) caused by El Niño. New strains of *V. cholerae* arise from genetic recombination and horizontal gene transfer.

Escherichia coli is now classified into five virotypes. *Pseudomonas aeruginosa* is an opportunistic pathogen in humans. *Helicobacter pylori* responsible for most human gastric and duodenal ulcers shows molecular mimicry between pathogen and the host protein sequences, suggesting horizontal gene transfer.

5 BACTERIAL ASSOCIATIONS OF PLANTS

Relationships between bacteria and plants are widespread and include mutualistic and parasitic associations (Preston et al., 1998). Bacteria constitute more than 90% of the total biomass of agricultural soils, followed by fungi and protozoa. Bacteria obtain energy and nutrients through decomposition of dead plant material and from exudates of living plants, and they may significantly influence interactions that take place between plants, including competition (Chanway et al., 1991). Most plant bacteria are facultative saprobes and contain plasmids in addition to large, circular chromosomes.

5.1 NITROGEN-FIXING SYMBIOSES

Plant growth is limited by how much nitrogen is in the soil. Nitrogen is an important element for living organisms and is used to make amino acids, proteins, and nucleic acids. Nitrogen is common in our atmosphere, but plants cannot use it in its elemental form and must absorb it from the soil principally in the form of nitrates. The supply of nitrogenous compounds in the soil is being continually replenished by bacteria that fix atmospheric nitrogen into ammonia and from decomposing organic matter. Other soil bacteria then convert ammonia into nitrites and nitrates. Nitrates absorbed by plant roots are converted back to ammonia, which is used to form the complex molecules needed by plants. Reactions involving nitrogen are part of a nitrogen cycle, which is an important part of every ecosystem.

Nitrogen-fixing bacteria commonly form mutualistic relationships with plants. The bacteria contain the enzyme complex *nitrogenase*, which can catalyze complex reactions, involving N_2, hydrogen ions, and free electrons, leading to the formation of ammonia. For these reactions to occur, the bacteria need energy-rich compounds, such as ATP, and also electrons, both of which are obtained from respiring sugars supplied by the plant. Bacteria need several factors to fix nitrogen: (1) access to atmospheric nitrogen; (2) a nitrogenase enzyme complex; (3) large amounts of ATP (16 moles ATP per mole N_2 reduced); (4) an anaerobic environment; (5) a supply of iron, magnesium, and molybdenum, which are required for nitrogenase activity; (6) temperatures below 30°C; (7) a means of regulating the amounts of ATP available for the system and for controlling fluctuating ammonia production.

Because oxygen readily denatures nitrogenase, nitrogen fixation must take place under low concentrations of oxygen. Symbiotic bacteria fix nitrogen in specialized structures such as nodules, heterocysts, and vesicles in which low levels of oxygen can be maintained. Another important constraint on bacterial nitrogen fixation is the high energy cost of the process. A symbiont's capacity for nitrogen fixation is controlled largely by the host because it regulates how much energy, in the form of sugar, the symbiont receives. In some cases as much as 30% of the photosynthetic products of a host plant may be used to support nitrogen fixation and assimilation.

As the bacterial symbionts fix nitrogen, the host inhibits their ammonia-assimilating enzymes, and they excrete much of the nitrogen they fix as ammonia. Host cells contain enzymes to convert the excreted ammonia into useful compounds. Such adaptations exist in the *Rhizobium*–legume and *Anabaena*–*Azolla* symbioses.

Rhizobium–Legume Symbiosis

All rhizobia (*Azorhizobium*, *Bradyrhizobium*, *Rhizobium*, *Sinorhizobium*) belong to the α group of the Proteobacteria(purple bacteria). *Rhizobium* is closely related to *Agrobacterium*, but not to *Bradyrhizobium* (Young and Johnston, 1989).

Rhizobium is a genus of rod-shaped, motile bacteria that live in the soil, usually in areas where legumes grow. These bacteria form symbiotic associations with the roots of legumes such as alfalfa, clover, peas, and soybeans. High levels of gene transfer might obscure the broad coevolution of bacteria with plants (Doyle, 1998). The bacteria stimulate the roots to form unique structures called *nodules*, within which nitrogen fixation occurs. In soil that is deficient in nitrogen, legumes that form nodules grow better than other plants. Root nodules can supply a plant

with all the nitrogen it needs. Plants without nodules must obtain their nitrogen from the soil, where it is often in short supply (Van Rhijn and Vanderleyden, 1995).

Signal molecules

Two main groups of *signal molecules* have been discovered: one for short-range interaction (e.g., cell–cell attachment), and the other for long-range signals (e.g., diffusible substances between rhizobia and host cells). Flavonoid molecules released by legume roots induce the transcription of rhizobial nodulation (*nod*) genes that code for enzymes needed for the synthesis of nodulation factors. Nod factors are lipo-oligosaccharide molecules that have a chitinlike sugar backbone and an N-linked long-chain fatty acid. These factors are released by rhizobia and act as long-range signals that cause root-hair deformation and cortical cell division in suitable legumes (Hirsch, 1992; Fisher and Long, 1992; Franssen et al., 1992; Verma, 1992; Denarie and Cullimore, 1993; Kannenberg and Brewin, 1994; Spaink, 1995) (fig. 5.1b).

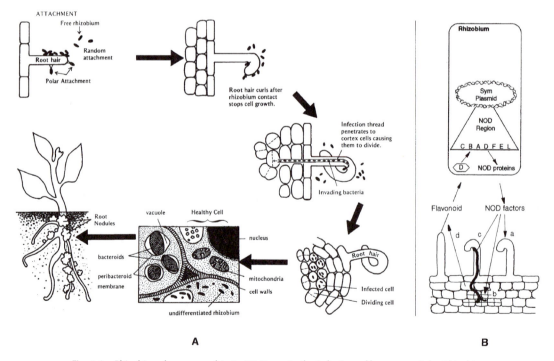

A   **B**

Fig. 5.1   *Rhizobium*–legume symbiosis. (A) Stages in the infection of legume roots by *Rhizobium* species. (B) Interaction between *Rhizobium* and legume roots. Host plant secretes flavonoids, which along with Nod D proteins regulate the bacterial nodulation genes (*nod*). The expression of nod factors is essential for root-hair curling, cortical cell multiplication, and other steps leading to nodule formation. (Redrawn from Van Rhijn and Vanderleyden (1995).

Infection process

Infection of a legume begins when bacteria attach, by one end, to newly formed root hairs of plant seedlings. Complementary chemical bonding may be involved. Proteins called lectins on the surface of root hairs bind to polysaccharides on the bacterial cell wall. In some instances, the binding is specific; that is, only certain lectins will bind to certain polysaccharides, which means that some strains of rhizobia infect only certain types of legumes. For example, *Bradyrhizobium japonicum* commonly infects soybeans; *Rhizobium trifolii* infects only white clover. Some strains have a broad host range. In some cases plant lectins may enhance nodulation rather than determine specificity (Kannenberg and Brewin, 1994). Recent studies, however, have shown that transferring a lectin gene from one plant to another can extend host range (Diaz et al., 1989; Van Rhijn et al., 1998).

When bacteria attach to a root hair, the hair deforms in response, some forming 360° curls called "shepherd's crooks." The exact mechanism of root hair curling is unknown. The bacteria then penetrate the cell wall and come in contact with the plasma membrane. Cell wall synthesis by the root hair is redirected toward the site of infection. As a result of this redirected cell wall synthesis, a tubular infection thread forms and grows inwardly toward the root cortex. The nucleus of the root hair doubles in size and directs the growth of the infection thread. The infection thread contains rhizobial cells, which are surrounded by a slimy substance (polysaccharide) in which they multiply. The infection thread is surrounded by cell wall except at its tip, where there is only naked plasma membrane. Specific antibodies and enzyme–gold-labeled probes have been used to observe the development of infection threads (Rae et al., 1992).

Bacteroids

As the infection thread grows through the root cells, cortical cells are activated to divide some distance away from the thread, possibly because of Nod factors released by the bacteria. These dividing cortical cells form nodules whose cells become colonized by bacteria delivered by the infection threads. After a thread penetrates a cortical cell, the bacteria become surrounded by a *peribacteroid membrane* (PBM), which originates from the host plasma membrane. They occur either singly or divide to form groups. Bacteria that are released from the infection thread are called *bacteroids*. Interactions between the plant and the bacteroids are regulated by transporters and channels on the PBM and the bacterial membrane (Udvardi and Day, 1997).

The cells of many strains of *Rhizobium* undergo radical changes in morphology and physiology when they become bacteroids. Some bacteroids are 40 times larger than the small rods they develop from, and their numbers, up to several thousand, can fill a host cell. Each legume host determines the size and shape of its enclosed bacteroids and also the number of bacteroids contained within a peribacteroid membrane. One strain of *Rhizobium* may form different types of bacteroids in different species of legumes. Bacteroids are nonmotile and assume a wide variety of shapes. They have thin walls that allow for easy passage of nutrients from the plant to the bacteroids and of nitrogen from the bacteroids to the plant. The transformation of rod-shaped rhizobia into bacteroids inside the legume cells initiates the process that leads to nitrogen fixation. Bacteroids behave as nitrogen-fixing organelles within the host cells. They cannot use the nitrogen they fix, but rather depend on the plant for their nitrogen and carbon compounds.

Role of oxygen-binding proteins

Legume root nodules contain large amounts of a proteinaceous red pigment called *leghemoglobin*. Leghemoglobin is located in the cytoplasm of the plant cell and forms only after the symbiotic association has taken place (O'Brian, 1996). The pigment binds oxygen to maintain a low concentration of free oxygen in the nodule. Nitrogenase is inactivated at high oxygen concentrations. Leghemoglobin is similar in structure and function to hemoglobin found in the red blood cells of mammals.

Nodules

Legume root nodules are generally spherical or club shaped, but lupine nodules are collar shaped. Nodules develop in a region behind the root tips in the emerging root hair zone. Nodules are outgrowths of plant tissue, and they consist of uninfected tissue as well as tissue that contains bacteroids (Nap and Bisseling, 1990; Hirsch, 1992; Brewin, 1993; Long, 1996; Hirsch and LaRue, 1997). The vascular system of the plant extends into the nodules and transports nutrients in and out of the nodule.

Some strains of *Rhizobium* that nodulate legumes can also form nodules with a non-leguminous plant, *Parasponia*, a member of the elm family. Nevertheless, whatever unique trait legumes have that enables them to fix nitrogen in association with rhizobia is so complex that it has rarely evolved in other plants. An example of such a unique trait might be isoflavonoid synthesis, which is rare in other plant families (Young and Johnson, 1989). As indicated earlier, the fixation of atmospheric nitrogen uses significant amounts of energy-rich compounds, which are supplied by the host plant. Such an energy-dependent process may explain why most plants do not form symbiotic associations with nitrogen-fixing bacteria.

Nodules have a limited life span. Only those that function at maximum capacity are maintained on the plant. When nodules decay, bacteria that have remained in the infection threads and the bacteroids present in the nodules are released into the soil. These bacteria most likely can reinfect other root hairs.

Leaf and stem nodules. Leaf nodules have been observed in more than 400 species of the family Rubiaceae, and leaf nodules of *Psychotria* and *Pavetta* have been studied in detail. The leaf nodules become covered with multicellular hairs and contain rod-shaped bacteria. The bacteria isolated from mature leaf nodules belong to the genus *Phyllobacterium* and do not fix nitrogen. *Ardisisia*, *Pavetta*, and *Psychotria* do not develop beyond the seedling stage without the presence of leaf nodule bacteria. It has been suggested that host plants are cytokinin-deficient mutants and depend on leaf nodule bacteria to provide them with plant hormones (Miller, 1990).

Azorhizobium produces nodules on stems of species in *Aeschynomene*, *Neptunia*, and *Sesbania*. In addition, some cultivated varieties of *Arachis hypogaea* and *Vicia faba* form nodules on the lower part of their systems (Fyson and Sprent, 1980). The bacterioids, when isolated from stem nodules and grown in culture, can fix nitrogen by using glucose, succinate, and lactate as energy sources (Trinchant and Rigaud, 1987). Nodules on stems develop at specific sites called *mamillae*, and bacteria occur both in the intercellular and intracellular spaces of the cortex. The bacteria multiply and stimulate host cell di-

vision. In contrast to the *Rhizobium*-legume symbiosis, *Sesbania* seeds contain large numbers of *Azorhizobium caulinodans* (Werner, 1992).

Root nodules. Root nodules formed on actinorhizal plants in response to *Frankia* or those on *Parasponia* roots following an infection with *Bradyrhizobium* are modified lateral roots. Though legume nodules have been considered homologs of lateral roots by some, others have suggested alternative developmental pathways. It has been hypothesized that nodules (1) originated as highly modified stems (Sprent, 1989), (2) evolved from carbon storage (Caetano-Anolles et al., 1993), (3) originated as a new organ formed from endomycorrhizae–root interaction (LaRue and Weeden, 1994), (4) derived from a defense mechanism used to protect the plant from pathogenic bacteria (Verma and Delauney, 1989), and (5) arose as an elaboration of a wound response (Sprent, 1997). Some experimental and anatomical evidence suggests that a nodule develops instead of a lateral root when meristematic cell divisions occur in the cortex instead of in the pericycle (Hirsch and LaRue, 1997).

Rhizobium *genes*

Legume mutualism is determined by many genes, among them genes called *nod* and *fix*, which code for nodulation and nitrogen fixation, respectively. Among the *fix* genes the most significant are the *nif* genes, which code for the synthesis of the enzymes of the nitrogenase complex. In most of the fast-growing *Rhizobium* species that have been examined, these genes are located on a large plasmid called pSym (symbiotic). Scientists have demonstrated a close linkage between the nitrogenase (*nif*) and nodulation (*nod*) genes on the pSym plasmid (Vlassak and Vanderleyden, 1997). The formation of a nodule in the host tissue is a complex, multistep process involving recognition (signals), binding receptors, infection, host cell multiplication, and cell enlargement (Hirsch, 1992). Mutants have been discovered for a number of nodular stages. These mutants have been used to discover how the process of nodulation is governed and controlled. Both bacterial and plant genes are involved in the formation of nodules. More than 50 different *nod* genes have been identified (Downie, 1994). *Rhizobium* genes controlling infection and nodule formation fall into two groups, one consisting of genes involved in the synthesis of exopolysaccharides (exogenes) and the other of *nod* genes (Verma et al., 1992; Gottfert, 1993; Van Rhijn and Vanderleyden, 1995). *Rhizobium* genetics has been advanced by transposon mutagenesis, recombinant cloning, and plasmid transfer experiments (Long, 1989, 1996).

Many aspects of the *Rhizobium*–legume symbiosis are not fully understood, and the information that is available comes from only a few species, which may not represent the situation in nature (Van Rhijn and Vanderleyden, 1995; Pueppke, 1996). Research on this symbiosis continues because of its agricultural significance and fundamental importance to natural plant communities. Efforts to increase food production to feed a growing world population require new supplies of fertilizer, of which nitrogen is a key element. It is more economical, as well as environmentally sound, to increase nitrogen production through biological systems, such as that of *Rhizobium* and legumes, than to produce nitrogen fertilizers by expensive chemical processes (Maier and Triplett, 1996).

Actinorhizal Symbiosis

Nitrogen-fixing nodules are also formed on nonleguminous plants by *actinomycetes*, which are filamentous soil bacteria (Pawlowski and Bisseling, 1996). Actinomycetes that form nitrogen-fixing symbioses belong to the genus *Frankia*. These associations, involving about 200 species of angiosperms (Huss-Danell, 1997), are called actinorhizae, and they develop on the roots of woody dicotyledonous plants. Symbiotic *Frankia* form numerous terminal swellings, called vesicles, within which nitrogen fixation occurs. Nodules of some actinorhizae also contain hemoglobin, which is similar to leghemoglobin in *Rhizobium*–legume root nodules. The stages of infection that lead to actinorhizal nodules are similar to those of *Rhizobium*. The sequence of events is generally as follows: root hair curling, infection through root hairs, growth of the bacterial filaments through the plant cell, rapid division of cortical cells to form nodules, invasion of the cortical cells by bacteria, and formation of hyphae, vesicles, and spores in the plant cells. Depending on the type of host plant, one bacterial strain may form club-shaped or spherical vesicles. In *Casuarina* the bacteria do not

form vesicles, but nitrogen fixation still occurs. Strains of *Frankia* have been isolated from more than 160 species of woody dicotyledonous plants, including species of *Alnus* (alders), *Casuarina* (beefwood), *Comptonia* (sweet fern), and *Myrica* (bayberry). Plants with actinorhizae are important sources of nitrogen in woods, fields, bogs, and roadsides, both from their leaf litter and decay of dead plants. *Frankia* strains have been cultured on artificial media. Their growth is much slower than that of most free-living bacteria, and they assume a variety of shapes. The molecular genetics of *Frankia* is being studied to determine if the symbiotic genes of this bacterium are similar to those of *Rhizobium* (Simonet et al., 1990; Franches et al., 1998). So far, the *nif* genes seem well conserved, but there is no evidence for other similarities in symbiotic genes.

Cyanobacteria and Plants

Cyanobacteria are another group of nitrogen-fixing bacteria. In addition to their probable contribution to the origin of plants, as chloroplasts, they form associations with a wide range of organisms that include fungi, algae, bryophytes, the water fern *Azolla*, cycads, and the angiosperm genus *Gunnera*. Cyanobacteria colonize a variey of plant organs such as roots, stems, and leaves (Bergman et al., 1996), and they fix nitrogen within heterocysts, which contain nitrogenase (Rai, 1990; Bergman et al., 1992b; Paerl, 1992). Recent advances in the molecular biology of cyanobacteria have increased our understanding of these organisms (Bryant, 1995).

Symbiotic associations between cyanobacteria and plants have a number of common features (Meeks, 1998). Symbiotic cyanobacteria usually move in a slow, gliding manner by means of short filaments called *hormogonia* and can penetrate cavities and spaces of plant organs and form colonies. After a population of cyanobacteria is established in a plant, growth of the endosymbiont either stops or slows down. The host plant may regulate the growth and population size of the endosymbiont, but exactly how this is done is not known. Symbiotic cyanobacteria have more heterocysts and fix nitrogen at higher rates than free-living cyanobacteria. The symbiotic forms have larger vegetative cells, fewer nutrient reserves, do not photosynthesize, and cannot assimilate the nitrogen they fix. The host plant inhibits the nitrogen-assimilating enzymes of the endosymbiont. As a result, the ammonia produced by the symbiont is excreted and passed to the host plant, which then converts it to other compounds. Symbiosis alters host plants. They produce hairs and extensions of cells that increase the surface contact between host and cyanobacteria.

Azolla

Azolla is a genus of small, aquatic ferns that float on the surface of freshwater ponds and marshes throughout the world (Peters and Meeks, 1989). The plant has tiny roots and a short, branched stem covered with small, overlapping leaves (fig. 5.2). The ferns multiply rapidly by vegetative reproduction. They can double their biomass in 2 days if growth conditions are optimal and quickly cover large bodies of water with a dense and sometimes impenetrable growth. Each leaf of *Azolla* is divided into a dorsal and ventral lobe. The ventral lobe floats on the water and contains filaments of the cyanobacterium *Anabaena azollae*, which lives in symbiosis with the fern. The bacteria are inside mucilage-filled cavities of the leaves. The cavities are sealed so that cyanobacterial filaments cannot escape and other organisms cannot enter. Sexual structures that the fern uses to overwinter contain *Anabaena* spores, so that when germination occurs in the spring the new individuals of *Azolla* are infected by the bacteria. Thus, no stage in the life cycle of the fern is free of cyanobacteria.

Anabaena fixes nitrogen in the leaf cavities of *Azolla* and supplies the fern with all the nitrogen it needs. With its own built-in fertilizer generator, the fern can grow in waters that are deficient in nitrogen. The fixed nitrogen, or ammonia, that is released by *Anabaena* is absorbed by specialized hairs in the leaf cavity. Each cavity has about 20 randomly located simple hairs that transport sucrose from the fern to the cyanobacteria, where it is used in nitrogen fixation. Each cavity also has two branched hairs that are always near a vein. These hairs are thought to assimilate the ammonia produced by the cyanobacteria, form amino acids, and transport them to the fern (Plazinski, 1990).

The leaf cavity that contains cyanobacteria acts as a greenhouse. Inside this space, the bacterial symbiont has a moist, protected environment, and its nutrient needs are provided by the fern. The extent to which *Anabaena azollae* contributes to the total photosynthesis of the symbiosis is not clear. The isolated *Anabaena* symbiont fixes carbon dioxide and undergoes normal photosynthesis, but this process is inhibited when the bacterium is in the leaf cavity. The fern inhibits the glutamine synthetase activity of the cyanobacterium. When this enzyme is inhibited, the *Anabaena* symbiont excretes rather than assimilates the ammonia it produces by nitrogen fixation. *Anabaena* filaments are attached to the growing tip, or meristem, of the *Azolla* plant. From this region the bacteria become incorporated into each new leaf as it develops. If the *Anabaena* colony on the meristem is sparse or absent, then bacteria-free leaves of *Azolla* will form. The bacteria on the meristem do not fix nitrogen because they lack heterocysts. Such bacteria receive nitrogenous compounds from the fern.

The development of the fern and *Anabaena* is synchronous. As each new fern leaf develops, the bacteria that become trapped inside the leaf cavity multiply, form many heterocysts (25–30% of total cell number), and

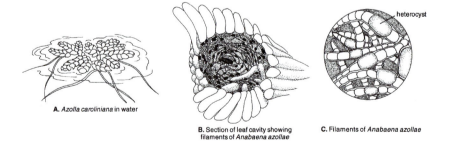

A. *Azolla caroliniana* in water

B. Section of leaf cavity showing filaments of *Anabaena azollae*

C. Filaments of *Anabaena azollae*

heterocyst

Fig. 5.2 *Azolla–Anabaena* symbiosis.

begin to fix nitrogen. Cells of the *Anabaena* symbiont enlarge and divide slowly when they are in the leaf cavity.

Coryneform bacteria also grow in *Azolla* leaf cavities, along with *Anabaena*. Whether these secondary bacteria have a role in the symbiosis is not known. They may produce some of the slime that is found in the leaf cavities, which is important for the growth of the *Anabaena* filaments (Braun-Howland and Nierzwicki-Bauer, 1990).

Experimental studies on the *Azolla–Anabaena* symbiosis have been hindered by the inability of bacteria-free (aposymbiotic) ferns to reestablish the symbiosis with isolated *Anabaena* symbionts. The specific recognition factors that are necessary to bring about the symbiosis are not known.

There are six species of *Azolla*, but whether they all contain *Anabaena azollae* as a symbiont has not been determined. There are no detailed studies to identify the cyanobacteria in the different species of *Azolla*. Morphologically, the species of fern differ according to their reproductive structures. Physiologically, the species differ only in their optimum growth temperature.

Much research is being conducted on the *Azolla–Anabaena* association. *Azolla* is used as a fertilizer in place of manure in Asian countries such as China, Thailand, and Vietnam because it grows well in the warm, stagnant waters of rice paddies and provides nitrogen for the rice plants. Farmers grow *Azolla* alongside rice plants or mix the fern with the soil of the paddies. The use of natural fertilizers such as *Azolla* has a long history in Asia. Natural fertilizers are an important and inexpensive alternative to commercial fertilizers (Nierzwicki-Bauer, 1990). *Azolla* is also used to feed animals and purify water (Wagner, 1997).

Bryophytes

Hornworts, such as *Anthoceros punctatus*, and thallose liverworts, such as *Blasia pusilla*, form symbiotic associations with the cyanobacterium *Nostoc*. Filaments of *Nostoc* live in mucilage-filled cavities on the undersurface of the bryophyte gametophytes. The bacteria fix nitrogen into ammonia, which is excreted into the cavities and absorbed by the bryophytes. The bryophytes provide the bacteria with carbon compounds such as sucrose and a protected place in which to live (Meeks, 1990).

Both symbionts are altered after they enter the association. The *Nostoc* symbionts virtually stop reproduction and do not photosynthesize. Their cells have less pigment and fewer storage granules than free-living cyanobacteria, and they appear to be nitrogen-starved, which suggests that they excrete most of the nitrogen they fix. As in the *Azolla* symbiosis, the host plant may inhibit the cyanobacterial enzymes that assimilate ammonia. The heterocyst frequency of *Nostoc* in the gametophyte cavities is much higher than that of the same symbiont growing outside the cavities (30–43% versus 3–6%, respectively). When *Blasia* becomes infected with *Nostoc*, its cavities form filamentous extensions of tissue that closely associate with the cyanobacteria. These extensions increase the area of contact between the symbionts and facilitate the exchange of nutrients (Kimura and Nakano, 1990).

Cycads

Cycads are primitive seed plants that were common and widespread during the Mesozoic era. Only about 9 genera and 100 species of cycads still grow in tropical and subtropical regions of the earth. About one-third of the known cycads contain symbiotic cyanobacteria in their roots. The bacteria are located in specialized branches of the roots that, unlike normal roots, are negatively geotrophic. Instead of growing down into the ground, these roots grow on and above the soil surface. The roots develop nodules that contain cyanobacteria. Because of their warty appearance, these specialized structures are called coralloid roots. The roots are loosely organized, and gases circulate easily in and out of them (Lindblad and Bergman, 1990).

The endosymbionts of cycads are strains of *Nostoc*. The cyanobacteria penetrate the coralloid roots through cracks or openings and move between the middle lamellae of cells until they reach the middle of the root cortex. Small host cells in this area degenerate and form mucilage-filled spaces where the cyanobacteria divide. Other host cells in the cyanobacterial zone elongate, possibly because of growth factors excreted by the cyanobacteria, and develop fingerlike extensions that may facilitate the exchange of nutrients between the symbionts (Caiola, 1992).

Cyanobacteria of cycad roots fix nitrogen, which they supply to the cycad in the form of ammonia. Cycads that grow in the forests of the Southern Hemisphere are important for the nitrogen economy of those forests.

Gunnera

Gunnera is an angiosperm genus with about 50 species, all of which are perennial herbs. The plants grow commonly in wet areas of the tropics and Southern Hemisphere. Near the base of the leaf petioles, there are secretory glands that contain the cyanobacterium *Nostoc punctiforme*. Filaments of *Nostoc* multiply in the mucilage that fills the glands and then penetrate cells at the base of the glands (Bonnett and Silvester, 1981). This association is unusual because the cyanobacterial symbionts are intracellular; that is, they are surrounded by the plasmalemma of the host cell. Each gland cell forms fingerlike extensions that increase the area of contact with the endosymbiont. Like other cyanobacterial symbionts, the *Nostoc* endosymbiont of *Gunnera* does not photosynthesize; it has the highest frequency of heterocysts recorded in any cyanobacterial population (Bergman et al., 1992a) and fixes nitrogen at a high rate. The *Nostoc* symbiont supplies *Gunnera* with all the nitrogen it needs and receives carbohydrates from the plant. Thus, in symbiosis the *Nostoc* endosymbiont behaves as a heterotroph (Bergman et al., 1996).

5.2 *AGROBACTERIUM*, GALLS, AND GENE TRANSFER

Galls are produced on plant roots and stems by species of *Agrobacterium*, *Corynebacterium*, and *Pseudomonas* and are composed of a disorganized mass of plant tissue. *Agrobacterium tumefaciens* causes crown gall disease on many woody plants, *A. rubi* forms cane galls on raspberries and blackberries, and *A. rhizogenes* induces development of hairy roots on apples. The type of symptoms that develop on the host are a function of the kind of plasmids the bacteria carry. An *Agrobacterium* with a tumor-inducing (Ti) plasmid produces crown gall symptoms; the hairy roots of apple are caused by a root-inducing (Ri) plasmid found only in strains of *A. tumefaciens* and *A. rhizogenes*. Crown gall disease is unique among plant diseases because it is malignant. Host plant cells once stimulated by *A. tumefaciens* continue to divide and no longer obey the host's hormonal commands.

Crown gall disease affects approximately 100 families of plants and causes millions of dollars worth of damage each year to agricultural and ornamental plants. *Agrobacterium tumefaciens*, a rod-shaped, Gram-negative bacterium is closely related to the *Rhizobium* bacteria that form nodules on the roots of legumes. *Agrobacterium tumefaciens* lives in the soil and infects plants through wounds that the plants suffer either during seed germination or from insects and nematodes. Infection is usually near the crown of a plant—that is, at the junction between the stem and the roots. The bacteria stimulate the plant cells around the wound to multiply and form a tumorlike growth, or gall. The disease generally affects only wounded plants because it is the new cells developing around the wound that are most susceptible to bacterial infection (Baron and Zambryski, 1995a). The wounded plant cells produce signal molecules such as acetosyringone and a hydroxyacetosyringone that turn on the virulence systems (*vir* genes) of *Agrobacterium*. These compounds are thought to develop from the wounding-induced degradation, or the repair, of cell wall lignin (Baron and Zambryski, 1995a). Galls generally appear several weeks after a plant is infected and may be a millimeter or more in diameter and may weigh up to 100 lbs depending on the plant host (fig. 5.3). Plants rarely die from crown gall disease, but they become weakened, grow more slowly, and produce fewer flowers and fruits. Infected plants are susceptible to other disease-causing agents, such as fungi.

The infectious process of crown gall disease has been studied intensively (Kado, 1991). Each cell of *A. tumefaciens* contains one or more large plasmids. The plasmid responsible for tumorigenesis, the Ti plasmid, contains about 100 genes. After the bacterium attaches to a plant cell, the plasmid passes through the host cell wall and part of the plasmid, called tDNA (transfer DNA), becomes integrated into the host chromosome (Hooykass and Schilperoort, 1992). Genes on the tDNA are expressed along with the plant's genes (Gelvin, 2000). The integrated bacterial genes cause the plant cell to multiply abnormally and to produce unusual chemical compounds. In effect, the plant cell is partly controlled by genes from the bacterium. Crown gall disease is an example of genetic parasitism in which the genes of *A. tumefaciens* are used to parasitize another organism. Once the plasmid is incorporated into the host

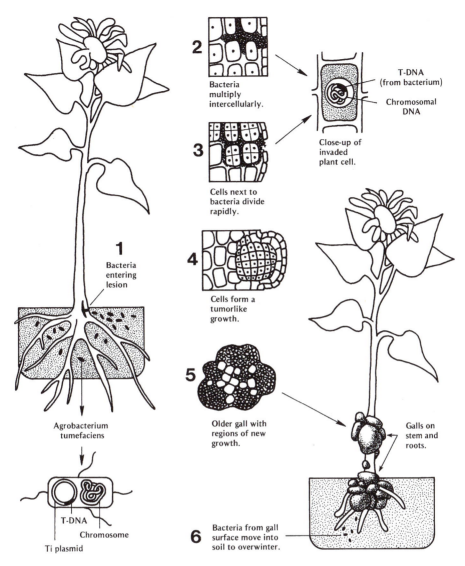

Fig. 5.3 Genetic colonization of a flowering plant by *Agrobacterium tumefaciens*.

chromosome, it replicates along with the plant's chromosomes. Thus, all the progeny of a cell with tDNA will have copies of the tDNA, and the bacterium no longer is necessary for the disease to progress. Crown gall is the only known example of a disease that persists in the absence of the inciting agent. The Ti plasmid, in part at least, may represent ancient plant genes that have become incorporated into *A. tumefaciens*.

Genes on the tDNA code for the production of *auxin* and *cytokinins*, plant growth hormones that stimulate growth of the tumor. Other genes on the tDNA segment code for the synthesis of opines, which are derived from amino acids and produced only by plant cells infected with tDNA. *Opines* are sources of carbon, nitrogen, and energy for *Agrobacterium* in the soil and they increase the rate of transfer of Ti plasmids to avirulent strains of *A. tumefaciens* that are in the soil near the crown galls. Opines also stimulate plasmid exchange between strains of *Agrobacterium* in the crown gall tissue.

Six different strains of *A. tumefaciens* are recognized, based on their ability to metabolize specific types of opines. For example, the octopine strain uses opines such as octopine, derived from the amino acid arginine; lysopine, derived from lysine; histopine, derived from histidine; and octopinic acid, derived from ornithine. The nopaline strain uses nopaline, derived from arginine, and nopalinic acid, derived from ornithine. Most strains have a wide host range, but a few are limited in the types of plants they can infect.

That a segment of a plasmid of *A. tumefaciens*, a prokaryotic cell, can fuse with a chromosome of a eukaryotic plant cell is surprising. It is equally intriguing that genes of a plasmid are expressed within a eukaryotic cell. This occurs because tDNA includes genetic sequences that can be read by enzymes of the eukaryotic cell. *Agrobacterium* is the only bacterium known to transfer DNA to eukaryotes as part of its life cycle (Winans, 1992).

The ability of *A. tumefaciens* to insert part of its plasmid into a plant chromosome has considerable implications for the improvement of crops (Vasil, 1994). Through the techniques of genetic engineering, *A. tumefaciens* has been used to introduce desired traits into plant cells. In fact, transformation brought about by *Agrobacterium* is the most popular and reliable method of creating transgenic plants in different species that can grow in tissue culture and regenerate. Such modified plants have been used to produce new products such as antibiotics, oral vaccines, and biodegradable plastic (Walden and Wingender, 1995).

An important event in the *A. tumefaciens* infection of plants is the binding of the bacterium to the host cell wall, a process that may be genetically controlled by the *A. tumefaciens* chromosome. A lipopolysaccharide constituent of the bacterial cell wall binds to a specific component of the plant cell wall. Only young, actively dividing plant cells, such as those that form around a wound, seem to have this component.

Agrobacterium tumefaciens infects a wide variety of plants. Dicotyledonous angiosperms and gymnosperms are susceptible to infection, but monocotyledonous plants appear to be resistant. The nature of this resistance is not understood. It may result from the inability of the bacterium to bind to the cell walls of these plants or from a failure of the tDNA to become integrated in the host DNA or even to notice such an integration. Suitable culture and regeneration techniques for these groups of plants have yet to be discovered (Walden and Wingender, 1995).

5.3 BACTERIA AS PLANT PATHOGENS

In 1878 an American plant pathologist, Thomas Burrill, discovered that bacteria (*Erwinia amylovora*) caused fire blight disease that killed apple and pear trees (Agrios, 1997). Although many plants are susceptible to bacterial diseases, mosses, ferns, gymnosperms, and monocots are generally resistant. Bacteria enter plants through wounds and natural openings and survive winter in plant debris.

The first avirulence gene was isolated in 1984 from *Pseudomonas syringae*, a bacterial pathogen of soybean, and since then more than 30 bacterial *avr* genes have been cloned. Most *avr* genes encode for hydrophobic proteins that require additional bacterial genes called *hypersensitive reaction and pathogenicity* (*hrp*). The gene products of *avr* and *hrp* confer resistance in the host, which is characterized by rapid localized cell death. The bacterial avr protein is translocated into the plant cell, where it affects nuclear gene expression. Cloned *avr* genes have been introduced into host plants. The *avr* gene activity in plant cells with an *R* gene is measured by induction of hypersensitive cell death (Ackerveken and Bonas, 1997) (fig. 5.4).

Several hrp proteins are homologous to proteins from human bacterial pathogens such as *Salmonella*, *Shigella*, and *Yersinia* that make up the type III secretion system. Several *Yersinia* proteins called YOP are translocated into the host cells. In *Pseudomonas*, type III protein called hrpA is associated with the formation of piluslike structures which are thin enough to pass through the host cell wall and allow the avr protein to enter the host cell cytoplasm. Hypersensitive cell death in resistant plants occurs when avr protein acts within the host cell (Lee, 1997). The following are some well-known examples of bacterial symbionts of plants.

Vascular Wilts

Several species of *Clavibacter*, *Erwinia*, *Pseudomonas* and *Xanthomonas* cause *vascular wilts* in cultivated plants. The pathogens multiply in the host's vascular tissues and destroy the cells using enzymes that cause leaves to wilt and die. Growth regulators secreted by the bacteria may cause hypertrophy and hyperplasia of xylem tissue.

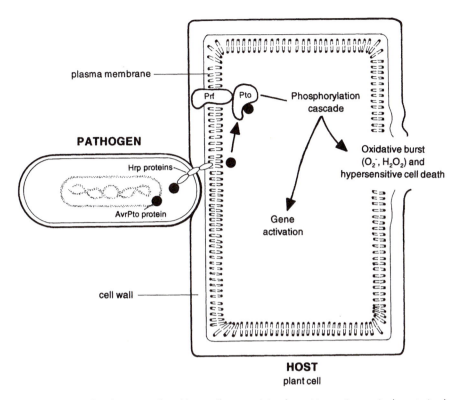

Fig. 5.4   Bacterial pathogen-mediated host cell apoptosis in plants. Host resistance is characterized by sudden, rapid death of the infected cell (hypersenstivity response). When a host resistance gene product (PTO) interacts with an avirulence gene product (Avr Pto), the resulting phosphorylation cascade leads to cell death. Adapted from Lamb (1996).

Soft Rots

Bacterial soft rots occur commonly in plants during storage, and the decomposing plants give off volatile gasses. The bacteria produce enzymes that dissolve the middle lamella in plant tissues. *Erwinia carotovor* and *E. chrysanthemi* cause soft rots and wilting and dwarfing of many fruits, vegetables, and flowers. The bacteria produce different pectinolytic enzymes by means of several gene regulatory systems (Hugouvieux-Cotte-Pattat et al., 1996). *Pseudomonas marginalis* causes pink eye disease in potatoes and slippery skin in onions.

Scabs

Streptomyces scabies causes the common scab of potato and is distributed worldwide. Bacteria multiply in the intercellular spaces of parenchyma cells, causing healthy cells around the lesion to divide and form layers of cork cells.

Cankers

Bacterial cankers destroy stems and twigs and cause economic losses. *Pseudomonas syringae* infects stone fruits and ornamentals such as roses and lilacs. Most strains of *P. syringae* produce a plasmid-encoded phytotoxin.

Citrus canker, caused by *Xanthomonas compestris* pv. *citri*, is one of the most feared citrus diseases in Florida because of the heavy economic losses it produces. The disease is endemic in Asia and was accidently introduced into Florida in the early 1910s with the import of infected nursery trees from Japan. It took more than 30 years of sustained eradication efforts to eliminate the bacterial pathogen that resulted in destruction of millions of fruit-bearing trees (Schumann, 1991).

Role of Phytotoxins in Bacterial Plant Diseases

Syringomycin is one of several types of phytotoxins that determine bacterial pathogenesis in a host plant. Purified toxin applied to young peach trees produces typical disease symptoms (DeVay et al., 1978). Recent studies on the chemical structure of syringomycin and other related toxins have provided significant insights into their role in disease physiology. The toxin lipopeptide molecule acts on host cell membranes, affecting electrical potential and pH gradients, thereby causing leaf stomata to close (Mott and Takemoto, 1989).

Rhizobitoxin is produced by *Bradyrhizobium japonicum*, *Pseudomonas aeruginosa*, and *P. andropogonis*. The toxin inhibits biosynthesis of methionine, which is needed to produce ethylene. *Phaseolotoxin*, a tripeptide toxin produced by *Pseudomonas syringae* pv. *phaseolicola*, causes chlorosis of leaves. *Tagetitoxin* is a phytotoxin produced by *P. syringae* pv. *tagetis* that interferes with chloroplast genes in developing tissues.

5.4 MYCOPLASMAS: BACTERIA WITHOUT CELL WALLS

Mycoplasmas are regarded as the smallest and simplest cellular organisms. Their genomes are the smallest of any organism, and the DNA sequences of *Mycoplasma genitatium* and *M. pneumoniae* encode for 470–500 proteins (Rottem and Naot, 1998). These prokaryotes lack the characteristic cell wall of other bacteria and were first seen in the phloem cells of plants with yellows disease. Yellows disease was long suspected to be caused by a virus because it was transmitted by insects, and the infectious agent passed through bacterial filters. Antibiotic treatment of sick plants, however, restored them to good health. Mycoplasmas are resistant to penicillin and other antibiotics that inhibit cell wall formation but are sensitive to antibiotics such as tetracycline. Today, we know of several different kinds of mycoplasmas and mycoplasmalike organisms (MLOs). Mycoplasmas are similar in many respects to Gram-positive bacteria (Dybvig and Voelker, 1996).

Plants infected with phloem-specific mycoplasmas are often stunted and have small flowers and fruits. Corn stunt and citrus stubborn are two diseases that are caused by the mycoplasma *Spiroplasma*. Aster yellows, lethal yellowing of coconut palms, elm phloem necrosis, peach X disease, and pear decline disease are other plant diseases suspected of being caused by mycoplasmas. These obligate symbionts occur intracellu-

larly in sieve-tube elements of the phloem. Sieve-tube cells are a highly specialized environment. A large volume of nutrients passes through these cells to areas of the plant where active growth is taking place or to storage areas of the roots. A unique feature of these cells is their high sugar concentration and hence high turgor pressure gradient. Additionally, sieve cells contain significant concentrations of K^+, Mg^+, Cl^- and PO_4^- ions, amino acids, proteins, and ATP molecules. It is this uniquely specialized cellular environment to which the symbionts have become adapted.

Leaf hoppers and other insects feed on phloem sap, and the chemical constituents of their hemolymph are strikingly similar to those of the sieve-tube cell. Insect hemolymph has high concentrations of organic and inorganic nutrients, which result in a high osmotic pressure. Leaf hoppers feeding on phloem cells acquire mycoplasmas that multiply rapidly in the alimentary canal, hemolymph, salivary glands, and ovaries of the insects. Mycoplasmas are eventually transported to the salivary glands, from which they are transferred to new phloem cells during a leaf hopper's feeding. The same mycoplasmas can reproduce in different hosts, such as flowering plants and insects, because the intracellular environments of the hosts are remarkably similar. The lack of cell walls in mycoplasmas may be viewed as an evolutionary consequence of intracellular symbiosis.

In recent studies, mycoplasmas have been observed in the leachates of flowers, suggesting that flowers may also be habitats for these prokaryotes. Possibly, they survive in nectar and are transmitted by pollinating insects. Indeed, insects could be the principal hosts of these prokaryotes, and the flowers may represent only transient habitats.

Animal-associated mycoplasmas first gained the attention of researchers when tissue culture cells obtained from respiratory and urogenital tracts developed large populations of these organisms. Scientists are beginning to realize that mycoplasmas may be a normal constituent of human and animal mucous membranes. Mycoplasmas have been reported from cells of the oral and nasal cavities, urogenital tracts, and lungs of cattle and humans. Some authors estimate that 20% of all human pneumonia may be caused by *Mycoplasma pneumoniae*. The significance of mycoplasmas in human and animal health should be better understood now that the complete nucleotide sequence of the *M. pneumoniae* genome is known (Krause, 1998).

5.5 SUMMARY

Some bacteria form mutualistic, nitrogen-fixing symbioses with plants. The nitrogen fixed by these organisms is an important part of different ecosystems throughout the world. The best known nitrogen-fixing symbiosis is that between *Rhizobium* and legumes. Intensive research is being conducted on this symbiosis to understand the genetics of nitrogen fixation and nodulation. The long-term goal of such research is to be able to introduce nodulating bacteria into crop plants and thus eliminate the need for artificial nitrogen fertilizers. Unfortunately, nitrogen fixation involves complex interactions between the host plant and its microbial symbionts, and therefore the symbiosis must be studied as an integrated whole to understand it fully. *Frankia* is an actinomycete that forms nitrogen-fixing nodules on the roots of many trees and shrubs and fixes nitrogen within root nodules in specialized vesicles.

Cyanobacteria also fix nitrogen and form associations with plants such as *Azolla*, bryophytes, and cycads as well as with lichen-forming fungi. Nitrogen is fixed in specialized structures called heterocysts, which are similar in function to vesicles of *Frankia*. In all nitrogen-fixing symbioses, the host inhibits the nitrogen-assimilating enzymes of the bacteria and thereby causes ammonia to be excreted from the symbiont to be used by the host.

The three principal features of the *Agrobacterium* symbiosis are tumor production, opine production, and Ti plasmids. Tumors occur when a plant cell with a tDNA segment derived from a bacterial Ti plasmid divides in an uncontrolled fashion and produces opines. Synthesis of these molecules is controlled by tDNA. Opines permit the symbiont to exploit the host's carbon and nitrogen energy sources. Ti plasmids contain genes that allow the bacteria to use opines as energy molecules, and they possess DNA that is transferred to the host genome. The infection process of this disease is a subject of much study because T-DNA is an important vector for carrying genes into plant cells using techniques of genetic enginering. The *A. tumefaciens* symbiosis may serve as a model for studying whether genetic information passes between symbionts. In other associations as well, including those of rhizobia and legumes, lichens, and mycorrhizas, there are morphological transformations of the symbionts and new chemical compounds produced, which future studies may show are caused by genetic colonization of a host by its symbiotic partner. At present, little is known about gene flow between the partners of a symbiosis, but considering that genetic exchange occurs between *A. tumefaciens* and plant cells as well as between the organelles of a eukaryotic cell, it appears likely that a similar situation may exist between partners of other symbioses, especially highly integrated ones. Gene flow would be one form of signal exchange.

Many species of bacteria cause plant diseases such as vascular wilts, soft rots, scabs, and cankers. Bacterial pathogens produce different phytotoxins that cause pathogenesis in plants. The gene-for-gene hypothesis explains the genetic basis of resistance and virulence in plants and pathogens. Mycoplasmas are the smallest known cellular organisms. They cause many plant diseases and also have been found in different animal tissues.

6

SYMBIOSIS AND THE ORIGIN OF THE EUKARYOTIC CELL

The cell theory, long a basic tenet of biology, states that all organisms are made up of cells and that cells are the basic units of living organisms. This traditional view of an indivisible cell has been challenged by the *serial endosymbiosis theory* (SET), which claims that the eukaryotic cell is not a single entity but rather a multiple of the prokaryotic cell (Margulis, 1992a). This controversial theory has slowly gained acceptance by biologists until now, according to Doolittle and Brown (1994), "the endosymbiont hypothesis for the origin of mitochondria and chloroplasts is as firmly established as any fact in biology" (p. 6721). Given the widespread acceptance of the SET, a revision or rejection of the cell theory is inevitable. Even the boundaries of an organism can be clouded over by symbiotic associations (Dyer and Obar, 1985).

Another firmly held biological tenet is the distinction between prokaryotes and eukaryotes, and this too is under attack. Critics contend that the definition of a prokaryotic cell is a negative one (it lacks what the eukaryotic cell has—namely, a nucleus and organelles) and therefore has no phylogenetic value (i.e., the absence of a particular feature does not mean that all prokaryotes are related) (Lake, 1991; Lake and Rivera, 1996). Prokaryotes should be defined on shared molecular traits, such as the structure of the small subunit rRNA. This challenge comes from Carl Woese and colleagues (Woese and Fox, 1977; Woese et al., 1990; Woese, 1993; Olsen et al., 1994), who from their phylogenetic analyses based on RNA structure and sequences, have found that archaebacteria are more closely related to eukaryotes than to the true bacteria. Margulis (1996) and others (Mayr, 1990), however, argue that despite their molecular differences, archaebacteria and eubacteria are prokaryotes and should be kept in one inclusive taxon, the Prokaryota (Prokarya). Nucleated organisms that evolved from symbiotic integration of several prokaryotes are placed in the Eukaryota (Eukarya).

6.1 KINGDOMS AND DOMAINS

The jump from a two-kingdom classification (Plants and Animals) to one that involves five kingdoms (Plantae, Animalia, Fungi, Monera, Protista) was a radical step a few years ago (Margulis and Schwartz, 1997). Despite the general acceptance of this new classification, it is now being challenged by new molecular findings (Woese et al., 1990; Sogin et al., 1996). Critics of the five-kingdom classification feel that it does not reflect the true evolutionary history of life on earth. Supporters of the five kingdoms (Margulis and Guerrero, 1991; Margulis, 1992a) are not persuaded.

The use of molecular information to determine evolutionary relationships has revolutionized taxonomy, especially among the bacteria, which do not display conspicuous morphological diversity as do large organisms, and for associations whose symbionts cannot be cultured independently of each other. Woese and colleagues (Woese et al., 1990) divide living organisms into three domains: Archaea, Bacteria, and Eukarya. Domains are taxonomic categories above the kingdom (Williams and Embley, 1996). Figure 1.2 shows the three domains with the Archaea being distantly related to the Eukarya (Keeling, 1998). Some molecular features shared by these two domains include N-linked glyco-proteins, absence of formylmethionine, shared resistance or sensitivity to certain antibiotics, and the presence of tRNA introns. There is also a similarity in eukaryotic and archaebac-terial transcription (Keeling and Doolittle, 1995).

The origin of eukaryotes was one of the most complicated events in the evolutionary history of life (Cavalier-Smith, 1991; Dyer and Obar, 1994; Vellai et al., 1998). The archae-bacteria and bacteria appear to have origi-nated in a hot environment (thermophiles) (Olsen et al., 1994; Monastersky, 1998). The term *progenote* has been used for primitive pre-organisms that were in the process of evolving a translation apparatus, and *genote* describes modern cells (Woese and Fox, 1977; Woese, 1998).

6.2 MOLECULAR VERSUS TRADITIONAL TAXONOMY

With the increased concern about biodiver-sity, the disappearance of tropical rainforests, and the concomitant loss of undescribed species, the importance of taxonomy has in-creased greatly. Unfortunately, in this resur-gence of popularity, a fierce debate has devel-oped as to whether the traditional (i.e., whole-organism based) taxonomy accurately reflects evolutionary relationships. Some scientists believe that molecular information is suffi-cient to create a phylogenetic tree using the pattern of genetic sequences from only one kind of RNA molecule (Woese et al., 1990). Discovery of extensive horizontal gene trans-fer between organisms, however, has under-mined confidence in the accuracy of trees of life based on molecular relationships (Doo-little, 1999; Pennisi, 1999). Molecular taxon-omists select a gene or its RNA or protein product in an organism and then study the se-quences of the same gene or product in other organisms. At present, the most popular genes are those that code for ribosomal RNA (16S rRNA). The similarity of these molecules in all organisms make them useful comparative markers of evolution (Sogin, 1989; Schlegel, 1994; Pace, 1997). Molecular taxonomists compare the chemical sequences of genes to see how they have changed during the course of evolution. The positioning of an organism in a family tree depends on how much its gene sequences differ from those of other or-ganisms. When one considers that all life on

earth probably arose from a common origin some 3.8 billion years ago, trying to determine phylogenetic relationships is a challenging task.

Many proteins and certain nucleic acid se-quences can be viewed as "living fossils" be-cause they have been conserved from the dawn of life. They have been found in all cel-lular life forms and are believed to have a common ancestor. Margulis (1992a, 1996) ar-gues that all living organisms have at least five different types of nucleic acids and at least 500 different proteins and that one alone is not an accurate reflector of phylogeny. She be-lieves that whole-organism biology (genetics, metabolism) should support phylogenetic classification.

6.3 THEORIES ON THE ORIGIN OF EUKARYOTIC CELLS

Several theories and hypotheses have been proposed to explain the origin of eukaryotic cells: these include the autogenous (direct fil-iation) theory, the SET, and the hydrogen hy-pothesis.

Autogenous Theory

The central assumption of the *autogenous theory* is that organelles evolved within the cell by progressive compartmentalization. Cell structures such as the nuclear membrane, endoplasmic reticulum, Golgi bodies (dic-tyosomes), vacuoles, and lysosomes are be-lieved to have evolved from elaborations of the cell and other internal membranes. The theory suggests a primitive ancestor, the *uralga*, evolved by gradual selection (Jensen, 1994). Furthermore, after the eukaryotic cell developed, endosymbiotic events could have occurred. The cytoskeleton of the cell most likely originated autogenously (Cavalier-Smith, 1991, 1992a).

Serial Endosymbiosis Theory

The SET best explains the origin of organelles such as mitochondria, chloroplasts, and mi-crotubular complexes. The possibility that peroxisomes, hydrogenosomes, and even the nucleus have a symbiotic origin is under in-vestigation (Palmer, 1997). These organelles may have arisen by successive symbioses about a billion years ago. An early step in this process was the invasion of anaerobic host cells (archaebacteria) by aerobic prokaryotes, which in time were transformed into mitochondria (Fenchel and Finlay, 1994). Later, eukaryotic heterotrophs acquired cyanobacteria and gained photosynthetic capabilities. These symbiotic cyanobacteria eventually developed into plastids, which include chloroplasts. Some researchers believe that the acquisition of mitochondria and chloroplasts by phagotropic hosts occurred more or less simultaneously rather than sequentially (Cavalier-Smith, 1987b). The *endosymbiont hypothesis*, first proposed by Margulis some 30 years ago, has recently been supported by several molecular sequence analyses (Brown and Doolittle, 1997; Martin, 1999).

The SET states that eukaryotic flagella, mitochondria, and chloroplasts were acquired by prokaryotes in that order. These organelles are viewed as transformed bacterial symbionts. This idea is not new: early scientists such as Ivan Wallin were convinced that mitochondria were bacteria that lived inside other cells. In 1927 Wallin used the term *symbionticism* to describe microsymbiotic complexes such as those of mitochondria with the rest of the cell. Wallin thought that symbionticism played an important role in the origin of species. As early as 1883, A.F.W. Schimper proposed that the chloroplasts of plants were cyanobacteria that lived symbiotically inside plant cells. These insightful ideas were not accepted by the general scientific community because they were impossible to prove experimentally. As with many scientific hypotheses, the ideas were eventually forgotten. The revitalization and expansion of these hypotheses is the work of Lynn Margulis, who has published books and articles promoting the symbiotic origin of eukaryotic cells. Her persistent and lucid arguments have resulted in the general acceptance of the symbiotic theory (Margulis, 1996).

The SET is strongly supported by findings from structural, molecular, and biochemical studies of eukaryotic cell organelles. The DNA of chloroplasts and mitochondria differs from that of the nucleus; organelle DNA is circular and histone-free, like that of prokaryotes. Ribosomes of these organelles are also similar to those of prokaryotes both in size and in the nucleotide sequence of their RNA. Chloroplasts and mitochondria synthesize proteins, divide by fission, and mutate. The mitochondria and chloroplasts of algal and plant cells are only semi-independent. For example, most of the proteins needed for the development and function of chloroplasts are coded for by genes in the nucleus and synthesized on ribosomes in the cytoplasm. The proteins are then transported into the chloroplasts. Some proteins are a joint product of genes found in the nucleus and chloroplast. For example, ribulose bisphosphate carboxylase is an important enzyme in photosynthesis that catalyzes fixation of atmospheric CO_2 into the organic matter of cells. Such interaction between the nucleus and chloroplast suggests that the symbiosis between chloroplasts and plant cells is an ancient one. Similar relationships also exist between mitochondria and the nucleocytoplasm.

Evidence suggests that during evolution, much of the genome of the bacterial ancestors of mitochondria and chloroplasts was transferred to the nucleus (Henze et al., 1995; Lake and Rivera, 1996; Martin, 1996). These organelles contain less genetic material than free-living bacteria, and segments of mitochondria and chloroplast DNA are identical to segments of nuclear DNA. The same type of DNA that occurs in more than one organelle has been called *promiscuous DNA* (Gray, 1989). DNA transfer between organelles appears to be an ongoing process (Stern and Palmer, 1984; Knoop and Brennicke, 1994).

The Hydrogen Hypothesis and Origins of Organelles

The recent *hydrogen hypothesis* of William Martin and Miklos Muller (1998) is based on the symbiosis between an α proteobacterium and an archaean (methanogen). A similar proposal, the syntrophy hypothesis, differs mainly in the types of eubacterial symbionts involved (Lopez-Garcia & Moreira, 1999). Hydrogen is key to the emergence of eukaryotes, and the classical endosymbiosis theory has been expanded to include new evolutionary insights (Martin and Muller, 1998; Doolittle, 1998a) (figs. 6.1, 6.2).

Glycolysis is central to the energy metabolism of eukaryotes, with 1 mol glucose generating a net gain of 2 mol ATP. In mitochondrial cell respiration, pyruvate is oxidized through the pyruvate dehydrogenase complex (PDH) to CO_2 and water with an additional 36 mol of ATP generated per 1 mol glucose. In amitochondriate eukaryotes, pyruvate is metabolized through pyruvate-ferredoxin oxidoreductase (PFO) Fig. 6.2. Type 1 amitochondriate eukaryotes lack organelles, and PFO decarboxylates pyruvate to produce

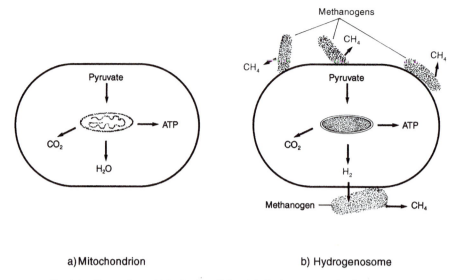

a) Mitochondrion   b) Hydrogenosome

Fig. 6.1   Comparison of (a) mitochondial and (b) hydrogenosomal cell respiration.

ferredoxin and acetyl-CoA. The acetyl-CoA is converted into ethanol and CO_2 or acetate with an additional gain of 2 mol ATP. Type II amitochondriate eukaryotes have hydrogenosomes, where PFO converts pyruvate into CO_2, acetate, and reduced ferredoxin. As hydrogenase reoxidizes ferredoxin, hydrogen is produced and an additional 2 mol of ATP are generated, with CO_2 and acetate as waste products. Molecular sequence analysis shows that PFO and PDH of hydrogenosomes and mitochondria share a common bacterial ancestry (Martin and Muller, 1998).

The proteobacterium excretes hydrogen and CO_2 as metabolic wastes of anaerobic fermentation, whereas the archaean host sym-

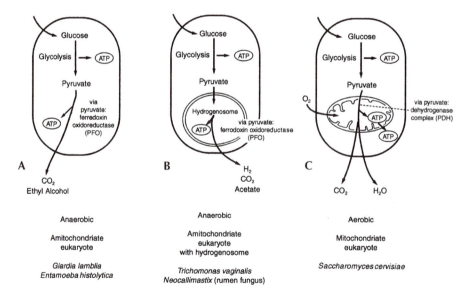

Fig. 6.2   Three types of eukaryotes according to their energy metabolism. (A) Amitochondriate eukaryotes that live anaerobically, (B) anaerobic eukaryotes with hydrogenosomes which produce hydrogen as a metabolic waste product, (C) aerobic eukaryotes with mitochondria that produce carbon dioxide and water. Adapted from Martin and Muller (1998).

biont consumes hydrogen and CO_2, thereby producing a mutually beneficial partnership. Thus, the first eukaryotes evolved from a methanogen that consumed hydrogen and CO_2 and produced methane. Its partner, the future mitochondrion, was a bacterium that produced hydrogen and CO_2. Because of the ever-present hydrogen supply, the archaean host became dependent on the α proteobacterium, and a tight genetic linkage between the two resulted. The degree of dependence increased following some gene transfer from the proteobacterial symbiont to its host genome. The host provided the membrane proteins for the import of substrates and enzymes for glycolysis, and ATP production began under anaerobic conditions. The new host with its completely enclosed symbiont changed its nutrition from simple molecules such as hydrogen and CO_2 to complex organic molecules. The symbiont in the course of evolution became (1) lost, as in amitochondriate archezoons (e.g., *Giardia*), or (2) a hydrogenosome, as in the protist *Plagiopyla*, and a mitochondrion leading to all complex cells (Martin and Muller, 1998) (fig. 6.3).

Ancestral host cell

What kinds of present-day organisms exist that might be similar to early host cells (uralga) whose symbiotic associations evolved into the modern eukaryotic cell? The primitive host cell appears to have been prokaryote-like and probably had the following features: anaerobic, possessed hydrogen-dependent metabolism, strictly autotrophic, and could perform endocytosis. Most modern prokaryotes are unsuitable models for the ancestral host cell because their cell wall is a barrier to phagocytosis. Early bacteria may have entered host cells in the same way as *Bdellovibrio*, a present-day parasitic bacterium that enters its hosts by forcibly penetrating the wall and membrane of the host cell. (The life history of *Bdellovibrio* symbionts is described in chapter 3.) Other organisms that provide insight into possible ancestral host cells include:

1. *Giardia lamblia*. This one-celled intestinal parasite causes dysentery in humans and is among the most primitive of the eukaryotes. From an evolutionary perspective, *Giardia* is "a bacterium in a eukaryotic cloak" (Upcroft and Upcroft, 1998, p. 256).

On the basis of rDNA phylogeny and ribosome size, *Giardia* resembles prokaryotic cells. *Giardia* is an obligate anaerobe that lacks mitochondria, peroxisomes, and endoplasmic reticulum. It can undergo endocytosis and so could have ingested prokaryotic symbionts that were the precursors of mitochondria (Cavalier-Smith, 1987a; Kabnick and Peattie, 1991; Sogin et al., 1989; Gupta et al., 1994). Unlike eukaryotes, which possess the enzyme PDH, *Giardia* has PFO, characteristic of prokaryotes. A 2.4-kb plasmid has been isolated from *Giardia*, and like that of some prokaryotes it carries genes for virulence and antibiotic resistance. The *Giardia* plasmid is unusual because it behaves like a transposable element and may be responsible for genomic recombination in an organism that lacks sexual reproduction (Upcroft and Upcroft, 1998). Some investigators think that *Giardia* and other primitive organisms may have had mitochondria at some point and then lost them (Hashimoto et al., 1998).

2. *Thermoplasma acidophilum*. This mycoplasmalike, free-living archaean lacks the characteristic cell wall of the eubacteria. *Thermoplasma acidophilum* has the smallest genome of any known free-living organism, and its genome resembles that of eukaryotic cells in having histonelike proteins associated with its DNA. In addition, actinlike filaments have been reported in this organism. According to Gupta and Golding (1996), the ancestral eukaryotic cell developed when a Gram-negative eubacterial host engulfed a thermophilic, sulfurmetabolizing archaean.

Mitochondria

The origin of mitochondria differs from that of chloroplasts and appears to be monophyletic (Gray, 1993; Gray and Spencer, 1996). Determining which living bacterium could represent the ancestor of mitochondria is difficult. Mitochondria cannot be cultured outside a cell and therefore cannot be compared directly with other bacteria. Comparisons can be made only through molecular and biochemical characteristics. Mitochon-

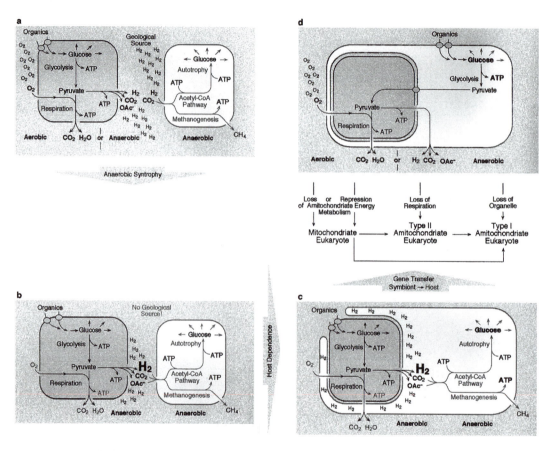

Fig. 6.3   The hydrogen hypothesis for the origin of the first eukaryotes. Symbiosis between an anaerobic, hydrogen-dependent, autotrophic archaean as a host cell and a bacterium that was able to respire and produce hydrogen as a waste product resulted in the first eukaryotes. From Martin and Muller (1998), with permission.

dria have double membranes and show greater integration into a eukaryotic cell than do chloroplasts. The outer membrane is believed to represent the original ancestor host cell vacuole membrane, and the inner membrane is thought to be that of the mitochondrial ancestor. New studies, however, have suggested that the outer membrane may be a remnant of the outer membrane of the Gram-negative endosymbiont which gave rise to the organelle (Maynard-Smith and Szathmary, 1995). Mitochondria not only have smaller ribosomes than those of the eukaryotic cell cytoplasm but also have an antibiotic sensitivity similar to that of prokaryotic cells. Mitochondrial DNA varies in size; there is two or three times more genome in plant mitochondria than in animal mitochondria. However, much of the plant mitochondrial DNA consists of noncoding sequences (introns) (Stern and Palmer, 1984; Brennicke et al., 1996; Franz Lang et al., 1999).

Fungal mitochondria have a unique genetic code. The codon UGA, which is a stop codon in prokaryotes and most eukaryotes, for example, is a second codon for tryptophan in mitochondria. Again, the sequences of tRNA in mitochondria resemble neither those of prokaryotes nor those of many eukaryotes. Mitochondrial mRNA genes, like the eukaryotic nuclear genome, have intervening sequences, or introns, along with post-transcriptional processing (removal of introns and splicing of exons); these processes are rare in prokaryotes.

The strongest evidence for the endosymbiotic origin of mitochondria comes from the sequencing of proteins and nucleic acids. The sequences of cytochromes, ferredoxin, and 5S rRNA are similar to those of photosynthetic prokaryotes such as *Rhodopseudomonas*. It has been demonstrated from sequencing rRNA from mitochondria of fungi, mice, and humans, from nuclei of animals, yeast, and corn chloroplasts, and from *Escherichia coli* that the greatest similarity in conserved sequences was between mitochondria and prokaryotes. Molecular evidence indicates that the nonsulfur purple bacteria (Proteobacteria) are the nearest present-day bacterial relatives of mitochondria (Gray and Spencer, 1996). More specifically, the relationship was closest to a subdivision of the α proteobacteria that includes intracellular parasites such as *Rickettsia*. Comparison of heat-shock proteins showed a similar picture (Viale and Arakaki, 1994; Gupta, 1995a). Mitochondria

of the mud-inhabiting protozoan *Reclinomonas americana* have characteristics of both free-living bacteria and host cells (Sogin, 1997). More than 1000 species of protists and a few fungi lack mitochondria (Cavalier-Smith, 1987a; 1997).

Chloroplasts

The chloroplast more closely resembles a prokaryotic cell than does the mitochondrion. The chloroplast probably arose when a phagotrophic host cell engulfed and retained a photosynthetic prokaryote, which gradually became transformed into an organelle (Mc Fadden and Gilson, 1995). This primary endosymbiotic event occurred at least once (Stiller and Hall, 1997). Other kinds of plastids arose through secondary endosymbiotic events involving fusions of eukaryotic cells (Gibbs, 1992; Jensen, 1994; Lewin, 1993; Lockhardt et al., 1992b; Valentin et al., 1992; Palmer, 1993; Douglas, 1994b). There are many different types of plastids, and they probably originated from multiple symbioses that involved cyanobacteria and algae. Most chloroplasts, like mitochondria, are surrounded by two membranes. The inner membrane represents the plasma membrane of the original prokaryotic symbiont, and the outer membrane either represents the vacuolar membrane formed by the host around the symbiont or is the remnant outer membrane of the Gram-negative cyanobacterium that was the endosymbiont (Maynard-Smith and Szathmay, 1995). The chloroplasts of green algae, plants, and red algae have two enveloping membranes. The chloroplasts of *Euglena* have three outer membranes, and those of brown algae and diatoms are surrounded by four membranes. It is clear that these different types of plastids could not have evolved from a common ancestor (Cavalier-Smith, 1992b). Cyanobacteria such as *Synechococcus* and *Gloeobacter* (Pace, 1997) are the most likely ancestors of the plastids of red and green algae, the latter group giving rise to green plants.

The chloroplasts of *Euglena* are thought to have arisen when a swimming protist ingested unicellular green algae that were not digested and subsequently became symbiotic. The first, or outer, membrane around the *Euglena* chloroplast is the vacuolar membrane that the host cell formed around the endosymbiont. The second and third membranes represent the plasma membrane and chloroplast membrane, respectively, of the original green algae. In brown algae and diatoms, the outer two of the four membranes around the chloroplast originate from the endoplasmic reticulum of the host cell. There are more than 27,000 species of algae with four-membraned chloroplasts, and they are placed in a separate kingdom, the Chromista which includes brown algae, diatoms, chrysomonads, cryptomonads, and haptophytes. The chloroplasts of chromistan algae are inside the lumen of the cell's rough endoplasmic reticulum.

Several unusual algae were once thought to be "missing links" in the cyanobacteria–chloroplast evolution. *Cyanidium caldarium*, *Cyanophora paradoxa*, and *Glaucocystis nostochinearum* are eukaryotic, unicellular algae that contain bluish green plastids, the cyanelles, which are equivalent to cyanobacteria without cell walls (Wasmann et al., 1987). How the host cell acquired the cyanobacteria is not clear. One possibility is that some algae lost their chloroplasts and ingested cyanobacteria, which were eventually transformed to cyanelles and assumed the role of chloroplasts. Having once had chloroplasts, such colorless eukaryotic hosts would be better equipped to maintain externally acquired cyanobacteria. Cyanelles cannot grow separately in culture, and the DNA content of some, such as *C. paradoxa*, is reduced, like that of true chloroplasts. It is not known whether there are genetic interactions between cyanelles and the host nucleus as there are in the case of true chloroplasts. Thus, although cyanelles have a superficial resemblance to chloroplasts and have reduced DNA, the present information is too limited to support the view that they are primitive chloroplasts (Seckbach, 1994). Similarly, *Prochloron*, a phototrophic prokaryote with chlorophylls a and b, was proposed as a missing link between prokaryotes and chloroplasts. However, studies have revealed species of *Prochloron* to be more closely related to cyanobacteria than to higher plant chloroplasts (Bryant, 1992; Lockhart et al., 1992a).

Evidence for the endosymbiotic origin of chloroplasts has been collected from molecular data, such as DNA–RNA hybridization of chloroplasts, nucleocytoplasm, and cyanobacteria and sequencing of plastocyanin, ferredoxin, and cytochrome c, as well as tRNA and 16S and 5S rRNA. All of these studies have indicated a close relationship between chloroplasts and blue-green prokaryotes. In particular, the 5S and 16S sequences of chloroplast rRNA show a greater similarity to those from cyanobacteria than to those from eukaryotic cytoplasm.

Complete chloroplast genome analyses from Cyanophora, *Odontella* (a diatom), *Euglena*, *Porphyra* (red alga), *Marchantia*, *Pinus*, *Nicotiana*, *Zea*, and *Oryza* indicate that of 210 different protein-coding genes, 45 are common to each and also to the cyanobacterial genome. Gene loss from chloroplast genomes occurs in all groups, and 44 different plastid-coded proteins are functional nuclear genes, thus providing strong evidence for endosymbiotic gene transfer to the nucleus in plants (Martin et al., 1998).

Microtubules

Another aspect of the serial endosymbiosis theory involves organelles that are used for motility (Margulis and McMenanin, 1990; Hinkle, 1991). The term *flagella* is restricted to motile structures found in prokaryotes (Margulis, 1991a). Following the old Russian and French literature, Margulis calls the organelles of motility found in eukaryotes undulipodia. Bacterial flagella are extracellular organelles that run by rotation of the motor at their base. The shafts are nonmotile, contain the protein flagellin, and lack microtubules. In contrast, eukaryotic flagella (undulipodia) are made up of microtubules, which consist of tubulin and are arranged in characteristic pairs, in a 9+2 arrangement. Margulis (1991b) suggests that eukaryotic flagella developed from tubule-containing spirochetes. Spirochetes are Gram-negative, highly mobile bacteria which are helically shaped and readily form ectosymbioses with other heterotrophs. They are thought to have formed an association with *Thermoplasma*-like host cells and eventually evolved into undulipodia. Symbiotic associations in the hindgut of termites and wood-eating roaches contain protistan symbionts such as *Mixotricha paradoxa* that are covered with a complex arrangement of spirochetes that are responsible for the swimming of the protist. Margulis (1991b) also proposes that symbiotic spirochetes gave rise, via mutations, to centrioles, chromosomal kinetochores, and the spindle of the mitotic apparatus. With the development of mitosis came an explosion of many new species of eukaryotic microorganisms, some of which eventually gave rise to fungi and animals.

Nucleus

The origin and evolution of the eukaryotic nucleus is uncertain. The following hypotheses have been proposed (Lake and Rivera, 1996):

1. The nucleus arose autogenously when prokaryotic DNA became enclosed within an invagination of the plasma membrane, thus creating two compartments, the DNA-containing nucleus and non-DNA cytoplasm. Each unit then evolved separately (Maynard-Smith and Szathmary, 1995).

2. The nucleus originated endosymbiotically from a prokaryotic cell that contained DNA and RNA. Gray (1992) observed the archaebacterial-like nature of the nuclear genome; Gupta et al. (1994) and Golding and Gupta (1995) proposed that the eukaryotic cell nucleus was formed by the fusion of an archaebacterium and a Gram-negative eubacterium. Lake and Rivera (1994) also posit the nucleus as an endosymbiont that took over control of the host cell. Sogin (1991) hypothesized an anuclear protoeukaryotic lineage separate from the archaebacteria and eubacteria that obtained a nucleus through engulfment of archaebacteria. The eukaryotic lineage, therefore, is a chimera (Doolittle, 1996; Gray et al., 1999). The presence of many eukaryotic nuclear genes of bacterial origin may be the result of a gene transfer rachet which occurs when a bacterial gene replaces a resident homolog of archeal origin (Doolittle, 1998b; Katz, 1998).

Peroxisomes

Peroxisomes may be yet another organelle that originated endosymbiotically (Cavalier-Smith, 1987a; Maynard-Smith and Szathmary, 1995). Their ancestors most likely were Gram-positive bacteria, which would explain the organelle's single membrane. However, the lack of DNA in peroxisomes needs explanation.

Hydrogenosomes

Hydrogenosomes are double-membraned organelles. They occur only in eukaryotes that do not have mitochondria and are facultative anaerobes. Hydrogenosomes produce hydrogen, CO_2, and acetate, which are substrates for methanogenesis (Brul and Stumm, 1994; Bui et al., 1996). Hydrogenosomes appear to have evolved independently from mitochondria in several different groups, including trichomonads, ciliates, and chytrids (Embley et al., 1997).

6.4 HOST CELL: AN INTRACELLULAR ECOSYSTEM

Ecosystems are ecological units of the biosphere, with identifiable boundaries and characteristic species of organisms. Each organism in an ecosystem generally forms colonies. From such a perspective, intracellular space represents an extreme in microhabitat. F.J.R. Taylor (1983) defines a cell as a *cytocosm* and identifies four features of this unique habitat: (1) a stable environment for the cell following symbiont intrusion (i.e., such a constant environment inhibits sexuality of the symbiont and encourages it to stay in one phase of its life cycle); (2) slow rate of succession; (3) exchange of nutrients and materials between the symbiont and host (the exchange is controlled by selective pemeability of the host membrane); and (4) colonization of cell cytoplasm by new symbionts, bringing forth processes for evolutionary change. The host cell as an intracellular ecosystem produces a mutational meltdown and genomic reduction of the resident symbionts (Andersson and Kurkland, 1999).

The plant or algal eukaryotic cell, therefore, is one of the oldest, smallest, and most intensive ecosystems. Individual organisms in the intracellular ecosystem should be genetically independent in order to reproduce successfully. Interdependent relationships between the individuals of the ecosystem involved division of labor in cellular functions. Modern-day eukaryotic cells have a variety of regulatory mechanisms, some in the form of negative feedback systems, that maintain and perpetuate the ecosystem.

The level of integration achieved by symbionts in the host cell is expressed in terms of its evolutionary history. Symbioses discussed in this book show a wide range of adaptations between the symbionts and their ecosystems. Intracellular parasitism can be viewed as a nutritional relationship in which the parasitic symbiont obtains its food from the host cell. Intracellular parasites may become lifelong residents in the host cell, but some can also live in the extracellular environment. An intracellular symbiont must enter the host cell, overcome host defense mechanisms, reproduce, avoid killing the host cell, and finally exit from the old cell to a new one. Several examples of intracellular parasitism are discussed in detail in this book.

6.5 SYMBIOSIS AND PARASEXUALITY

Symbiosis is a form of *parasexuality*, a special type of reproduction found in some fungi that does not involve the usual stages of the sexual process. Two fungi may fuse, and one may transfer its nuclei to the other. The nuclei remain together in the hyphae. Each hyphal cell, therefore, may contain two genetically different nuclei, a condition known as *heterokaryosis*. The different nuclei may fuse occasionally and undergo a mitotic reduction that produces nuclei with new genetic traits. In a symbiotic association the genes of both partners are close together but never actually fuse. Natural selection acts on both genomes as a unit. Many symbioses are more fit than their individual partners in given environments, and some have cycles during which the individual and symbiotic states alternate (Margulis, 1993b).

6.6 SUMMARY

Basic tenets in biology such as the cell theory (as applied to symbiogenetic cells) and prokaryote/eukaryote distinction are being questioned as new findings in molecular evolution further confirm the serial endosymbiosis theory and reveal new relationships between the archaea and the eukaryotes. The five-kingdom classification of organisms has been criticized as unrealistic, and new domains (superkingdoms) have been suggested (Archaea, Bacteria, Eukarya) as well as six kingdoms of Eukarya, i.e., Archezoa, Protozoa, Chromista, Fungi, Animalia, and Plantae. The serial endosymbiosis theory has been used to explain the origin of six different cell organelles (flagella of eukaryotic cells, mitochondria, chloroplasts, nucleus, peroxisomes, and hydrogenosomes). The autogenous theory explains cell membrane systems. A new hydrogen hypothesis is based on the symbiosis between an α proteobacterium and a methanogen. Ancestral prokaryotic host cells that may have evolved into modern eukaryotic cells could have been similar to *Giardia lamblia* and *Thermoplasma acidophilum*. Cyanelles are remnants of cyanobacteria that were ingested by primitive algae that had lost their chloroplasts. Eukaryotic cells have been compared to complex ecosystems. Parasexuality is a form of symbiosis and is a type of reproduction common to some fungi.

7

FUNGAL ASSOCIATIONS OF PROTOZOA AND ANIMALS

More than 1 million species of fungi are estimated to exist, but only about 69,000 species have been described. Most of the undescribed fungi occur in tropical and subtropical regions of the world (Hawksworth and Rossman, 1997). Some fungi are large, like mushrooms, but most of them are microscopic. *Armillaria bulbosa*, a fungal parasite of the roots of conifers and hardwoods, may be one of the largest and oldest living organisms. One individual strain of this fungus occupies more than 30 acres of Michigan forest, weighs more than 10 tons, and is estimated to be 1500 years old (M.L. Smith et al., 1992). Fungi grow practically everywhere and produce large numbers of asexual spores that float in the air, are buried in soil and water, and cling to plants and animals. Under favorable conditions, the spores germinate and produce colonies. About two-thirds of all fungal species live in symbiotic associations with a wide variety of organisms. The major groups of filamentous fungi are ancient in origin, with a fossil record from the Precambrian and early Paleozoic (Sherwood-Pike, 1991).

Fungi have exploited a large number of ecological niches in the biosphere. As saprophytes, they play a key role in the decomposition of dead plants and animals and thus in the recycling of carbon. Fungi are an important part of human affairs (Hurdler, 1998). Some species are beneficial to humans, but others are extremely harmful. Penicillin, an antibiotic obtained from a fungus, has saved countless lives; yeasts such as *Saccharomyces cerevisiae* are fungi that people throughout the world use to make bread and alcoholic beverages. Some fungi have caused destructive plant diseases. *Phytophthora in-*

festans (Oomycota) was responsible for the potato blight disease that led to the Irish famine in 1845–1851, during which more than 1 million people died and another million emigrated to North America. The Bengal famine of 1945–1946 was caused by an infection of rice crops by the leaf blight fungus *Helminthosporium oryzae*. The infection reduced rice yields by 75–90% and resulted in the deaths of an estimated 2 million people. Between 1870 and 1890, the British switched to drinking tea because *Hemileia vastatrix* caused a rust infection that destroyed the coffee trees in Sri Lanka. In the United States large populations of two common trees have been destroyed by fungal infections. The American chestnut has been virtually eliminated by the chestnut blight fungus, *Cryphonectria parasitica* (Anagnostakis, 1995); the American elm has been decimated by the fungus *Ophiostoma novo-ulmi*. Fungi cause many chronic illnesses in humans and other animals. Ringworm and athlete's foot are familiar fungal infections of skin tissue. Many types of allergies and food poisoning are caused by common soil and airborne fungi (Pirozynski and Hawksworth, 1988; Alexopoulos et al., 1996).

7.1 CHARACTERISTICS OF FUNGI

Fungi, like animals, are heterotrophic, eukaryotic organisms. For their nutrition they break down and absorb organic molecules produced by other organisms. They store food as glycogen or lipid. This mode of nutrition differs from that of animals, which ingest food, and from plants, most of which can manufacture food through photosynthesis. Fungi are biotrophic, necrotrophic, or saprophytic, depending on whether they obtain nutrients from living cells, dead cells of organisms they kill, or dead organisms, respectively. Fungi were earlier classified with plants, but the unique features of these organisms have been recognized and they are now placed in a separate kingdom.

Many fungi form branching filaments, the *hyphae*, which grow and usually produce a circular colony, a *mycelium*. A fungal hypha grows only at its tip and may be partitioned by crosswalls to form a row of cells. Some fungi lack crosswalls, and a few, such as yeasts, are unicellular. Fungal cell walls contain chitin.

The process of mitosis and meiosis is unique for most fungi. When the nucleus divides, its membrane does not break down, as in other organisms, but rather pinches in and forms two daughter nuclei. Further, fungal chromosomes do not align along the equatorial plate of the cell during metaphase.

Many fungi reproduce sexually. Their gametes are not motile, and the male gametes are carried to the female gametes by wind or water. In some fungi the male and female sex organs grow toward each other and fuse. Nuclei from these sex organs form pairs but

do not fuse. Each pair of nuclei ($N + N$) is called a *dikaryon*. Fungi commonly have a dikaryotic phase, which may be a major portion of their life cycle. Eventually, the cell nuclei fuse to produce a diploid ($2N$) nucleus, which undergoes meiosis. Spores are produced in distinctive fruiting bodies, which are used in fungal identification.

7.2 CLASSIFICATION

Not all organisms that have been called fungi are closely related and may not share a common evolutionary history based on morphology, nutrition, and ecology (fig. 7.1). The following is a brief synopsis of the major groups of fungi as they are recognized today. This list is intended to familiarize the reader with the groups from which we have selected examples of fungal symbioses.

Kingdom Stramenopila

Phylum Oomycota (zoosporic fungi): blights, downy mildews, and aquatic fungi that attack rotifers, nematodes, and arthropods

Kingdom Fungi (Mycota)

Phylum Zygomycota (zygosporic fungi)

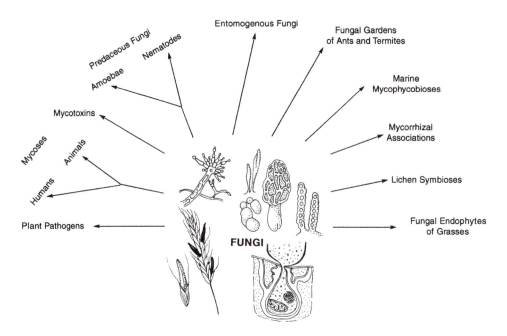

Fig. 7.1   Diversity of fungal associations, from mutualistic to parasitic, discussed in this chapter and chapter 8.

Class Zygomycetes: predatory fungi of insects, vesicular–arbuscular mycorrhizal fungi

Class Trichomycetes: obligate symbionts of arthropods

Phylum Ascomycota (sac fungi)

Class Hemiascomycetes: yeasts, fungi that cause leaf curl and witches-broom disease

Class Plectomycetes: saprophytes and plant pathogens such as powdery mildews

Class Laboulbeniomycetes: obligate symbionts of arthropods

Class Pyrenomycetes: lichen-forming fungi, plant pathogens that cause Dutch elm disease, ergot of cereals, and plant wilts and rots

Class Discomycetes: lichen-forming fungi, saprophytes and plant pathogens that cause brown rot of fruits and gray molds of vegetables

Phylum Basidiomycota (club fungi)

Class Teliomycetes: rust fungi, smut fungi

Class Hymenomycetes: toadstools, jelly fungi, ectomycorrhizal fungi

Class Gasteromycetes: puff balls, stink horns, ectomycorrhizal fungi.

Phylum Deuteromycota (imperfect fungi)

Class Blastomycetes: yeastlike fungi

Class Hyphomycetes: conidial fungi, nematode-trapping fungi

Class Coelomycetes: plant pathogens causing anthracnose of cotton, beans, and flax

7.3 SYMBIONTS OF PROTOZOA AND ANIMALS

Protozoa

Soil amoebae and testaceous (with shells) rhizopods are common hosts of specialized groups of zygomycetes. The fungi use adhesive hyphae to capture their hosts. Some fungal species do not penetrate their host, whereas others are internal symbionts. Each species of fungus produces a characteristic assimilative organ that digests the host cyto-plasm. *Stylopage rhabdospora* traps soil amoebae by means of thin, sticky strands of mycelia. At the point of contact with the amoeba, the fungus produces penetration pegs that enter the protist and branch to form assimilative organs. *Stylopage anomala* is commonly found in dung, where it parasitizes amoebae. It reproduces by spores that are dispersed by mites, which in turn are carried by dung beetles (Blackwell and Malloch, 1991). Species of *Cochlonema* and *Endocochlus* are obligate endosymbionts of soil amoebae. *Cochlonema explicatum* produces adhesive spores that attach to amoebae, germinate, and then produce germ tubes that enter the host cytoplasm. Once inside the host, the tip of the germ tube swells and elongates. The amoeba gradually becomes sluggish, and after it dies fungal spores are released from the host. A few hyphomycetes, such as *Dactylella tylopaga*, also capture rhizopods. The morphological adaptations of these fungi, such as adhesive hyphae and assimilative organs, are similar to those of predaceous zygomycetes.

Fungal attacks on soil amoebae are widespread in nature. These fungi are obligate symbionts whose axenic culture has not been successful because their spores will germinate only in the presence of a host. It is not known what specific nutrients the fungi obtain from the amoebae (fig. 7.2).

Nematode-trapping Fungi: Predatory Fungi

An American mycologist, Charles Drechsler, pioneered the study of fungi that trap microscopic organisms. From 1933 to 1975, Drechsler published papers about fungi that attacked nematodes and amoebae (Drechsler, 1934). In England, Charles Duddington's book "*The Friendly Fungi: A New Approach to the Eelworm Problem*," described the curious behavior of these soil fungi and captured the imagination of scientists and amateurs. Nematologists have explored the possibility of using carnivorous fungi to control nematode pests of plants. A soil fungus, *Paecilomyces lilacinus*, has been used successfully to control plant disease caused by nematodes. The fungus invades and destroys the egg masses of the root-knot nematode, *Meloidogyne arenaria*.

Nematodes are roundworms that are common inhabitants of soil (Barron, 1992). In addition to feeding on bacteria and fungi, nematodes parasitize plants and animals. Most

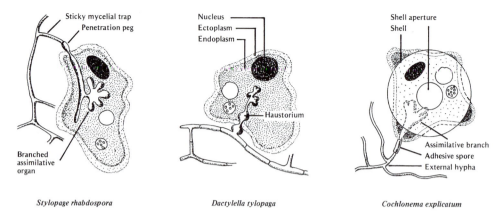

Fig. 7.2   Examples of fungal parasitism of soil amoebae. Adapted from Cooke (1977).

fungi that trap nematodes are hyphomycetes, but some are zygomycetes. These fungi are common in decaying organic matter (Gray, 1987). When a wandering nematode touches a hypha of the zygomycete *Stylopage hadra*, the fungal cytoplasm moves rapidly to the point of contact, where a drop of viscous fluid is released. The nematode sticks to the fluid and, after a brief struggle, becomes immobilized. At this time, the fungus penetrates the nematode's body by means of a slender peg, which grows and branches and fills the host body cavity with hyphae. The hyphae of *S. hadra* become sticky only after they are touched by a moving organism.

Hyphomycetes that capture nematodes produce specialized hyphal traps, which may or may not be adhesive depending on the fungal species. Following are examples of common types of traps (fig. 7.3).

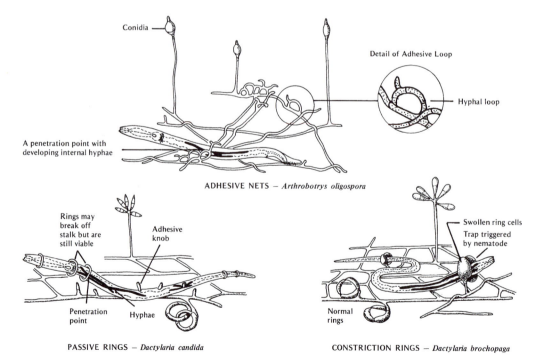

Fig. 7.3   Three types of mechanisms used by hyphomycetes to capture soil amoebae. Adapted from Barron (1977).

Adhesive nets and branches

Nets and branches are characteristic of many nematode-trapping fungi. Arthrobotrys oligospora produces a complex of three-dimensional nets and loops. The fungus produces short lateral branches that grow, curl, and then fuse with the parent hypha. A branch may arise from the loop hypha and produce another loop. Repeated loop production results in a network of adhesive hyphae. When a roundworm contacts a trap, it becomes caught, and as it struggles it entangles itself in other traps until it is no longer able to move. A penetration peg may develop from any point of the fungal network. In the body cavity of the nematode, the peg produces a bulbous structure from which the assimilative hyphae arise. Most of the cell contents of the host are digested and absorbed by the fungus.

Passive rings

Passive rings are typical of Dactylaria candida. Each ring consists of three or four cells and is supported on a stalk. The diameter of a ring is barely wide enough to accommodate the head of a nematode. If a nematode tries to pass through the ring, it gets caught and is unable to retract. The ring often becomes detached from the mycelium during a nematode's struggle to free itself, but the ring remains viable, and ultimately a peg from one of its cells enters the host. The killing action of this type of fungus is slow, and frequently a nematode will have several rings around its body as it moves through additional traps. In addition to passive rings, D. candida also captures nematodes by means of short, adhesive knobs. The late Harvard University mycologist William Weston dubbed these fungi the "nefarious nematode noose" fungus and the "lethal lollipop" fungus.

Constriction rings

Constriction rings are the most sophisticated trapping mechanisms produced by fungi such as Arthrobotrys anchonia and Dactylaria brochopaga. A typical ring consists of three cells supported on a one- or two-cell hyphal stalk. As the nematode's head moves through the ring, it contacts the ring cells and causes them to swell instantly, thereby trapping the nematode by constriction. Once a nematode is captured, a penetration peg from one of the ring cells grows into the body cavity and forms as-

similative hyphae. Only the inner surface of the ring swells, and the process is completed in 0.1 sec. The mechanism of ring constriction is poorly understood. Hot water, dry heat, and glass microneedles can trigger the constriction response.

What stimulates a nematode-trapping fungus to form traps or how it attracts and kills its prey is not fully understood. Some nematophagous fungi grown in axenic culture form traps only in the presence of nematodes. In one study, a culture broth in which nematodes had grown contained substances, collectively called nemin, which induced trap formation in Arthrobotrys spp. Amino acids such as valine and leucine, carbohydrates such as arabinose and ribose, and glycerol have stimulated trap formation in some fungi. Some scientists believe that the fungus produces chemicals that attract wandering nematodes to the traps. It is also possible that fungal spores may appear as food to nematodes. Toxins produced by the fungus may be the cause of death of captured nematodes. Fungi use nematodes to supplement nutrients they obtain from dead organic matter.

Endosymbionts of Nematodes

Endosymbiotic fungi of nematodes have several common features. These include (1) almost total absence of any hyphal structure outside the nematode, (2) exit tubes for the release of spores outside of the host, (3) resting spores that allow the fungus to survive in the absence of a suitable host, and (4) host infection resulting from ingestion of a spore. Following are examples of fungi that are internal symbionts of nematodes.

Meria coniospora (hyphomycete)

Meria coniospora produces tear-shaped spores that are sticky at their pointed end. The spores attach to the head of a nematode and produce penetration pegs from which infection hyphae develop. Young hyphae with a characteristic wavy appearance grow in the host body cavity. When mature, some hyphae break through the body wall of the host and produce spores (fig. 7.4).

Harposporium anguillulae (hyphomycete)

Spores of H. anguillulae are crescent shaped and have a pointed end. When the spores are ingested by a nematode, they become embed-

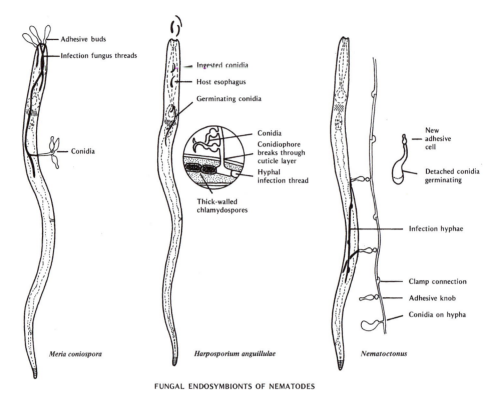

FUNGAL ENDOSYMBIONTS OF NEMATODES

Fig. 7.4   Fungal endosymbionts of nematodes. Adapted from Barron (1977).

ded in the muscular esophagus and the nematode is unable to feed. The spores germinate and produce hyphae, which grow throughout the body cavity of the host. At maturity the fungal colony develops erect hyphae, which produce numerous spores or conidia. The fungus also produces thick-walled chlamydospores that can withstand unfavorable environmental conditions (fig. 7.4).

Nematoctonus (basidiomycete)

Some species of *Nematoctonus* capture nematodes by means of adhesive knobs or aerial hyphae; other species are endosymbiotic. In the trapping species, each trap consists of a secretory cell that is swollen at both ends, constricted in the middle, and surrounded by a viscous, sticky fluid. When a nematode touches the fluid, a bond forms that is so strong that it rips open the cuticle of a struggling host. The fungus penetrates the captured nematode and fills its body cavity with hyphae. Endosymbiotic species produce sausage-shaped conidia on small, stalked hyphae. After dispersal, a conidium germinates

and produces an adhesive cell at the tip of the germ tube. The infection process begins when the cell becomes attached to the cuticle of a wandering nematode. A toxin produced by the germinating conidium immobilizes the nematode, which dies soon after it contacts the conidium (fig. 7.4).

Fungal Pathogens in Aquatic Ecosystems

Water molds play a significant role in decomposition and recycling in aquatic ecosystems. Species of *Achlya*, *Aphanomyces*, and *Saprolegnia* attack fish and crayfish. *Aphanomyces astaci* has devastated freshwater crayfish populations in Europe (Alderman and Polglase, 1988). *Lagenidium callinectes* parasitizes the eggs and larvae of marine crustaceans such as shrimp, lobsters, crabs, and barnacles. The shrimp embryos become colonized with a bacterium that produces an antifungal substance called *isatin*. Shrimp embryos that are surrounded by isatin are protected from *Lagenidium* infection (Gil-Turnes et al., 1989). *Lagenidium giganteum* parasitizes mosquito larvae and has been considered for biological control of mosquitos (Federici, 1981).

Arthropods

The phylum Arthropoda includes more than 70% of all the known species of animals. Insects, with an evolutionary history of 300 million years and more than 800,000 species, are the largest class of animals and represent one of the great success stories in evolution. Insects have occupied almost every conceivable ecological niche in terrestrial and aquatic habitats and have developed associations with fungi that range from casual to intimate. Fungal symbionts and insect hosts show a variety of morphological, physiological, and ecological coadaptations, and they also have some common characteristics. Both contain chitin as a structural component, which is present in the walls of fungi and in the exoskeleton of insects. Many fungi and insects are small and produce large numbers of offspring, and both groups play an important role in the recycling of organic compounds (Samson, 1988; Wilding et al., 1989).

Pathogenic fungi of insects

Species of *Beauveria*, *Cordyceps*, *Entomophthora*, *Metarhizium*, and *Verticillium* are well-known examples of fungi that are pathogens of insects. About 750 species of fungi infect insects. About 10 species of these fungi have been used successfully in the biological control of insect pests (mycoinsecticides). The fungi infect the insect hosts, assimilate their hemolymph, digest soft tissue, and produce mycotoxins, which kill the host (Hajek and St. Leger, 1994).

Many characteristics of insect–fungus symbioses can be generalized. Fungal spores usually attach to the insect cuticle in a specific manner, possibly through physical and chemical compatibilities between the fungus and its host. High humidity (>95% in some species) is necessary for spore germination. The spore produces a germ tube whose tip swells and attaches to the cuticle. Penetration pegs then develop and enter the host. On agar medium, a germinating spore produces protease, lipase, and chitinase, enzymes believed to be responsible for the digestion of the host integument. The fungus may take several days to reach the insect hemocoel, a blood-filled space in the tissues. If the insect molts

before the fungus reaches the hemocoel, the symbiont is discarded along with the old cuticle. The fungus multiplies in the hemocoel by small, budding bodies, the *blastospores*. The developing fungus produces a toxin that kills the host within 48 h. The host suffers from tremors and loss of coordination before death. The fungus then enters the mycelial phase and develops hyphae, which consume all the organs of the host. The insect body becomes mummified, and hyphae grow out of the body. These hyphae produce conidia that are dispersed by wind or water.

Molecular mechanisms of fungal pathogenesis in insects

Insect pathogenic fungi produce conidia that germinate and form *appressoria* as the first essential step in the infection process. *Metarhizium anisopliae* can produce appressoria on any hard hydrophobic surface, and their growth is stimulated by nitrogenous compounds. Intracellular second messengers such as Ca^{2+} and cAMP are involved in the formation of appressoria. Enzymes such as endoproteases, aminopeptidases, and PR1, a serine protease, rapidly degrade the insect cuticle. Molecular genetic analysis of PR1 and similar proteins has shed new light into how chitin barriers are breached by fungal pathogens. Within the host tissue, *M. anisopliae* produces toxins called *destuxins* (DTXs) which cause paralysis of muscles by activating their Ca^{2+} channels. Destuxinlike secondary metabolites are believed to be widespread among fungi and may also be involved in their growth and development (Clarkson and Charnley, 1996). Following are some well-known examples of entomogenous (insect-inhabiting) fungi.

Entomophthora (zygomycete). Species of *Entomophthora* have been used to control West African black flies, which are carriers of a nematode that causes blindness in humans. The fungus is an obligate parasite of the insect. Aphids, muscoid flies, caterpillars of butterflies and moths, and grasshoppers are common hosts of other species of *Entomophthora*. In Russia, pea aphids of legumes have been suppressed by *Entomophthora* infections. One major limitation in the commercial use of this fungus as a biological control agent is the relatively short life of its spores, about 2 weeks. One curious aspect of *Entomophthora* is that it produces singlespored sporangia that are "shot" from the tips of specialized hyphae. An infected insect becomes surrounded by a halo of white sporangia that are discharged from hyphae that grow out of the insect's body (fig. 7.5).

Cordyceps *(ascomycete)*. *Cordyceps* attacks many different types of insects and spiders (fig. 7.6). The infected hosts shrivel and dry after death but resist decay because of an antibiotic compound, *cordycepin*, produced by

Fig. 7.5   Conidial life cycle stages of *Entomophthora muscae*, parasite of the onion fly. (A) Infected female fly. (B, C) Lateral view of fly, showing fungal sporulation (arrows). (D) Conidiophores emerging from insect abdomen. (E) Single conidium attached to insect thorax. (F) Two conidia attached to an abdominal seta of the host fly. Note the mucilage attached to the conidia. (Courtesy Raymond I. Carruthers, USDA, Boyce Thompson Institute.)

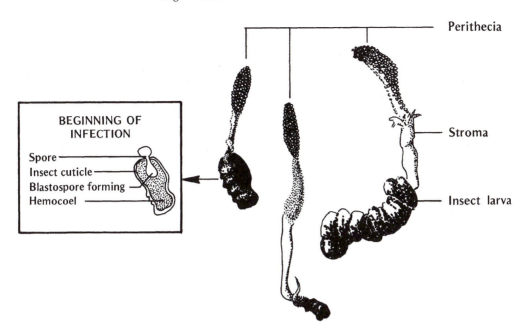

Fig. 7.6   *Cordyceps* infection of an insect larva.

the fungus. After the host dies, the fungus produces a cylindrical structure (*stroma*) that grows out from the host and produces many fruiting bodies (*perithecia*), which contain spores. The ancient Chinese were familiar with the mummified silkworms and cicadas produced by *Cordyceps* infections. Caterpillars infected with *Cordyceps* have been used in Chinese medicine for curing many illnesses ranging from drug addiction to the effects of overeating. A Chinese custom was to place precious stone carvings of the infected insects in the mouths of the deceased in the hope of immortalizing them.

Beauveria bassiana *(hyphomycete). Beauveria bassiana* is commonly isolated from dead, overwintering insects. In Russia, large quantities of the fungus are prepared commercially and used to control the potato beetle. The fungus reduces the longevity of the adult insects and also causes high larval mortality. *Beauveria* conidiospores remain viable in soil for up to 2 years. Efforts are being made in several other countries, including the United States, to use *Beauveria* to suppress potato beetle infestations.

Metarhizium anisopliae *(hyphomycete). Metarhizium anisopliae* shows promise for controlling the rhinoceros beetles that have de-

stroyed many coconut trees in the South Pacific. The beetle was accidentally introduced into the region in the 1930s. Beetle-infested trees do not produce fruit, and new plantings are destroyed in the seedling stage. The application of fungus around the base of an infected tree dramatically improves its health, as the fungus kills the beetles. Similar efforts to control rhinoceros beetle larvae with *M. anisopliae* in southern India have been successful. The fungus is frequently isolated from soil-inhabiting insects and is also being used to control mosquito larvae and the sugar cane spittlebug. In Brazil, spittlebug control by *M. anisopliae* application has been successfully extended to many acres of sugar cane and pasture land.

Symbionts of insects: nonpathogenic associations

There are many fungus–insect associations in which the host is not harmed. The associations are obligatory for one or both partners. The following are examples of nonpathogenic symbioses.

Laboulbeniomycetes. The *Laboulbeniales* are a group of microscopic ascomycetes that are ectosymbionts of insects and other arthropods. The fungi appear as black or darkish bristles to the naked eye and do not harm the insects. The fungi are obligate symbionts, and although some species have a wide host range, most live only on certain insect species. The laboulbeniomycetes became known to the world through the life-long scholarship of Ronald Thaxter at Harvard University, whose dedication to these insects resulted in a multivolume monograph (Thaxter, 1896–1931).

Incredibly, a fungus may be restricted to specific locations on the insect such as mouth parts, legs, or wings. Species of *Laboulbenia* and *Stigmatomyces* are common on flies and beetles (fig. 7.7). *Stigmatomyces ceratophorus* is the only species that has been grown in axenic culture, on fly wings kept on nutrient agar. The nutritional requirements of these fungi have yet to be investigated. A typical fungal life cycle begins with the germination of an ascospore on the host integument. The

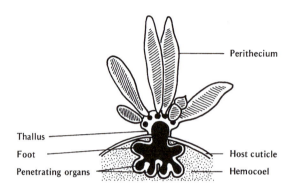

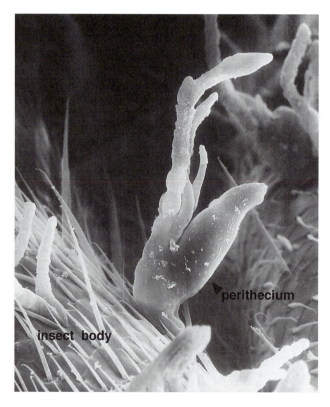

Fig. 7.7   *Laboulbenia* parasitism of insects. (Top) Schematic drawing of the fungus with its penetrating organs and perithecia. (Bottom) Scanning electron micrograph of the fungus perithecia emerging from the insect cuticle. (Courtesy Gerald Van Dyke, North Carolina State University.)

fungus produces a foot cell that anchors the symbiont to the host cuticle and then produces a thallus that may consist of only a few cells. The thallus produces hairlike appendages, which may be unicellular or multicellular. Male and female reproductive structures and the fruiting body, or perithecium, occur interspersed with the appendages. In most Laboulbeniales, the symbiont is anchored to the host by a foot cell that does not appear to penetrate the cuticle. How these fungi obtain nutrients from the host, without visible signs of penetration, is not clear. The fungus may produce minute penetrating structures, or it may be that the cuticle contains sufficient nutrients for the fungus. In a few species, such as *Herpomyces stylopage* and *Trenomyces histophthorus*, penetrating structures develop from the foot cell and grow through the host integument to the hemocoel, where fungal assimilative organs are formed. In *T. histophthorus*, when the fungus reaches the hemocoel, it produces spherical swellings with nodular branches that penetrate the fat bodies of the host. In no case do the fungal symbionts cause apparent ill effects in the host (Tavares, 1985).

Trichomycetes (hair fungi). Trichomycetes are obligate symbionts that usually live attached to the intestinal wall of marine, freshwater, and terrestrial arthropods (Lichtwardt, 1986). The fungus receives nutrients from the contents of the host's digestive tract but does not harm or benefit the host; thus, the symbiosis is commensalistic (Moss, 1979). The host becomes infected when it swallows fungal spores. The spores germinate and the germ tubes attach to the gut lining of the host by a specialized cell known as the *holdfast*. The fungal thallus consists of either branched or unbranched hyphae. Mosquito larvae, when deprived of sterols and vitamin B, survive better if they are infected with the trichomycete *Smittium culisetae* (Horn and Lichtwardt, 1981). Internal trichomycetes can be identified only by dissecting the insect host. *Amoebidium parasiticum* is an exceptional trichomycete in that it attaches only to the outside of aquatic crustaceans and insects (fig. 7.8).

Fungi and ambrosia beetles. Associations between fungi and beetles have been known for more than 160 years. In 1836, J. Schmidberger observed an association of beetles with a glistening white material, which he could not identify, in the galleries that beetles excavated in wood. He thought the material was sap from the wood and called it ambrosia. In 1844, Theodor Hartig discovered that the "ambrosia" was actually a fungus that was cultivated by the beetles and used as food.

Ambrosia beetles make up about 36 genera in 2 families of weevils. The beetle genera *Platypus* and *Xyleborus*, each with more than 200 species, are important groups of ambrosia beetles. Other common genera of ambrosia fungi include *Ambrosiella*, *Cephalosporium*, *Endomycopsis*, and *Fusarium*, all ascomycetes. A few basidiomycete ambrosia fungi are also known. The beetles inhabit the sapwood of weakened trees. Healthy trees and dead trees are usually not infected. Some beetle species attack only trees in a particular plant family, whereas other species infest

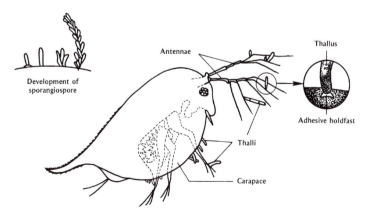

Fig. 7.8 The trichomycete *Amoebidium parasiticum* attached to *Daphnia* sp. Adapted from Moss (1979).

only certain regions of a tree, such as the terminal branches. The beetles excavate galleries in the wood and feed on the fungus that grows on the gallery walls. The beetles are well adapted to live in the galleries, being small and dark with elongated bodies and short legs. Their mouth parts are delicate and covered with flexible setae.

Each species of ambrosia beetle forms its own characteristic tunnel pattern. For example, in hickory trees, the *Xyleborus* beetle produces branched galleries from a single entrance into the bark. The female beetle carries fungal spores into the tunnel and deposits them on a substrate of chewed wood prepared by the beetle. The spores germinate to produce abundant mycelia and new spores all over the tunnel walls. The beetles graze on the chains of fungal spores. New fungal growth appears at regular intervals. The fungal garden, in general, is remarkably free from bacteria, yeast, and other contaminants. It is not known how the beetle maintains an almost pure fungus culture. The female beetles tend the fungus gardens, carefully preparing new beds into which they incorporate larval excrement. The female beetles also tend their offspring. Larvae with poor survivability are eaten or entombed in a branch tunnel. Larvae quickly die after the death of the mother beetle. The larvae are unable to maintain the fungus garden, and bacteria and other contaminants soon overgrow the galleries. Male beetles, which are wingless and short-lived, have no role in the ecology of a fungus garden, their main function being to inseminate the females. Some species of ambrosia beetles raise

their offspring in communal galleries; among others, each larva develops singly in a tunnel, excavating as it grows. The larvae have strong mandibles and can excavate wood. Female beetles carry the fungus inoculum in an externally attached saclike *mycetangium*, which is located on certain parts of the beetle, such as the mandible, elytra, or thorax (fig. 7.9). The fungus grows within the mycetangium and obtains nutrients from the host's secretions and excretions. The fungus is protected from drying out throughout the beetle's life and is thus kept available for seeding new sites during tunnel excavation (Norris, 1979).

Germinating spores of an ambrosia fungus produce germ tubes that penetrate the wood parenchyma cells adjoining the tunnel wall. The cells become filled with mycelium, and the tunnel walls become covered with a velvety mass of hyphae that bear spores. Most ambrosia fungi are dimorphic, producing a yeastlike ambrosial phase and a mycelial phase. The ambrosial phase commonly occurs in the mycetangium, in the beetle galleries, and sometimes in laboratory culture. The mycetangia of beetles often contain axenic colonies of ambrosia fungi, but many beetles emerging late in the season may carry secondary, foreign fungi that develop in the tunnels. Oak wilt fungus, *Ceratocystis fagacearum*, and the Dutch elm fungus, *Ophiostoma novo-ulmi*, are two important contaminants carried by ambrosia beetles.

The symbioses between ambrosia fungi and beetles are mutualistic. The fungal symbiont obtains several advantages from the as-

Tunnels of Ambrosia Beetle

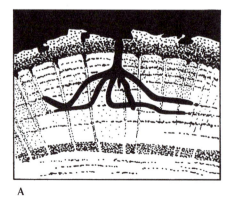

A

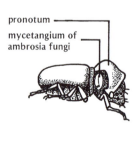

SCOLYTID BEETLE

pronotum ———
mycetangium of ———
ambrosia fungi

B

Fig. 7.9   Ambrosia symbiosis. (A) Tunnels produced by the ambrosia beetle, *Xyleborus*, in hickory tree. (B) Mycetangium of scolytid beetle. From Norris (1979).

sociation (Beaver, 1989): its range is increased by the beetle; its penetration of woody tissue is facilitated because of the mechanical injury caused by the beetles; it uses insect remains and excretory products as nitrogen sources; and it multiplies and obtains nutrients from the host within the mycetangium. The beetle host also benefits from the symbiosis. The fungus produces enzymes that decompose cellulose and thus break down woody tissues, which makes the excavation process easier for the beetles. The larvae feed exclusively on fungal conidia, and adults of many species also feed on the hyphae. The fungus produces steroids which the beetle uses for its development and for beetle pheromones.

Both the beetle and the fungal symbiont show the consequences of coevolution. The behavior of the female beetle is the key to the success of the symbiosis. Survival of the larvae depends on her ability to maintain the fungal garden and to suppress the growth of foreign contaminants in the galleries.

Wood wasps are similar to ambrosia beetles in their association with fungi. Female wood wasps lay eggs in the wood of conifer trees that have been weakened by disease or other insects. While the eggs are moving through the insect's ovipositor, they become coated with spores of a wood-rotting basidiomycete of the genus *Amylostereum* or *Stereum*. The fungal spores are contained in mycetangia located near the base of the ovipositor. The spores germinate in the wood and produce a mycelium that breaks down the wood and serves as a food source for the insect larvae. The larvae bore through the decaying wood, form tunnels and chambers,

and develop into pupae and adults (fig. 7.10). The wasp is an obligate symbiont, but the fungi involved in these symbioses also occur free-living.

Fungus gardens of termites and ants. Some species of termites and ants cultivate fungi for food. These termites (Termitidae; higher termites) use the fungi as a supplemental food source, for vitamins, whereas the ants rely entirely on the fungi for their nutrients. In termites the ingested fungi also provide enzymes that remain active in the gut and help the insect by digesting cellulose and other plant polysaccharides (Martin, 1992). The fungi are grown in specially prepared beds that contain a mixture of plant material and insect excrement. Termites and ants are similar in many aspects of their symbioses with fungi. Both insects have a rigid caste system that includes workers and soldiers, and both build nests that may contain a million or more insects. Termite nests are called *termitaria* and they are architectural wonders, with built-in temperature and humidity controls. Nests are produced in soil and may be up to 20 feet high (fig. 7.11). Fungus-cultivating termites are common in Africa and Asia and differ from other termites in their lack of intestinal symbiotic protozoa. The fungi cultivated in termitaria belong to the basidiomycete genus *Termitomyces*. White nodules in the fungus gardens are aggregates of conidiophores and conidia (Wood and Thomas, 1989).

Leaf-cutting ant species of the genera *Acromyrmex* and *Atta* maintain fungus gardens in nests throughout Central and South America. The nests consist of chambers in

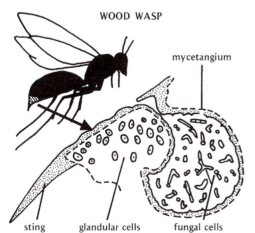

WOOD WASP

mycetangium

sting   glandular cells   fungal cells

Fig. 7.10 Mycetangium of a wood wasp, located next to the ovipositor. Adapted from Cooke (1977).

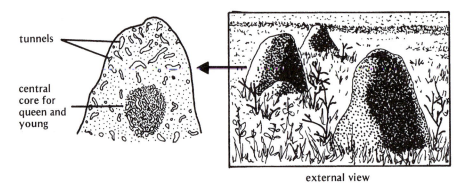

tunnels

central
core for
queen and
young

external view

TERMITE MOUNDS

Fig. 7.11   Termite nests with characteristic tunnels, where fungal gardens are maintained.

which the fungus is cultivated. The fungal symbionts of leaf-cutting ant species belong to the basidiomycete *Attamyces bromatificus* (Cherrett et al., 1989), an obligate fungus that has never been found ouside of a nest. Genetic studies have revealed that the ants have propagated the same fungus for 23 million years (Chapela et al., 1994; Hinkle et al., 1994). Numerous cultivars of the fungus have been identified (Mueller et al., 1998). New nests are started by fertilized females (queens) that carry some of the fungus from their old nest. Ant fungus gardens are developed by workers who cut leaf blades into large segments (10 mm in diameter) and carry them to the nest (fig. 7.12). Young leaves are preferred over old ones, and flower parts are also used. Well-worn trails are evident between ant nests and the bushes and trees that supply leaves. The leaf segments are cleaned, cut into small pieces (1–2 mm in diameter), and then chewed to a pulpy consistency. The ant worker mixes the spongy plant material with drops of its own exudates, which are rich in ammonia and amino acids. A small fragment of vegetation that contains a fungus from an old garden is then placed on the newly prepared bed. As the new fungus garden grows, ant workers continue to add drops of anal exudates to the bed. Fungal hyphae, which become swollen under these culture conditions, and spores are harvested periodically and used as food by the ant colony. The worker ants continuously remove secondary invaders from the nest, keeping bacteria and yeast populations at low levels. After the gardens reach an age of several months, they are discarded and replaced with new ones. Abandoned fungus gardens quickly become overrun with foreign contaminants. Experimental evidence has shown that the fungus may pass on chemical signals to the foraging ants regarding the suitability of some substrates as food (North et al., 1997).

Septobasidium and scale insects. Septobasidium (teliomycete) is a basidiomycete that commonly occurs in warm areas of the world. Most species are obligate symbionts of scale insects. The fungus produces brown-black, flat thalli that adhere firmly to the leaves and bark of living trees. The thalli have many chambers within which the host insects live. The chambers are interconnected by tunnels, and each chamber contains one insect. The fungus penetrates the insect by a few hyphae that form coiled haustoria within the hemocoel. The infected scale insect inside the fungal chamber inserts its stylets into the plant tissue and feeds on plant sap (fig. 7.13). Nutrients the insect obtains from the plant are also assimilated by the fungus in the hemocoel. Infected insects are small and sterile but are seldom killed. A scale insect colony also contains many uninfected individuals. In the spring, as the fungus produces new mycelial growth, the uninfected female insects in the fungal chambers produce nymphs. At the same time, the fungus forms erect fruiting bodies, which produce basidiospores that divide by yeastlike budding. The spores stick to the bodies of the wandering nymphs. The spores germinate and the germ tubes penetrate the body hemocoel and form haustoria. Many nymphs settle in the same fungal colony, but some leave and start new colonies elsewhere. In this mutualistic association, the scale insects provide nutrients to the fungus and the nymphs disseminate the fungus, while the fungus protects the insects from temperature extremes and from predation by birds and other organisms (Couch, 1938). Symbiosis between scale insects and *Septobasidium* represents an example of kin selection, where the reproduction of some members of the insect colony is sacrificed when they become infected with the fungus. The scale insects that reproduce are protected from parasitoid wasps (Seifert, 1981).

Yeasts as endosymbionts of insects. Many insects whose diet is blood, plant sap, or wood possess yeast or yeastlike endosymbionts of the genera *Candida*, *Taphrina*, and *Torulopsis*. These fungi occur in the intestines, gastric cecae, fat bodies, and Malpighian tubules of in-

sects. The symbionts are housed in specialized cells, *mycetocytes*, which are grouped together in *mycetomes*. An intestinal mycetome has an epithelium that consists mainly of mycetocytes, which are made up of enlarged fungus-filled cells with irregularly shaped nuclei. The cells lack microvilli on their absorptive surface, do not divide, and often are polyploid (fig. 7.14). The mycetocytes of some insects harbor fungi only, but others contain a mixture of fungi and bacteria. Some of the symbionts can be cultured, but most are obligate. The endosymbionts produce vitamins and essential amino acids that are used by the host insect and in return receive a stable and favorable environment in which to live. The endosymbionts are transmitted to new individuals by one of the following ways: licking and stroking of the offspring by the female parent;

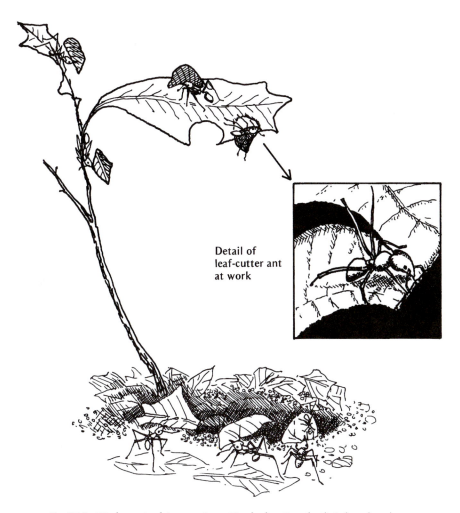

Detail of
leaf-cutter ant
at work

Fig. 7.12   Worker ants of *Atta* species cutting leaf sections for their fungal gardens.

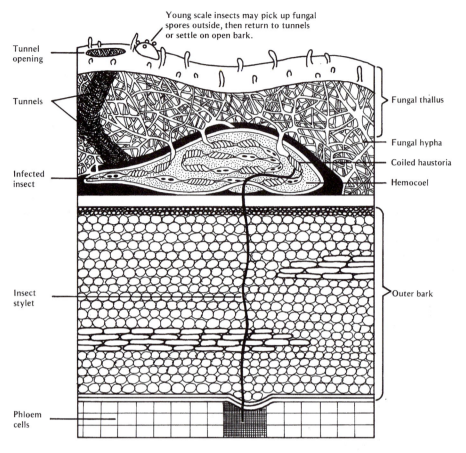

Fig. 7.13 Mutualistic symbiosis of the fungus *Septobasidium* with scale insects. Adapted from Cooke (1977).

the outer surface of the eggs becoming coated with fungi during the egg-laying process by means of fungus-containing structures attached to the ovipositor; or by the egg cell cytoplasm carrying the fungus (Nardon, 1988).

The symbionts in a coccid insect *Stictococcus diversisetae* enter the ovary and infect some of the eggs produced by the insect. These eggs develop parthenogenetically and give rise to females; uninfected eggs develop into males. Loss of a chromosome is involved in the production of males. Some scientists believe that endosymbionts of infected eggs, in a manner not understood, prevent the chromosomal loss.

Symbionts of Humans and Other Vertebrates

Associations of fungi and warm-blooded animals are almost universal. In nature, most an-

imals have low-grade fungal infections. Some fungi are well known because they cause mycoses in domesticated animals as well as in humans. Fungal and protozoan infections are now a serious public health hazard, especially for those who are immunocompromised because of AIDS, cancer therapy, or drugs taken to prevent organ rejection (Kasper and Buzoni-Gatel, 1998).

Endogenous fungi make up the normal biota of the skin and mucous membranes and often attack the host through injured tissue. Exogenous fungi live as saprophytes in the soil and sometimes produce diseases in animals. Fungal infections of vertebrates are of two main types: systemic infection, of deep internal tissues and vital organs such as lungs and brains; and superficial infection, of scalp, skin, or nails from fungi known as *dermatophytes*. Fungal parasites of vertebrates have a wide host range and are easily transmitted between humans and animals. Some fungal infections have become associated with certain occupations and lifestyles of humans. For example, histoplasmosis is prevalent among poultry farmers and cave explorers because bird droppings are a common source of fungus. Farmers, gardeners, and agricultural workers often have mild fungal infections.

Recently, an increasing number of plant pathogenic fungi have become implicated in immunosuppressed patients. *Alternaria alternaria*, *Aspergillus flavus*, and *Fusarium oxysporum* are common pathogens of strawberries, tomatoes and other plants that infect the lungs of immunocompromised patients. Opportunitistc fungal infections in immunosuppressed patients makes the development of new antifungal drugs critical (Miller, 1986).

Fungi that attack vertebrates have been reported from all major fungal groups, although only a few infections are caused by basidiomycetes. Following are some examples of fungal diseases of humans.

Candidiasis

Candidiasis is caused by a yeastlike fungus, *Candida albicans*. In humans, *Candida albicans* as well as true yeasts are residents of the mouth, alimentary canal, and vagina. The fungus normally lives as a saprophyte receiving nutrients from sloughed-off epithelial cells of the mucous membranes. *Candida albicans* causes *thrush*, in which the fungus invades the superficial tissue and produces soft, gray-white lesions on the gums, tonsils, and tongue. Candidiasis frequently occurs in individuals undergoing antibiotic therapy, on prolonged use of birth control pills, during pregnancy, and with diabetes mellitus. The fungus has a hyphal form during the invasive phase of the infection. The mechanisms controlling fungus morphology are not known. *Candida* can also cause serious systemic infections.

Aspergillosis

Species of *Aspergillus* cause several types of mycoses. *Aspergillosis* is increasingly being recognized as an important complication in individuals suffering from malignant diseases, such as leukemia and lymphoma, and in those receiving immunosuppressive drugs. *Aspergillus fumigatus* is a virulent pathogen that causes pulmonary aspergillosis and can spread to other organs. Asthmatic symptoms and fever are common in these mycoses. *Aspergillus fumigatus* may invade the nasal sinus tissue and can cause endocarditis in patients with heart valve replacements. *Fumigatotoxin*, a high molecular weight toxin, has been isolated from the fungus.

Several species of *Aspergillus* cause avian

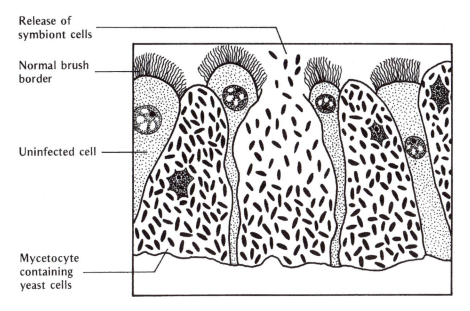

Fig. 7.14 Intestinal epthelial cells of insects with mycetome-containing yeast cells. Adapted from Cooke (1977).

aspergillosis. Young birds die from appetite loss, high temperature, and diarrhea. In 1940, many herring gulls in the Boston Bay area died from infections caused by *A. fumigatus* that had developed first on rotting seaweed. Aspergilli have also been involved in equine gutteral pouch mycosis, mycotic abortion, and mycotic mastitis.

Histoplasmosis

Histoplasmosis is a serious and often fatal infection of the lymphatic system. The disease is widespread in warm, humid regions of the world and is caused by *Histoplasma capsulatum*, a fungus that reproduces like a yeast in host tissue and produces mycelial growth in culture. The infection begins when spores of the fungus are inhaled into the lungs. After a prolonged exposure to the fungus, the individual develops chronic, tuberculosislike symptoms of cough, slight fever, and pulmonary cavitation. Progressively, the disease spreads to the spleen, liver, adrenals, kidneys, nervous system, and other organs. Histoplasmosis is frequently associated with soils that become contaminated with fecal material from bats and birds, including chickens. In city parks and plazas, where birds often roost, the soil and the buildings frequently become contaminated with *H. capsulatum*. There is no satisfactory treatment for this disease.

Coccidiomycosis

Coccidioides immitis causes a mild respiratory infection that is called the *San Joaquin Valley fever*. The fungus causes lesions in the upper respiratory tract and progressively develops in bone, joints, and other tissues, including the central nervous system. Symptoms include fever, chills, sweating, and general weakening. *Coccidioides immitis* infections are widespread in arid parts of the United States and throughout Central America. The fungus is soilborne and produces spores that remain viable for long periods. Rodents, domesticated animals, and humans become infected from inhaling the airborne spores. The fungus is dimorphic; in the invasive phase it produces spherical cells that function as sporangia at maturity; in the saprobic phase the fungus produces arthroconidia. Most people and animals living in areas where the fungus is common have mild coccidiomycotic infections. The chance of becoming infected increases with the length of stay in the area. The disease is rarely fatal.

Dermatophytic diseases: ringworms

The fungi collectively known as dermatophytes are limited to the keratinized layers of skin, nails, hair, or feathers of humans and animals. Diseases from dermatophytic fungi include various types of ringworm. Dermatophytic fungi belong to three genera: *Epidermophyton*, *Microsporum*, and *Trichophyton*. Species of *Epidermophyton* are confined mostly to humans and infect only the skin. These fungi are widely distributed in soil, and most infections are acquired through intimate contact. Humans contract ringworm disease from infected animals and also from contaminated combs, hairbrushes, caps, and furniture. *Microsporum canis* infection is usually acquired from dogs and cats. *Trichophyton* species are transmitted to humans from horses, cattle, dogs, and other animals.

Dermatophytic fungi flourish under warm, humid conditions. Most people during their lifetime become infected with at least one dermatophytic fungus. Infections are favored under conditions caused by continuous wet skin (swimmers), moist and oily skin, tight clothing and shoes that prevent evaporation of perspiration, and obesity, which results in increased folding of the skin where oil and moisture accumulate.

It is assumed that dermatophytic fungi can digest and assimilate keratinized tissue, although the mechanism of keratin digestion is not understood. Dermatophytic fungi can be grown in axenic culture, but in nature most species need a host to grow.

Mycotoxins

It is only since 1960 that we have learned about the significance of mycotoxins in illness and even death of humans and domesticated animals. In 1960, the deaths of more than 100,000 turkeys within a 100-mile radius of London led to the discovery of the mycotoxin *aflatoxin*, which was isolated from *Aspergillus flavus*, a fungus that had contaminated the peanut feed of the turkeys. Mycotoxicoses are now recognized as occurring worldwide, and a number of mycotoxin-producing fungi have been identified. Scientists in the former Soviet Union were first to appreciate the significance of mycotoxin-produced diseases. Perhaps the prolonged storage of grains and other animal feed during the long Russian winters increased the possibility of mold growth. Consumption of moldy hay was found to be responsible for the deaths of many horses in the Ukraine in 1937. In 1954, A.K. Sarkisov published a detailed monograph in Russian on mycotoxicoses, but the work remained unknown in Western countries until the mid-1960s.

All mycotoxin-produced diseases are characterized by noncommunicability, unresponsiveness to antibiotic drugs, and geographically limited outbreaks usually linked to moldy food. The toxic response in the affected individuals occurs through organs such as the liver, kidney, lung, and nervous system. The chief mycotoxin-producing fungi belong to the genera *Aspergillus*, *Fusarium*, and *Penicillium*. Following are some examples of fungi and their mycotoxins.

Aspergillus

Aspergillus flavus and *A. parasiticus* are two aflatoxin-producing fungi that are prevalent in hot, humid climates. Aflatoxins have been isolated from edible nuts such as peanuts, walnuts, almonds, pecans, Brazil nuts, and pistachio nuts and from grains such as wheat, rice, corn, sorghum, cottonseed, and millets. In a 1960 British study, aflatoxin types B_1, B_2, G_1, and G_2 were identified. Since then it has been established that dairy products from cows that ingested B_1 toxin contaminated hay contained M_1 and M_2 types of aflatoxins. The M aflatoxins are firmly bound to casein molecules. The G_1 and M_1 toxins are carcinogenic. Aflatoxin-susceptible hosts such as rabbits, guinea pigs, rainbow trout, rats, and monkeys develop liver damage with necrotic lesions and swollen kidneys. The first signs of the disease are loss of appetite and weight, followed by poor muscular coordination. A high incidence of human liver cancer in parts of Africa, Southeast Asia, and Japan is correlated with aflatoxin-contaminated food (Bilbrami and Sinha, 1992).

Claviceps purpurea

Ergotism is caused by the ascomycete *Claviceps purpurea* and it is one of the oldest recurring epidemics in European history (Schumann, 1991). Brownish-black sclerotia are formed in infected ovaries of grasses such as rye. It has been suggested that women who displayed signs of "bewitchment" in Salem, Massachusetts, during 1692 may have been suffering from convulsive ergotism (Matossian, 1989). Ergot mycotoxins are a complex mixture of alkaloids, and the two main types are lysergic acid and agroclavine. Disease symptoms of ergotism include gangrenous necrosis of fingers or toes and convulsions when ergot-contaminated grains are ingested by humans or cattle. Ergot mycotoxins cause vasoconstriction and have found widespread medical applications (Marasas and Nelson, 1987).

Penicillium

Penicillium viridicatum is an important mycotoxin-producing species. It usually infects stored grains and causes illness and death in humans and domesticated animals. *Penicillium citrinum* is widespread in the tropics and subtropics, and its toxin produces a lethal kidney disease. Mycotoxins of *P. islandicum* have received attention from Japanese scientists because consumption of rice contaminated with this fungus produces the disease rice toxicosis. As many as four potent hepatotoxic carcinogens have been isolated from *P. islandicum*. One of these toxins, *islanditoxin*, causes severe liver damage, hemorrhage, and rapid death. *Penicillium cyclopium* is commonly associated with food grains and produces penicillic acid, cyclopiazonic acid, and two tremor-producing compounds. These fungi usually inhabit soil as saprobes and develop on the seeds only after harvest.

Fusarium

Fusarium graminearum infects corn and produces trichothecene mycotoxins that induce vomiting in pigs. *Fusarium poa* and *F. sporotrichioides* produce the fescue food syndrome in cattle. The animals show symptoms of lameness, weight loss, arched back, swelling in the hind legs, separation of the hoof from the foot, abnormal horn and hoof growth, and dry gangrene in the extremities. If not checked, the disease can be fatal. Trichothecenes are produced only by *Fusarium* and related fungal species. In several Brazilian plant species of *Baccharis* the female plants produce trichothecenes which are concentrated in the flowers and fruits. It is speculated that this represents an example of horizontal gene transfer between the fungal symbiont and *Baccharis* host plant (Kuti, et al., 1990).

Mycotoxins are potent carcinogens that form a DNA adduct, often resulting in mutations as a result of G to C transversions (Baertschi, et al., 1989). Recently, a genetic mechanism has been reported that suggests how these mycotoxins are a causal factor in human hepatocarcinogenesis (Aguilar et al., 1993). Aflatoxin has been demonstrated to cause critical transversions in the p53 tumor-suppressor gene, and such a change has been found in more than half the liver tumors examined in regions of the world where aflatoxin contamination of foods is high.

Mycotoxin research is still in its infancy. There are a number of challenging questions that need additional research, such as the prevalence of mycotoxins in food crops, their relationship with carcinogenicity, and ways to prevent their occurrence.

7.4 SUMMARY

Fungi have successfully formed associations with all forms of life in most ecosystems and play a significant role in decomposition. Fungal diseases in plants, animals, protozoa, and even other fungi have received considerable attention from mycologists. Unique features of fungi, such as the way they obtain nutrients, their mycelial growth and spore production, and the chitinous nature of their cell walls have convinced scientists that fungi belong in a separate kingdom. Fungal taxonomy is changing and organisms traditionally called fungi (chytrids) have been placed in other kingdoms. In this chapter fungal associations with soil amoebae, nematodes, insects, and animals were examined in detail.

Some fungi have evolved morphological and physiological adaptations to trap and capture nematodes. Species of *Laboulbenia* are microscopic ascomycetes that are obligate ectosymbionts of insects and arthropods. Some species of these fungi are so highly specialized that they grow only on certain parts of an insect host. Trichomycetes form commensalistic relationships with arthropods. Ambrosia beetles and wood wasps form mutualistic associations with wood-rotting fungi, and some species of termites and ants establish fungal gardens for food. The basidiomycete *Septobasidium* is an obligate symbiont of scale insects, with which it forms a unique mutualistic relationship. Yeasts are endosymbionts of many insects. Fungal parasites of nematodes and insect pests have provided opportunities for scientists to develop strategies for biological control. Some fungi cause diseases of humans, such as candidiasis, histoplasmosis, coccidiomycosis, and ringworms. These opportunistic fungi attack hosts with weakened immune systems. Mycotoxins such as aflatoxin and islanditoxin produce diseases in humans and animals. Fungal disease epidemiology is receiving increasing attention from the medical community.

8

FUNGAL ASSOCIATIONS OF FUNGI, ALGAE, AND PLANTS

Symbiotic interactions between fungi and phototrophs were instrumental in the colonization of land by algae and plants. Fungi may have helped the early phototrophs cope with the drying conditions of the terrestrial environment. Such interactions led to permanent associations now recognized as lichens, mycorrhizas, mycophycobioses, and endophytes of grasses (Atsatt, 1988, 1991; Selosse and LeTacon, 1998).

8.1 MYCOSYMBIONTS OF FUNGI

More than 1000 species of fungi obtain nutrients by infecting other fungi. Most of these mycosymbionts are conidial fungi (Hyphomycetes), and they have been used as biological control agents of plant pathogens such as *Phytophthora*, *Pythium*, and *Sclerotinia*. Necrotrophic mycosymbionts kill their host fungus before the infection process is established, but most fungal associations are biotrophic and do not appear to injure their hosts. The terms *hyperparasitism*, *mycoparasitism*, and *fungicolous* are often applied to interfungal associations (Elad, 1995; Jeffries, 1997).

Necrotrophic Mycosymbionts

Common soil-inhabiting fungi such as *Ampelomyces quisqualis*, *Gliocladium roseum*, *Rhizoctonia solani*, and *Trichoderma viride* are necrotrophic mycosymbionts. They can live saprobically on dead organic matter and also attack a wide range of fungi (fig. 8.1). Many mycosymbionts grow well in laboratory cultures. When a host fungus is introduced into a culture, the mycosymbiont grows toward it, and, in some cases, its hyphae coil around the host hyphae. The mycosymbiont produces diffusible toxic substances that change the host cell's permeability. Nutrients leak out from the infected cell and it dies. *Trichoderma* spp. produce hydrolytic enzymes such as β-glucanase, chitinase, and other volatile metabolites. *Rhizoctonia solani* is a well-known root pathogen that kills young seedlings. In culture, *R. solani* can destroy susceptible fungal hosts such as *Rhizopus nigricans* and species of *Mucor* by penetrating the host mycelium.

Gliocladium roseum attacks a wide range of fungi in nature and in laboratory culture. The fungus curls around the host hyphae by means of short branches and kills the host on contact. The fungus then penetrates the cells and absorbs their nutrients. *Gliocladium roseum* hyphae penetrate cells of *Mucor* sp., *Rhizoctonia solani*, and *Rhizopus nigricans* while they are still alive.

Ampelomyces quisqualis attacks powdery mildew fungi. At first, the mycosymbiont grows harmlessly among the host hyphae, but later it kills them. The fungus is believed to overwinter as a saprobe on mildew-infected leaves. In one study, clover powdery mildew was controlled by artificial inoculation with *A. quisqualis*.

Darluca filum attacks only spores and mycelia of rust fungi. Because of this limited host range, some scientists have speculated that the ancestor of *D. filum* was a leaf pathogen that lost its ability to infect leaf tissue and instead parasitized rust fungi.

Tuberculina maxima is often described as the purple mold of rust fungi, and it is a nat-

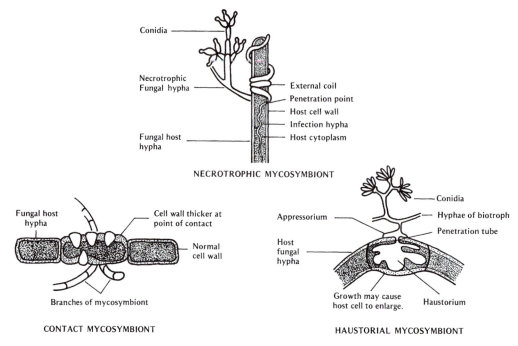

Fig. 8.1   Three types of interfungal parasitic symbiosis. Adapted from Cooke (1977).

ural enemy of pine blister rust. This mycosymbiont infects mostly the pycniospores of the host, which reduces aeciospore production. *Tuberculina maxima* also destroys rust mycelium in infected tissues of white pine. Spores of the fungus remain viable for several years. There has been interest in using *T. maxima* as a biological control agent of pine blister rust in the western United States and Canada (Schumann, 1991).

Biotrophic Mycosymbionts

Biotrophic fungal symbioses are highly specialized, long-lived relationships, and many fungal species are obligate symbionts of their host fungi.

Contact mycosymbionts

Contact mycosymbionts were not recognized until 1958 but are now believed to be common. About six species of hyphomycetes from genera such as *Calcarisporium*, *Gonatobotrys*, and *Stephanoma* are involved in contact mycosymbioses. Extensive observations have been made on *C. parasiticum* and *G. simplex*.

The mycosymbiont produces short, lateral hyphae that contact the host hyphae and cause them to release nutrients. Except for arrested growth, the infected fungus shows no adverse effects. Mycosymbionts will grow in culture only if a growth factor, *mycotrophein*, extracted from the host fungus, is part of the culture medium. Contact mycobionts obtain the vitamins biotin and thiamine from their hosts (Moore-Landecker, 1996).

Haustorial mycosymbionts

Infection of species of the order Mucorales (zygomycete) by other members of the same order is common. Spores of the mycosymbiont germinate in reponse to substances that diffuse from a host mycelium. The germ tube grows toward the host hyphae, and upon contact, the tip of the germ tube swells to form an *appressorium*, a flat, hyphal cell. Fine hyphal branches from the appressorium penetrate the host cell and then enlarge to form inflated *haustoria*. An internal mycelium develops from the haustoria, and on maturity hyphae emerge from the host and form sporangia.

Piptocephalis virginiana attacks a large number of *Mucor* species. Electron microscopic studies have shown that the haustorium of *P. virginiana* is similar in structure to that of fungal symbionts of plants.

8.2 FUNGAL–ALGAL ASSOCIATIONS

Mycophycobioses

A *mycophycobiosis* is an obligate symbiosis between a filamentous marine fungus and a large marine alga in which the alga is the dominant partner (Kohlmeyer and Kohlmeyer, 1979). The fungus grows between cells of the algal thallus but never penetrates or damages the cells. The effect of the fungus on the alga is not clear. The fungus may protect the algae from drying during low tides when the algae are exposed, and the fungus may be necessary for algal development. Sporelings of some marine algae, such as *Ascophyllum nodosum*, will not develop a thallus unless they are infected by a fungus (Garbary and London, 1995).

Fungi that form mycophycobioses include species of *Blodgettia* and *Mycosphaerella*. *Blodgettia bornetii* is a hyphomycete that grows in the walls of tropical species of *Cladophora*. The fungus produces spores and is absent only from the growing tips of the alga. The ascomycete *Mycosphaerella ascophylli* is always associated with the brown algae *Ascophyllum nodosum* and *Pelvetia canaliculata*. Hyphae grow between cells of the cortex and medulla of the host, and the fungus undergoes sexual reproduction and produces fruiting bodies while still within the host.

The ascomycete *Turgidosculum complicatulum* forms associations with the green algae *Prasiola borealis* and *P. tesselata*. Hyphae of the fungus grow throughout the algal thallus and separate cells of the thallus into groups of four or into rows. These algae can also live without the fungus. *Prasiola* thalli infected with fungus are more common than uninfected forms in exposed or drier parts of the intertidal zone. The fungus protects the alga from drying. *Turgidosculum ulvae* is a common symbiont of the green alga *Ulva vexata*. The fungus grows so extensively that it separates the layers of the algal thallus.

Lichens

A *lichen* is an association between a fungus (mycobiont) and a photosynthetic symbiont (photobiont) that results in a stable thallus, or body, of specific structure. The photobiont is either an alga or a cyanobacterium. A remarkable feature of a lichen is the transformation that the symbionts, in particular the fungus,

undergo during the association. A new entity, the thallus, is formed, and unique chemical compounds are synthesized. The physiological behavior of the symbionts also changes in symbiosis (Honegger, 1991; Hill, 1994). There are about 15,000 species of lichens, an indication that this type of symbiosis has been highly successful and has involved many species of fungi (Nash, 1996). Surprisingly, only about 30 different types of algae and cyanobacteria have been reported as photobionts. Taxonomically, only the fungus and alga of a lichen have Latin names, although commonly the name of the fungus is also used for that of the lichen. For example, *Cladonia cristatella* is the name of a mycobiont but, unofficially, the name of a lichen as well. The photobiont of this lichen is *Trebouxia erici*.

Types of lichens

A lichen thallus usually consists of layers such as an upper and lower cortex, algal layer, and medulla. The layers differ in thickness and are better developed in some species than in others. Fungal hyphae make up most of a thallus; the photobiont cells are only a small percentage (about 7%) of the total volume (Ahmadjian, 1993).

There are three main types of thalli: crustose, foliose, and fruticose. A *crustose* thallus lacks a lower cortex and is generally considered to be the most primitive type. Thalli of *Lepraria* species do not have layers but consist only of powdery granules. There are more species of crustose lichens than other types, and most of them belong to the genera *Lecanora* and *Lecidea*. Many crustose lichens stick tightly to the substratum and appear to be painted on it. Some species grow inside rock crevices and bark and still manage to produce separate layers. *Squamules* are typical of many species of *Cladonia*. Squamules are a specialized type of crustose thallus and are attached at only one end to the substratum.

A *foliose* thallus has an upper and lower cortex, an algal layer, and medulla and is usually loosely attached to the substrate by hairlike structures called *rhizines*. The thallus has many different sizes and shapes and is often divided into lobes. Common foliose genera include *Anaptychia*, *Cetraria*, *Parmelia*, *Physcia*, and *Xanthoria*. Some foliose lichens, such as *Umbilicaria* (rock tripe), have thalli that are attached to the substrate by only one central point.

Fruticose thalli are upright or hanging, round or flat, and often highly branched. Thalli of *Usnea* are hairlike and can reach a length of 5 m, whereas those of *Evernia* are shorter and strap-shaped. The layers of a fruticose thallus may surround a central thick cord, as in *Usnea*, or a hollow space as in some *Cladonia* species.

Some lichens, such as *Collema*, which have *Nostoc* as a photobiont, do not form a well-organized thallus. In these cases, fungal hyphae grow inside the thick gelatinous sheaths of the photobiont, which makes up much of the thallus. *Geosiphon pyriforme*, once thought to be a primitive lichen with a *Nostoc* photobiont, is now considered to be a vesicular arbuscular mycorrhizal fungus (Gehrig et al., 1996).

Most lichens have an ascomycete as the fungal partner, and only about a dozen are formed by basidiomycetes. The lichenized ascomycetes are one of the largest groups of the phylum Ascomycota in the kingdom Fungi. The lichenized habit obviously has had great selective benefits for fungi.

Distribution

Lichens grow practically everywhere—on and within rocks, on soil and tree bark, on almost any inanimate object. They grow in deserts and in tropical rainforests, where they occur on living leaves of plants and ferns. They have been found on the shells of tortoises in the Galapagos Islands and on large weevils in New Guinea. In the dry valleys of Antarctica, endolithic lichens, such as *Buellia* and *Lecidea*, grow inside sandstone crevices (Hively, 1997). *Dermatocarpon fluviatile* and *Hydrothyria venosa* grow in freshwater streams, and species of *Verrucaria* are common in the intertidal zones of rocky, ocean shores. *Verrucaria serpuloides* is a permanently submerged marine lichen that grows on stones and rocks 4–10 meters below mean low tide off the coast of the Antarctic Peninsula. *Verrucaria tavaresiae* is another unusual marine lichen with a brown algal photobiont (Moe, 1997). Douglas Larson has estimated that about 8% of the earth's terrestrial surface is dominated by lichens (Ahmadjian, 1995a). Lichens abound in areas with high annual humidity, such as the fog belt zones of Chile and Baja California. Extensive lichen populations also grow in the cool, northern forests of the world, where hundreds of miles of forest floor are covered with thick carpets of reindeer lichens (*Cladonia*). Trees along the coasts of the northwestern United States may be blanketed with beard lichens such as *Alectoria* and *Usnea*.

Lichens with organized thalli do not grow well in areas that are continuously wet, such as tropical rainforests. Only poorly organized species of *Lepraria* and leaf-inhabiting lichens are found in these regions. *Lecanora conizaeoides* and *Lecanora dispersa* colonize trees and gravestones in industrial cities and towns, but most lichens cannot tolerate the polluted atmosphere and persistent dryness of urban areas. The sensitivity of lichens to atmospheric pollutants such as sulfur dioxide, ozone, and fluorides has made them valuable indicators of pollution in cities and industrial regions. Lichens have been used to map pollution zones around many of the world's major cities (Richardson, 1992).

Dispersal and reproduction

Lichens are dispersed by thallus fragments and vegetative diaspores such as isidia and soredia. Each diaspore consists of a few algal cells and fungal hyphae. *Soredia* are powdery granules that originate inside the thallus, as localized overgrowths of algae and hyphae, and break through the upper cortex. *Isidia* are cylindrical extensions of the thallus. About 39% of all foliose and fruticose species of lichens produce isidia. Diaspores are dispersed by water, wind, insects, and birds.

Lichen fungi produce the same type of reproductive structures as other ascomycetes. Only some aspects of the sexual process have been seen in lichens, such as the fusion of microconidia to the tips of trichogynes. Dispersal of photobionts occurs by means of motile (zoospores) and nonmotile (autospores) spores.

Interactions between symbionts

Physical relationships. The basic unit of a lichen thallus is one algal cell with enveloping hyphae. Fungal hyphae adhere to the surface of an algal cell by means of a mucilage produced by both symbionts. As the fungus envelops the algal cells, it forms two types of specialized cells, appressoria and haustoria. These structures are common features of pathogenic fungi. The appressorium fastens the mycobiont tightly to the photobiont and gives rise to hyphae that grow into the algal cell and form haustoria. Hyphae penetrate the algal cell by enzymatic and physical means; that is, they partially dissolve their way through the algal wall and also push their way through. The plasma membrane of the algal cell always remains intact, no matter how deeply the hyphae grow inside the cell. In some lichens, hyphae barely penetrate the algal cell wall; in others they extend more deeply. Lichens with small haustoria are thought to be more highly evolved than those with large haustoria. Presumably, as lichens evolved, fungal penetrations into the algal cells were inhibited by stronger algal defenses. Small haustoria are difficult to see because they are obscured by the chloroplasts that fill many algal cells. An algal cell may have multiple haustoria (Honegger, 1992).

The frequency of haustoria in lichens shows great variation. *Trebouxia* photobionts, the most frequent type of photobiont, commonly contain haustoria. In *Cladonia cristatella* more than 50% of the algal cells have haustoria. Crustose lichens generally have the largest haustoria. Large foliose lichens such as *Sticta* and *Peltigera* possess algal cells that are closely associated with fungi but that are not penetrated by hyphae. This absence of haustoria is unusual among lichens and is thought to result from the presence of *sporopollenin* in walls of the *Coccomyxa* photobionts of these lichens. Sporopollenin forms a rigid layer in the algal cell wall and appears to stop fungal penetration. Sporopollenin does not occur in the cell walls of *Trebouxia*, a fact that may explain why these algae have many haustoria. The function of haustoria is not clear, but most likely they absorb nutrients from the algal cells. Details of the interfaces between the symbionts of lichens have been observed with low-temperature scanning electron microscopy (Scheidegger, 1994; Honegger et al., 1996).

Recognition mechanisms. Whether or not organisms recognize each other within a symbiotic system is a subject of much interest to scientists. In the *Rhizobium*–legume association, symbiont specificity is close, and it may be regulated by a recognition mechanism that involves complementary binding of lectins and polysaccharides. The binding occurs on the external surfaces of the symbionts. In lichens, the evidence for a similar recognition mechanism is inconclusive. Although lectins have been isolated from lichens, there are

conflicting reports on whether they function in symbiont interactions (Molina and Vicente, 1995). If there is a recognition phase in lichens, it may occur later in the development of the thallus. Initial contacts between fungi and algae are nondiscriminatory, and mycobiont hyphae will bind to any object, even glass beads. Similarly, it is not known if there is any type of signal exchange between lichen symbionts (fig. 8.3).

Secondary compounds. Lichens contain unique secondary compounds, which are commonly called *lichen acids*. These compounds were thought to be products of the symbiosis, but studies of isolated mycobionts growing in culture with high concentrations of sucrose have revealed that the fungus alone produces these compounds (Hamada et al., 1996). In lichens secondary metabolism appears to be connected to desiccation and aerial growth of the mycobiont (Armaleo, 1995).

Secondary compounds of lichens may have important ecological roles (Nash, 1996). Many have antibiotic activity and may prevent the microbial decay of lichen thalli, which may live for hundreds and even thousands of years. Lichen substances also inhibit bryophytes, other lichens, as well as seed germination and seedling development in vascular plants. The compounds protect thalli from grazing by invertebrate herbivores, such as slugs and insect larvae, and act as light screens to protect the photobionts from high light intensity. Lichen substances are also chemical weathering compounds that have a role in soil formation because of their chelating properties.

Synthetic lichens. Some lichens have been produced in axenic laboratory cultures by recombining their separate fungal and algal symbionts. The most successful artificial syntheses have been with *Cladonia cristatella* ("British soldiers"; fig. 8.2) and other similar species of *Cladonia*, and with *Usnea strigosa* (beard lichen). The synthetic lichens formed in the laboratory have been identical to their natural counterparts in morphology and, to some extent, in chemistry.

Synthetic lichens have provided useful information concerning the nature of the lichen symbiosis and the range of specificity that exists among symbionts. Synthesis studies carried out with *Cladonia cristatella* have revealed that the mycobiont will lichenize species of *Trebouxia* other than its natural

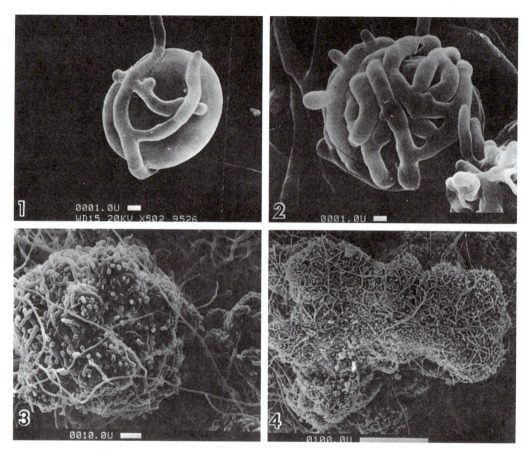

Fig. 8.2 Early synthesis stages of the lichen *Cladonia cristatella*. (1) Algal cell enveloped by fungal hyphae. The hyphae adhere to the surface of the algal cell by means of a gelatinous matrix around both symbionts. (2) One algal cell enveloped by hyphae and penetrated by a fungal haustorium (arrow). (3) A group of algal cells completely enveloped by hyphae (a presquamule stage). (4) Fungal hyphae are forming a network (upper cortex) over the algal groups and are starting to differentiate into individual squamules.

phycobiont; however, it will not accept "foreign" algae (i.e., those belonging to different genera). In several experiments, the mycobiont partially lichenized *Friedmannia israeliensis*, a free-living alga, i.e. formed soredia. Other algae that were foreign to the *C. cristatella* mycobiont were parasitized and killed before the initial stages of lichenization took place. Compatible algae were also penetrated by the fungus, but they were not killed, possibly because they possessed a defensive mechanism against the fungus. The relationship between fungus and alga in a lichen is thought to be a controlled parasitism. The fungus parasitizes the alga, but under natural conditions the parasitism is slow and infected algal cells may live for years (Ahmadjian, 1995b).

Many lichens have been cultured in the laboratory by means of the Yamamoto method (Yoshimura et al., 1993), which involves grinding thalli and then passing the suspension through different sizes of sieves. Pieces of the residue are then transferred to culture media and incubated. Experimental cultivation of lichens has helped clarify our understanding of the developmental stages of lichens (Stocker-Worgotter, 1995).

Physiology of the symbiosis

Carbohydrate transfer from photobiont to mycobiont. Most of the studies that have dealt with the physiological interactions between lichen symbionts have focused on the passage of nutrients from the photobiont to the mycobiont.

The reason for this is the availability of radioactive isotopes, such as $^{14}CO_2$, that can be fixed photosynthetically by the photobiont and then traced in other compounds as they move within the thallus. Studies using radioactive isotopes have revealed much information about the physiology of the lichen symbiosis (D.C. Smith and Douglas, 1987).

In a lichen thallus the photobiont excretes more than 90% of the carbon that it fixes photosynthetically as a polyol or a sugar such as glucose. The polyol excreted by green symbionts is ribitol, erythritol, or sorbitol; blue-green photobionts excrete glucose. The fungus may control the rate of polyol excretion by the photobiont, according to the urease theory. This theory states that urea in a lichen thallus is hydrolyzed by the enzyme urease to produce CO_2 and NH_3. Carbon dioxide stimulates photosynthesis of the photobiont, while NH_3 increases its respiration and carbohydrate release. When the lichen fungus is actively growing, it can increase the flow of

nutrients from the photobiont cells by producing more urease. Lichen acids and certain proteins act as a feedback control because they inactivate urease (Perez-Urria et al., 1989). Thus, during periods of fungal growth there is a greater release of nutrients but also an increase in lichen acids, which inactivate urease and slow down photosynthesis.

What happens to the large amounts of polyols and glucose released by the photobionts? These compounds are absorbed by the mycobiont and converted to mannitol, which is a fungal storage product. Such a conversion creates a sink to which algal nutrients continue to flow. The fungus uses some of the mannitol for growth and development, but the rest is used to help it withstand the extreme conditions of its habitat. Some scientists believe that the polyol reserves that accumulate in the fungus are used up during resaturation respiration. This occurs each time a dry lichen is rewetted. Lichens cannot control their water content and therefore undergo

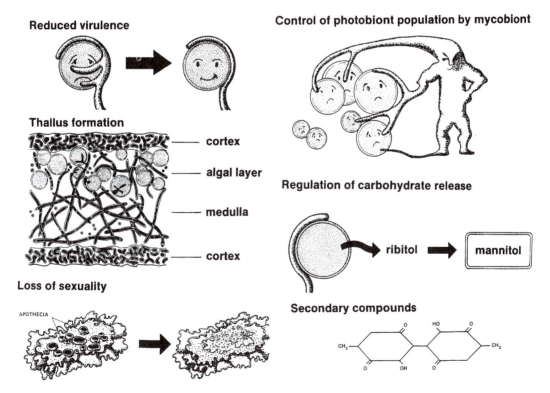

Fig. 8.3   Lichen stages that may involve signals between lichen symbionts.

frequent cycles of wetting and drying. Each time a dry lichen is wetted, its respiration increases, remains elevated for several hours, and then returns to normal. In species of *Umbilicaria* resaturation respiration lasts for 40 min, but in *Cetraria cucullata* it may persist for 5 h. Presumably, during resaturation respiration, the fungus respires polyols instead of proteins and other vital compounds. A lichen has to compensate for losses resulting from resaturation respiration as well as for losses that occur when carbon compounds leak out of the photobiont cells during the first few minutes of wetting.

Photosynthesis and respiration. In a lichen, photosynthesis reflects the activity of the photobiont, whereas respiration is mostly that of the mycobiont, which makes up the bulk of the thallus. In *Xanthoria parietina*, the fungus constitutes about 43% of the thallus volume, while the alga makes up only 6.7%; extracellular substances and air spaces make up the rest of the thallus. This wide difference in the proportions of symbionts in a thallus makes measurements of metabolic processes difficult. Moreover, the close physical relationship between the symbionts makes it impossible to obtain pure fractions of the algal symbiont, thereby limiting the types of studies that can be performed.

The metabolic rate of a lichen is influenced by light, temperature, day length, and water content. Some lichens can adjust their metabolic responses to different environments and seasons. For example, *Caloplaca trachyphylla* and *Peltigera rufescens* can acclimate their photosynthetic rates to winter and summer temperatures, thereby achieving near maximum rates throughout the year. Similar adaptive responses may occur with light intensity, water content, day length, and season (Ahmadjian, 1993).

The net photosynthetic rate of many lichens depends on the amount of thallus water. If a thallus is saturated with water, diffusion of CO_2 to the phycobiont is much slower than when a thallus has air spaces. In some lichens, crystals of chemical substances coat the outer surface of the fungal hyphae and prevent a thallus from becoming waterlogged. Many chlorolichens (those with green algal photobionts) can achieve maximum rates of photosynthesis by absorbing only water vapor from the atmosphere. Lichens kept at 95–98% relative humidity will reach water contents of up to 70% of their dry weight, and

their rate of CO_2 uptake will be about 90% of the rate achieved when their thalli are wetted with liquid water. In contrast, cyanolichens are at an advantage when wetted with liquid water (Green et al., 1993). Populations of desert coastal lichens, such as those in Baja California and in Chile, are never wetted by liquid water and depend on daily fog to carry out photosynthesis and respiration. Photosynthesis also depends on the amount of light that reaches the algal cells. Most lichen thalli have an upper cortex, which covers the algal layer. The cortex frequently contains lichen acids and pigments that shade the photobiont.

A CO_2-concentrating mechanism has been found in cyanolichens with *Nostoc* photobionts and in lichens with *Trebouxia* photobionts, but not in lichens whose photobionts (e.g. *Coccomyxa*) lack pyrenoids (Palmqvist, 1993).

Nitrogen fixation. About 8% of the known lichen species have cyanobacteria as their photobiont, usually species of *Calothrix*, *Fischerella* (*Stigonema*), *Nostoc*, or *Scytonema* (Rai, 1990). Cyanobacteria may be primary symbionts, as in *Collema* and *Peltigera*, or secondary symbionts as in *Lobaria*, *Stereocaulon*, and *Sticta*. As secondary symbionts the photobionts are housed in cephalodia, which are gall-like structures, or in separate regions of a thallus that occur on or inside a primary thallus that has a green photobiont. Most lichen-forming cyanobacteria fix atmospheric nitrogen inside specialized cells, the *heterocysts*. The percentage of heterocysts to total vegetative cells is much greater in cephalodial cyanobacteria (i.e., 20–30%) than in those that are primary photobionts (i.e., 4%). The reason for this wide difference is not known.

Rates of nitrogen fixation in lichens are affected by light intensity, darkness, thallus moisture, temperature, pH, desiccation of the thallus, and season. Seasonal variation in nitrogenase activity has been reported for *Peltigera canina* and *Stereocaulon paschale*.

Cyanobionts of lichens release more than 95% of the nitrogen they fix as ammonia, which is converted by the mycobiont into amino acids and proteins. Hyphae near the photobiont layer of *Peltigera canina* have been found to contain high concentrations of glutamic dehydrogenase, an enzyme that assimilates ammonia. The specific activity of nitrogen-assimilating enzymes of the cyanobionts, such as glutamine synthetase, is inhibited by compounds produced by the mycobiont (Ahmadjian, 1993). Thus, the cyanobionts are unable to use most of the nitrogen they fix, and their growth is slowed because of nitrogen deficiency. This deficiency, in turn, causes an increase in the number of heterocysts.

Lichens with nitrogen-fixing cyanobionts are important contributors to the nitrogen economy of different ecosystems. *Lobaria oregano* is a large, foliose lichen that contains a blue-green photobiont in cephalodia. The lichen is abundant in the Douglas fir forests of the Pacific Northwest of the United States, and when its thalli fall off the trees onto the ground, they decay and release significant amounts of nitrogen (2 lb N per acre per year). Organic nitrogen compounds may also leach out from lichen thalli during heavy rains.

Molecular biology of lichens

Molecular studies have provided new and important information toward understanding lichens. The focus of these studies has been on evolution and systematics. Some of the findings from molecular studies include the following: (1) the variable occurrence of group I introns in the *Cladonia chlorophaea* complex suggests that lichen fungi have evolved rapidly, a view that is in sharp contrast to the traditional belief that lichen evolution has been static (DePriest, 1993, 1995); (2) at least five independent origins of the lichen habit occurred among different groups of fungi (Gargas et al., 1995); and (3) fungi that adopted a mutualistic lifestyle, (e.g., those that formed lichens) evolved faster than nonmutualists (Lutzoni and Pagel, 1997). Species relationships in *Trebouxia*, the most common lichen photobiont, are being determined (Friedl and Rokitta, 1997).

8.3 FUNGI AND PLANTS

Mycorrhizas

Mycorrhizas are symbiotic associations between fungi and the roots of terrestrial plants (Ruehle and Marx, 1979). These associations were first described in 1885 by the German botanist A.B. Frank, who named the infected roots *mykorrhizen*. Seven types of mycorrhizas are recognized: vesicular-arbuscular mycorrhiza, ectomycorrhiza, orchid mycorrhiza, ericoid mycorrhiza, ectendomycor-

rhiza, arbutoid mycorrhiza, and monotropoid mycorrhiza. The first four types are the most common. In each type of mycorrhiza the fungal hyphae grow extensively through the soil around the infected roots and provide a greater area of absorption of nutrients than nonmycorrhizal roots.

Mycorrhizas occur in practically all terrestrial plants and are an important reason for the well being of plants, especially those growing in nutrient-poor soils (Allen, 1991; Varma and Hock, 1995; S.E. Smith and Read, 1997). Mycorrhizal plants are better able to withstand drought, disease, pests, high soil temperatures, toxic metals, and transplantation than plants without mycorrhiza. Fungi benefit from the association by receiving organic compounds from the plant.

Vesicular-arbuscular mycorrhizas are the most common type of mycorrhiza. They occur in most families of angiosperms, especially in herbaceous plants, in all gymnosperms except the Pineaceae (pine family), and in ferns and liverworts (Varma, 1995). The next most common type, the ectomycorrhizas, occur only in trees and shrubs, in about 3% of the known gymnosperms and angiosperms (fig. 8.4).

Vesicular-Arbuscular mycorrhiza

Fungi that form vesicular-arbuscular mycorrhizas are zygomycetes that belong to the order Glomales and to the genera *Acaulospora*, *Gigaspora*, *Glomus*, and *Sclerocystis*. About 80 species of these fungi are found throughout the world. The species have wide host ranges, although some sporulate better with some hosts than others. The fungi are present in the soil or on nearby roots, and they infect the developing roots. This type of mycorrhiza is difficult to recognize because there are no structural changes in the root. The fungus grows between the cortical cells of the root and also penetrates the cells. The fungus does not form an outer sheath around the root, but the hyphae do grow out from the root into the soil.

Fungal hyphae that penetrate the cells of the cortex form shrublike growths called *arbuscules* that fill most of the host cell. Arbuscules are elaborate, branched haustoria that remain surrounded by a plasma membrane of the host (fig. 8.5A). These structures develop in response to fungitoxic compounds produced by the plants. An arbuscule may live for up to 15 days before it degenerates.

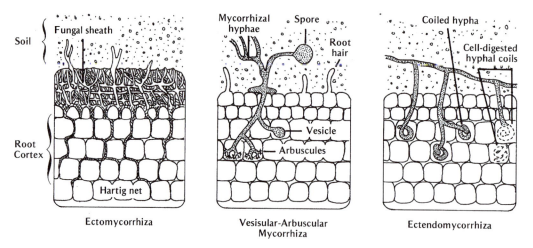

Fig. 8.4   Types of mycorrhizas. Adapted from Deacon (1980).

The host cell may then become infected by new arbuscules. The fungus also produces thick-walled swellings, the vesicles, both within and between the plant cells (fig. 8.5C). Arbuscules and vesicles are so characteristic of these fungi that the associations they form are called vesicular-arbuscular mycorrhizas. Species of *Gigaspora* and *Scutellospora*, however, do not form vesicles, leading some to suggest that the name "arbuscular-mycorrhiza" would be more appropriate for this group (Walker, 1995). Vesicles store lipids, whereas arbuscules are structures through which nutrients pass between the fungus and plant cell. An infected host cell has an enlarged nucleus and lacks starch granules. Host cells that digest the arbuscules return to their preinfected condition; that is, their nuclei return to a normal size and starch granules appear.

Vesicular-arbuscular mycorrhizas differ widely with respect to the extent of their infection in the root and the types of structures they form within the host cells. Many of these fungi form spores and fruiting bodies of different kinds. The fungi infect only young, living root cells, generally in an area directly behind the growing root tip. As the root matures, its cells are no longer susceptible to outside fungal infection. Once infection occurs, new root tissue may be infected by hyphae that grow out from the infected plant cells or by new fungi from the soil.

Fungi that form vesicular-arbuscular mycorrhizas are obligate symbionts. They cannot live independently, and they receive their organic compounds from the host plant. The fungi lack enzymes, such as cellulase and pectinase, that are necessary to degrade vegetable litter and humus. The obligate nature of these fungi may explain in part why they do not grow in axenic laboratory cultures in the absence of plant roots. Modern techniques of molecular biology are providing new insight into the cellular dynamics of the mycorrhizal symbiosis (Harrison, 1998; 1999).

Curious structures that resemble bacteria have been found in the hyphae of some vesicular-arbuscular mycorrhiza. The origin and function of these structures is not known (Perotto and Bonfante, 1997).

The earliest known vascular plants, such as *Rhynia* and *Asteroxylon*, contained vesicular-arbuscular mycorrhizas, according to a study of the fossil remains of these plants. The presence of these associations in such early vascular plants and their widespread occurrence among present land plants indicate that mycorrhizas played an important role in helping plants colonize the terrestrial environment. The fungi may have protected the early plants from drying and helped them obtain nutrients and water from the soil. An increase in the concentration of atmospheric CO_2 stimulates not only plant growth but also that of its mycorrhizal fungi (Staddon and Fitter, 1998).

Vesicular-arbuscular mycorrhizas are extremely common in tropical forests and in agricultural crops. These associations have immense potential as biological fertilizers and can be used to reclaim vast areas of barren soils in the world, particularly in the tropics, which have low levels of available phosphate. These mycorrhizas interlink different plants and are important in the transfer of nutrients between plants.

Some plant families, such as the Cyperaceae (sedges), Caryophyllaceae (carnation), Cruciferae (mustard), and Juncaceae (rush), lack mycorrhizal infections. It is not clear whether this absence of mycorrhiza results from the environmental habitats of these plants (many grow in wet habitats), which suppress fungal growth, or from a type of root epidermis and cortex that prevents fungal penetrations.

Ectomycorrhiza

In ectomycorrhizas the fungus forms a sheath or *mantle* around the root. The mantle gener-

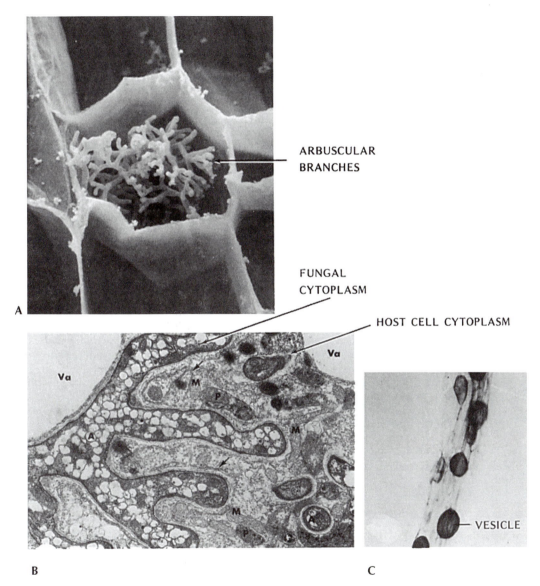

ARBUSCULAR
BRANCHES

FUNGAL
CYTOPLASM

HOST CELL CYTOPLASM

VESICLE

A

B   C

Fig. 8.5 Vesicular-arbuscular mycorrhizal symbiosis. (A) Scanning electron micrograph showing arbuscules of *Glomus mosseae* in root cell of *Liriodendron tulipifera*. (B) Electron micrograph showing part of a vacuolate arbuscule of *Glomus mosseae* inside the root cell of yellow poplar. (C) Vesicles of *Glomus fasiculatus* in stained soybean roots. (Parts A and C, courtesy Merton F. Brown, University of Missouri; part B, courtesy Gerald Van Dyke, North Carolina State University.)

ally is 20–40 μm thick, completely surrounds the root, and consists of either a tissuelike mat or a loose web of hyphae. From the mantle, individual hyphae or strands of hyphae grow outwardly into the soil and inwardly into the root cortex. Hyphae grow through the middle lamella that separates the cortical cells and form a network between the cells called the *Hartig net*. Hyphae do not penetrate the cortical cells or the inner core of the root where the vascular system is located. Fungal infection generally inhibits the development of root hairs and produces structural changes in the root. The changes are a result of growth hormones, auxin and cytokinins, produced by the fungus. Infected roots often assume a stumpy, coral-like appearance, which is typical of many ectomycorrhizas.

In contrast to vesicular-arbuscular mycorrhizas, which involve relatively few species of zygomycetes, more than 5,000 species of basidiomycetes, mostly hymenomycetes and gasteromycetes, and even a few ascomycetes, form ectomycorrhizas. Ectomycorrhizas are common in temperate and tropical forests. A fungus may be specific to one type of tree, or, more usually, it may form associations with different trees. For example, a fungus may associate with a beech tree as well as a spruce tree. Conversely, one tree can associate with different fungi. A Douglas fir tree can form mycorrhizas with as many as 100 different fungi. Different fungi may form mycorrhizas simultaneously on the roots of one tree without competing with each other. Fungi present in the roots of a seedling are often replaced by other fungi as the tree matures.

Ectomycorrhizas vary considerably in shape. They may be club-shaped or nodular, simple or branched. They start to develop 1 to 3 months after a seed germinates. Differences between ectomycorrhizas include the color of the outer fungal sheath or mantle, shape of the infected roots, and the abundance and texture of the hyphae that are in the soil around the root (fig. 8.6).

Ectomycorrhizas protect the root from other parasitic organisms. The fungal sheath around the roots is a barrier through which pathogenic organisms cannot penetrate. Mycorrhizal fungi also secrete antibiotics that inhibit the growth of potential parasites. Fungitoxic compounds produced by cortical cells as a result of mycorrhizal infection also inhibit growth of pathogens.

Ectomycorrhizas develop best when plants are growing under suboptimal nutritional conditions. The fungi absorb and make available the nutrients the plants need. Because the absorbing surface of the root is surrounded by a fungal sheath, all nutrients must pass through the fungus before they enter the root. In this way the fungus controls the amount of nutrients that pass into the root. The fungus provides phosphorous, nitrates, potassium, and other minerals to the plant. The most important mineral is phosphorus, which moves more rapidly through the fungal hyphae than through the soil. The sheath that a fungus forms around the root stores carbon compounds and minerals that can be used by both symbionts during adverse periods. If abundant nutrients, such as nitrogen, are present in the soil, the development of ectomycorrhizas is inhibited, and established associations revert to their nonsymbiotic condition. High nitrogen concentrations inhibit the fungus from producing auxin, which is necessary for mycorrhizal formation.

The plant provides sucrose to the fungus, which converts it into carbohydrates such as trehalose, mannitol, and glycogen. Because the plant cells cannot use these carbohydrates, the sheath around the roots becomes a sink to which there is a steady flow of sucrose from the plant.

Other mycorrhizas

There are several variant forms of ectomycorrhiza, one of which is the *ectendomycorrhiza*. In this type of association, which is common in pine seedlings, the outer fungal sheath is greatly reduced or absent. The Hartig net is present and the hyphae penetrate the host cells. Another variant form is the *arbutoid mycorrhiza*. This type occurs in members of the heather family (Ericaceace), such as trees and shrubs of the genus *Arbutus* (Madrone), in *Pyrola* (wintergreen), and in the small, creeping plant *Arctostaphylos uva-ursi* (bearberry). This mycorrhiza has a sheath, a well-formed Hartig net, and extensive penetrations of the host cells. Ectendomycorrhizas and arbutoid mycorrhizas are produced by fungi that also form ectomycorrhizas. The kind of association produced depends on environmental conditions and the type of host and fungus (Brundrett, 1991).

The *ericoid mycorrhiza* and monotropoid mycorrhiza occur in the heather family and related families. Most species of these families produce ericoid mycorrhizas, in which neither a sheath nor a Hartig net is formed.

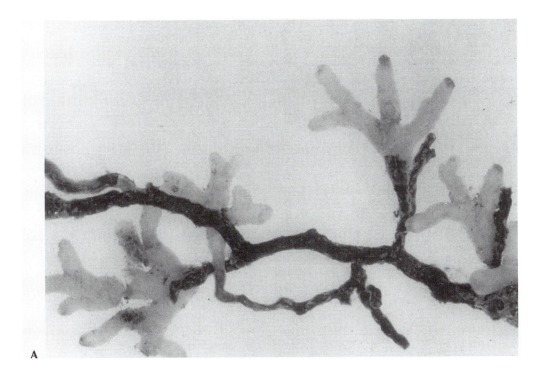

A

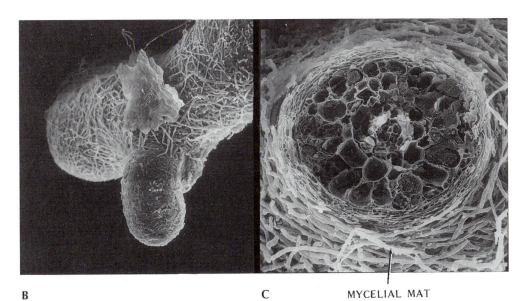

B   C   MYCELIAL MAT

Fig. 8.6   Ectomycorrhizal symbiosis of the fungus *Pisolithus tinctorius*. (A) Ectomycorrhiza on slash pine. Note characteristic dichotomous branching. (Courtesy Donald Marx, USDA, Forest Service, Athens, Georgia.) (B, C) Scanning electron micrographs of mantle of fungal hyphae around roots of *Eucalyptus nova-anglica*. (Courtesy Gerald Van Dyke, North Carolina State University.)

The fungus penetrates the root cells and forms hyphal coils. When the hyphae in a host cell degenerate, the cell dies, a situation that is unlike that of other mycorrhizas in which the host cells remain alive even after their internal hyphae degenerate or are digested. The fungus that forms ericoid mycorrhizas is usually an ascomycete, such as *Pezizella*, although the basidiomycete *Clavaria* may also be involved.

Monotropoid mycorrhizas occur in the genus *Monotropa* (Ericaccae), in plants known commonly as Indian pipe and pinesap. These plants lack chlorophyll and thus cannot manufacture their own food. They live as heterotrophs and also receive nutrients from nearby trees by means of mycorrhizal connections. *Boletus*, a common forest basidiomycete, simultaneously forms ectomycorrhizas with the roots of forest trees such as beech, oak, pine, and spruce, and monotropoid mycorrhizas with species of *Monotropa*. Organic compounds pass from the tree roots through connecting fungal hyphae to the chlorophyll-less plants, which some scientists regard as epiparasites on the host trees. The monotropoid mycorrhiza has a thick sheath and Hartig net, and the host cells are penetrated by haustoria.

Orchid mycorrhizas differ from those of other plants. The fungi that associate with orchids are basidiomycetes, and they are unique because they can break down complex carbohydrates, such as cellulose and lignin, into simpler carbon compounds, which they transport to the orchid seedlings. Other mycorrhizal fungi can use only simple carbohydrates and depend on the plant to provide them with these compounds. Orchid seeds have very little stored food, and when they germinate the embryos rely on mycorrhizal fungi to supply them with carbohydrates and other nutrients during their early development. The fungi that infect the cortical cells of adult orchids are often digested by the plant cells. In this way, the plant obtains nutrients and also controls the extent of the fungal infection. Some tropical orchids that are epiphytes and some that lack chlorophyll are epiparasites on neighboring plants by means of mycorrhizal fungi, which connect the orchids to the trees.

The relationship between the fungus and the cells of a root is similar to that between a fungus and the photobiont of a lichen. From a casual inspection, mycorrhizas appear to be mutualistic associations, one in which both partners benefit (F.A. Smith and Smith, 1996). The fungus absorbs water for the plant, provides it with minerals, and also protects it from pathogenic organisms. In return, the fungus receives simple organic compounds such as sucrose and also growth factors from the plant. What appears to be an idyllic association, however, may conceal a different situation. Elias Melin, a pioneer of mycorrhizal research, has proposed that the relationship in ectomycorrhiza is one of controlled parasitism. Melin (1962) considers the mycorrhizal fungus to be a pathogen whose intrusion into the plant root is checked by fungitoxic compounds produced by the plant cells. Thus, a mutual standoff between fungus and plant results in a long-lasting symbiosis. If conditions change (e.g., if the plant is given an optimal nutritional supply from external sources), then the plant cells kill the fungus and end the mycorrhizal association. Controlled parasitism also describes the symbiosis of lichens, where the parasitism of the photobiont cell is slow enough to allow for the development of a symbiotic relationship (Ahmadjian, 1993).

Identifying the fungi that form mycorrhizas is difficult. Ectomycorrhizal fungi that are isolated and grown in axenic cultures develop slowly, produce compact colonies similar to those of lichen fungi, and do not have any of the characteristics of natural mycorrhizas. The cultured fungi have not revealed any unique nutritional needs or differences from other fungi to explain their symbiotic habit. The best way to identify ectomycorrhizal fungi is to see if fruiting bodies of basidiomycetes growing near the trunks of trees are connected by means of hyphal strands with the mycorrhizas of the trees.

Plant Pathology: The Study of Plant Diseases

Fungi cause millions of dollars worth of damage each year to agricultural crops and ornamental plants. Principal fungal diseases of plants include wilts, rots, rusts, smuts, and cankers. Since the days of Anton de Bary, plant pathologists have studied life cycles, taxonomy, biochemistry, and now the molecular biology of host–fungal relationships. In collaboration with plant geneticists, they have successfully developed disease-resistant varieties of plants. Other control measures include disease-free seeds and seedlings, fungicides, plant protection quarantine, and manipulation of agricultural practices. How to control fungal pathogens of plants has been a persistent problem for humankind, one that will never be completely solved because of the changing nature of fungi. As resistant strains of plants evolve or are developed, new strains of the pathogen arise naturally to infect them. Understanding the life cycle of a fungus and the physiological and genetic basis of pathogenesis is the first step in any effort to control plant diseases (Kombrink and Somssich, 1995; Valent, 1999). Pathogen-triggered host cell death in a variety of plant–microbe interactions has become a focus of current research in plant pathology (Dangl et al., 1996; Gilchrist, 1998).

Fungal phytotoxins

Fungi produce toxins in infected plants and in culture medium. *Toxins* are cellular poisons that are effective at low concentrations. Host-selective toxins and non–host-selective toxins are two classes of toxic substances that pathogens produce in plants. *Host-selective toxins* affect only the host of the toxin-producing pathogen and are required to determine pathogenicity. Mutants for host-selective toxins lose their ability to infect and cause disease. *Nonselective toxins* cause disease symptoms in host plants and in other plant species that are not normally attacked in nature. They are not essential for a pathogen to cause disease.

Host-selective toxins. Victorin, or HV toxin, is produced by the fungus *Cochliobolus* (*Helminthosporium*) *victorae* in oat plants. The fungus infects the basal parts of oat seedlings, and its toxin is carried to the leaves, causing leaf blight and plant death. HV toxin is a cyclic pentapeptide that binds to several host cell membrane proteins. Its production is controlled by a single gene.

T-toxin causes southern corn leaf blight and is produced by *Cochlibolus heterostrophus*. The disease first appeared in 1968 and spread rapidly throughout the corn belt in the United States. The pathogen attacks only corn varieties with Texas male sterile (Tms) cytoplasm. Resistance and susceptibility to *C. heterostrophus* and its toxin are maternally inherited. T toxin destroys mitochondria of susceptible cells and inhibits host ATP synthesis. Corn varieties with Tms cytoplasm contain a mutation in their mitochondrial DNA for URF13 protein, which is absent in corn with normal cytoplasm. In the presence of T toxin, protein URF13 causes pores to be formed in the mitochondrial membrane of corn with cytoplasmic male sterility.

HC toxin is produced by *Cochilobolus* (*Helminthosporium*) *carbonum*, a pathogen of corn, and is responsible for a leaf spot disease. HC toxin affects some corn varieties, but the biochemical and genetic mechanisms of resistance are not understood.

AM toxin is produced by the fungal pathogen *Alternaria alternata*, which is responsible for alternaria leaf blotch of apples. The toxin is highly selective for susceptible apple varieties and causes structural changes in the host cell membrane and chlorophylls and a loss of electrolytes.

Other host-specific toxins are known, and new discoveries are giving plant pathologists more insight into the molecular mechanisms that govern pathogenicity.

Non–host-selective toxins. Tenotoxin is produced by *Alternaria alternata* and causes chlorosis in plant seedlings. Tenotoxin binds to the chloroplast protein and inactivates ATP synthesis. Examples of other non–host-specific toxins include fumaric acid produced by *Rhizopus* sp. and oxalic acid produced by *Cryphonectria parasitica*.

Hydrophobins and fungal infection

Hydrophobins are structural proteins that can aggregate into rod-shaped structures. Hydrophobins facilitate mycelial growth and spore dispersal. Mutants lacking hydrophobins are easily wettable and do not disperse easily. Large amounts of hydrophobins are produced by *Cryphonectria parasitica* (chestnut blight). Virulent strains of *Ophiostoma ulmi* (dutch elm disease) accumulate the hydrophobin cerato-ulmin as the disease progresses (Wessels, 1997).

Magnaporthe grisea, the fungus that causes rice blast disease, develops appressoria, which give rise to hyphae that penetrate the host cells. Appressorial formation is accompanied by increased levels of cAMP and MPG1 hydrophobin. Hydrophobins may also play an important role in root colonization by ectomycorrhizal fungi. For example, *Pislothus tinctorius*, a fungus associated with *Eucalyptus*, undergoes a complex developmental process that includes hydrophobin gene expression (Ebbole, 1997).

Floral mimicry by plant pathogens

Pseudoflowers are fungus-infected leaves that occur among regular flowers and mimic them in form, color, smell, and sugar content. The bright yellow surface of the infected leaves with sugary fluid forms patterns under visible and ultraviolet light that closely resemble those of natural flowers. Insects are attracted to pseudoflowers and contribute to the reproduction and dissemination of their fungal pathogens. Floral mimicry deceives insects as well as naturalists, who unknowingly collect pseudoflowers (Roy, 1993). Rust pseudoflowers attract bees, butterflies, and especially flies, which bring about sexual reproduction in the fungal pathogen by transporting the rust spermatia to compatible hyphae (Roy, 1993).

On blueberry plants, *Monilinia* sp., known as the mummy berry fungus, causes blight on leaves, flowers, and twigs and mummifies berries into white, hard structures. Suzanne and Lekh Batra described the deceit used by the fungus to attract bees to infected leaves. Mummified berries overwinter in moist debris and in the following spring form cups on their surface which release ascospores that infect young leaves. The infected leaves become discolored in a uniform fashion, with leaf edges remaining green while the middle of the leaf turns bluish brown with a central gray patch of fungal conidia. The diseased leaf looks and smells like a flower to an insect, which is rewarded with sugar-coated conidia. The mummy berry fungus succesfully modifies its host plant to trick its pollinators to spread the pathogen (Batra and Batra 1985).

Fungal castration of plants

Rusts, smuts, downy mildews, and endophytes are fungi that castrate (suppress sexual reproduction) plants. Fungi that sterilize the plants they infect prevent the host's coevolutionary response (i.e., to produce resistant progeny). In addition to affecting sexual reproduction, these fungi stimulate clonal reproduction and dispersal of their hosts, thereby producing more individuals which their spores can infect (Clay, 1991).

Rust diseases in plants

Rusts are basidiomycetes that cause many plant diseases (Yamamoto, 1995). The fungi get their name from the brownish red or black spores that cover the infected areas of the plant. There are about 5,000 species of rust fungi, which infect many flowering plants, conifers, and ferns. Human concern over rusts is due to their ability to infect important crops such as wheat, corn, barley, oats, and other cereals. Up to 10% or more of a crop may be lost because of these destructive pests.

Rusts are unusual basidiomycetes because they lack fruiting bodies and have a complex life cycle during which they produce five different types of spores. Each type of spore has a particular function in the life cycle and is produced at a different time of the year. Some rusts complete their life cycle on only one plant, but many need two host plants, usually unrelated to each other. For rusts needing two hosts, eradication of the least important plant is the best way to control the spread of the disease. For example, *Puccinia graminis* (black stem rust of wheat) infects cultivated wheat and has the common barberry as its alternate host. The sexual process of the fungus takes place on the leaves of the barberry, and the spores produced can infect wheat. If barberry plants are eradicated in wheat-growing areas, the life cycle of the fungus is broken, and the wheat will not become infected. Unfortunately, fungal spores are light and can travel on air currents for hundreds of miles. Thus, even a single infected barberry plant miles away from cultivated wheat crops may serve as a source of infection. On barberry plants, the rust pathogen can evolve virulent races that can break down resistance in existing wheat varieties.

The life cycle of *P. graminis* begins in the spring, when basidiospores of the fungus infect the leaves of a barberry plant (fig. 8.7). The basidiospore produces a hyphal filament that penetrates the epidermis and forms branches in the leaf tissues. Hyphae grow inside the cells as well as between the cells. Within 1–2 weeks, the fungus produces pycnia on the upper leaf surface. Pycnia produce pycniospores that serve as gametes. The spores are contained in a thick, sugary fluid that covers the top of the pycnia. Small insects, attracted to the nectarlike fluid, pick up the sticky spores and transfer them to receptive hyphae that grow out from the pycnia. Rust fungi have plus and minus mating strains. Sexual compatibility occurs when pycniospores of one strain fuse with receptive hyphae of another strain. After fusion, the nucleus of a pycniospore migrates through the receptive hypha to the lower surface of the leaf, where the fungus has started to form ae-

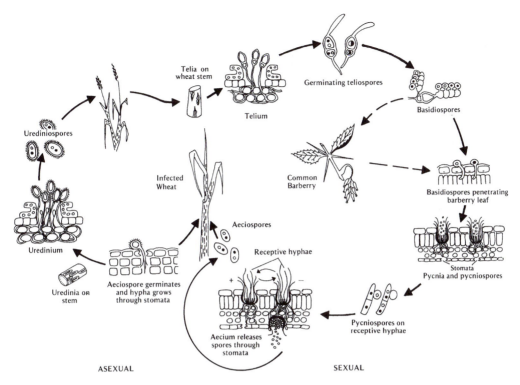

Fig. 8.7   Life cycle of wheat stem rust fungus, *Puccinia graminis tritici.*

cia, structures that produce another type of spore. The pycniospore nucleus pairs with a nucleus in the aecial primordium, and the resulting hyphae contain both types of nuclei. These dikaryotic (two-nuclei) hyphae give rise to aeciospores, each of which has two nuclei. Numerous aeciospores are released from the lower surface of the barberry leaf and are carried away by the wind. If an aeciospore lands on a wheat plant, it germinates and gives rise to a dikaryotic mycelium that infects the wheat tissue. Several weeks later, the fungus produces another type of spore, the urediniospore. These spores are formed in such large numbers that they rupture the plant epidermis and produce rusty streaks on the stems and leaves of the infected wheat. Urediniospores infect other wheat plants and, if unchecked, can infect an entire field of wheat. Continued urediniospore production in the wheat-growing regions can provide a source of inoculum for new wheat seedlings. In the autumn, the fungus produces a fifth type of spore, the teliospore. These black spores have thick walls and allow the fungus to survive the winter. In the spring, teliospores germinate and form basidiospores,

which then infect the barberry plant and complete the life cycle of this rust.

Rust infections occur in many plants other than wheat. The infections usually result in abnormal growth of the host plants and the formation of galls. "Cedar apples" are galls on juniper plants and are caused by the rust fungus *Gymnosporangium juniperi-virginianae.* The galls are made up of plant tissue that contains fungal hyphae and teliospores. Apple trees serve as the alternate host, where the fungus reproduces sexually.

The fungus that causes white pine blister rust, *Cronartium ribicola,* produces urediniospores and teliospores on the leaves of currants and gooseberries. The teliospores form basidiospores, which infect white pine trees. In the bark of the pine tree the fungus produces pycnia and aeciospores in such large numbers that they rupture the bark surface and form blisters. The quality of pine wood is affected adversely by the rust fungus.

Ustilago maydis

Smuts are basidiomycetes that lack fruiting bodies. Unlike some rusts, they do not need an alternate host to complete their life cycle. Smuts produce large numbers of brown chlamydospores, which are often packed together into large spore balls. Smuts also produce basidiospores that behave like yeast cells; that is, they divide by budding and often fuse to form a dikaryotic stage. Each spore eventually gives rise to a dikaryotic mycelium, which infects the host plant. There are more than 1,000 species of smuts. They infect many plants including agricultural crops such as corn, wheat, and oats (Hargreaves et al., 1995).

Corn smut disease is characterized by yellowing of leaves, stunting, and the formation of tumors in the leaves, stems, tassels, and ears. Infected ears of corn are called *huitacoche* in Spanish and are considered to be a culinary delicacy. Some corn growers cultivate corn smut for gourmet food markets (Schumann, 1991). The fungus *Ustilago maydis* is dimorphic; its life cycle includes a yeastlike form that multiples by budding and is haploid and saprobic, and a filamentous form which is dikaryotic and requires a corn plant for its growth. The filamentous form results from fusion of two compatible haploid cells. In the *U. maydis* life cycle, dimorphism and pathogencity are controlled by two incompatibile loci, *a* and *b*. The gene *a* locus controls a pheromone response pathway; gene *b* locus codes for a regulatory protein. Three other genes called *fuz1*, *fuz2*, and *rtf1* are involved in the development of the filamentous form (Banuett, 1992, 1995). *Ustilago maydis* can synthesize auxin and cytokinin and cause tumor induction in the infected corn plant. In this respect it is similar to *Pseudomonas* olive knot disease and *Agrobacterium* crown gall disease. One gene (*fuz7*) from the *b* locus is required for tumor formation. Some interesting questions on the molecular nature of the *U. maydis*–corn symbiosis include how karyogamy (nuclear fusion) is regulated in the life cycle, how pheromones encoded by one mating type stimulate filamentous growth, how the fungus transforms the host growth, and how the host plant regulates fungal meiosis.

Mildews

Downy mildews are obligate pathogenic symbionts of flowering plants. They attack young leaves, twigs, and fruits, and their development depends on the film of moisture on the plant surfaces. The epidemic downy mildew of grapes (*Plasmopara viticola*) during the late nineteenth century in France led to the discovery of the first successful fungicide, Bordeaux mixture, in 1885. In 1979, *Peronospora tabacina* caused the downy mildew of tobacco throughout the United States and Canada, with losses to growers in millions of dollars.

Powdery mildews are ascomycetes that are obligate parasites of many plants, including lilacs, apples, grapes, and roses. These fungi form a thin mycelial layer on the surface of leaves and penetrate the epidermal cells by means of haustoria (fig. 8.8). The powdery mildews produce large numbers of asexual spores called conidia and these spores, together with the mycelium, form a powderlike coating over the infected leaves (Kunoh, 1995).

The fungal haustorium

Haustoria are specialized hyphal cells that are produced inside living cells of other organisms by parasitic fungi. In the case of plants, fungi penetrate the cell walls by enzymatically digesting the outer wall and then mechanically pushing through the inner wall. After the fungus has entered a cell, it forms a haustorium. Haustoria have many different shapes. Those of lichen fungi are club-shaped, and those produced by mycorrhizal fungi are branched. Haustoria of powdery mildews have fingerlike lobes and are associated with specialized structures (fig. 8.8). The fungal haustorium is a branch of an extracellular hypha and terminates in the host cell. In some fungi the haustorium develops from an appressorium, another specialized cell of the fungus that attaches closely to the outer surface of the host cell by means of a mucilaginous substance. Haustoria of rust fungi have a narrow neck region at the site of penetration and a swollen region inside the host cell; most contain a nucleus, mitochondria, and other organelles and are surrounded by a dense, granular matrix. The host cell forms a new plasma membrane, different in structure from that of the uninfected cell, around the penetrating fungus.

Although there is little direct experimental evidence to reveal the function of haustoria, it is presumed that they absorb nutrients from the penetrated cells. Whether haustoria kill the infected cells depends on the type of relationship the fungus has with the host. In rust and powdery mildew diseases, infected cells

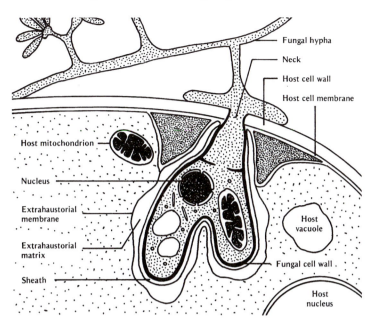

Fungal hypha

Neck

Host cell wall

Host cell membrane

Host mitochondrion

Nucleus

Extrahaustorial membrane

Extrahaustorial matrix

Sheath

Host vacuole

Fungal cell wall

Host nucleus

Fig. 8.8   Schematic diagram of haustorium of powdery mildew fungus. Adapted from Spencer (1978).

may be more active and live longer than un-infected cells. In lichens, haustoria kill the cells of the photobiont but only after a period of time, because the parasitism is gradual and under control.

An infected plant cell may have one or many haustoria inside its protoplast, depending on its resistance to the invading fungus. Some plant cells produce wall material around haustoria and limit their penetration into the cell.

Fungal Endophytes of Grasses

Many species of grasses, both wild and cultivated, are infected with ascomycetes of the genera *Atkinsonella*, *Balansia*, *Epichloe* (*Acremonium*), and *Myriogenospora* (Ball et al., 1993; Latch, 1995; Redlin and Carris, 1996; Schardl, 1996). *Acremonium coenophialum*, the fungal endophyte in tall fescue, has coevolved with its host to their mutual benefit. The fungus produces only mycelia, spreads by infected seeds, and does not sporulate (fig. 8.9). The infected grass is resistant to insects and nematodes and is also toxic to grazing animals (Ball et al., 1993; Saikkonen et al., 1998).

Symptoms of toxicity in livestock that eat the grass are caused by ergot alkaloids produced by the fungus in the leaf sheaths, where

the fungus concentrates, and translocated throughout the plant. Nitrogen-fixing microorganisms (diazotrophs) have been reported as endophytes of sugarcane and rice and are believed to significantly contribute to the nitrogen requirement of the host plant.

Hypovirulence

Some strains of pathogenic fungi are less virulent to their hosts than are normal pathogenic strains. Hypovirulent strains of *Cryphonectria parasitica* have reduced the blight of the European chestnut (*Castanea sativa*) in many parts of Europe, and similar strains offer the first glimmer of hope of controlling the devastating fungal blight of the American chestnut (*Castanea dentata*).

Hypovirulent strains can spread by means of hyphal fusions among a normal pathogen population and transmit to the normal strain an agent that may be a dsRNA virus. This virus somehow diminishes the virulence of the pathogen. In effect, the hypovirulent strains cause a disease of the normal strains. Because the virus is present in the asexual spores (conidia) of the hypovirulent strains, the disease can spread rapidly and over great distances. The effectiveness of hypovirulent strains depends on the ability of their hyphae to fuse with those of normal strains. If there

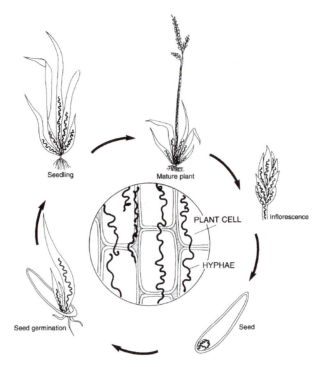

Fig. 8.9 Life cycle of *Acremonium coenophialum,* an endophytic fungus of grasses. The fungus only produces hyphae that grow between plant cells. Adapted from Ball et al. (1993).

are genetic incompatibilities among the strains that prevent hyphal fusions, then the virus cannot be transmitted and the pathogen is unaffected. Thus, a disease may be controlled in some geographical areas, but not in others.

8.4 SUMMARY

Mycosymbionts are fungi that parasitize other fungi, either killing their hosts or obtaining food from living hosts. These fungi are of interest to researchers who want to develop biological controls for fungal pathogens. Mycophycobioses, associations between marine fungi and marine algae in which the alga is the larger partner, are being studied by scientists interested in the early stages of symbiotic evolution.

Lichens are associations of fungi that are mostly ascomyetes and photobionts (photosynthetic partners), which are generally unicellular green algae and cyanobacteria. The lichen thallus is the result of a morphological transformation of the fungal symbiont and to a lesser extent the photobiont. The main types

of thalli are crustose, foliose, and fruticose. Lichens produce secondary compounds that have a variety of roles in the symbiosis. Lichens are dominant species in some parts of the world, except in cities, where air pollution inhibits their growth. Studies on synthetic lichens have provided information about symbiont specificity, thallus development, and the physiological nature of the symbiosis. Many studies have examined carbohydrate transfer between lichen symbionts, photosynthesis and respiration of lichen thalli and nitrogen fixation and metabolism. The photobiont excretes most of its photosynthetic products and, in the case of cyanobacteria, much of the nitrogen they fix. The fungus exerts some control over the photobiont and can regulate the amount of excreted compounds. Molecular studies are revealing important information about lichen evolution and systematics.

Mycorrhizas are mutualistic associations of fungi and the roots of terrestrial plants. There are seven different types of mycorrhizas. The most common type is the vesicular–arbuscular mycorrhiza, formed by zygomycetes, which produces arbuscules and vesicles inside root cells. Basidiomycetes form ectomycorrhizas, the second most common type, which are characterized by a sheath or mantle of fungal tissue around the roots. Other types of mycorrhizas are ectendomycorrhiza, arbutoid mycorrhiza, ericoid mycorrhiza, monotropoid mycorrhiza, and orchid mycorrhiza. Mass-scale inoculations of pine seedlings with the ectomycorrhizal fungus *Pisolithus tinctorius* have been successfully carried out at several locations around the world. The prospect of developing a means for the host plant to obtain minerals without having to depend on chemical fertilizers has encouraged reforestation efforts on marginally suitable lands.

The role of fungi in causing diseases of agricultural crops has been a major factor in the growth of the science of plant pathology. Rusts and smuts are obligate parasitic symbionts of plants. These fungi have complex life cycles and inflict heavy crop losses in many parts of the world. Efforts to control mildew and blight infestations of plants have played a historic role in the development of fungicides. Fungal endophytes are common in wild and cultivated grasses. Fungus-infected leaves appear as flowers to some pollinating insects (floral mimicry), which helps to disperse the fungal spores. Some fungi can suppress sexual reproduction of their plant hosts, thereby limiting their coevolutionary response and the production of resistant progeny. Fungi produce different types of toxins in infected plants.

Hypovirulence, a reduced pathogenicity in strains of some fungi, may be caused by a virus. Hypovirulent strains of pathogenic fungi are being used to control blights of chestnut trees.

Axenic culturing of obligate parasitic fungi has had limited success. Some species of rusts and smuts have been cultured, but powdery mildews have not yet been grown in axenic culture, nor have fungi that form vesicular–arbuscular mycorrhizas and many lichen fungi. Some lichen fungi and ectomycorrhizal fungi can be cultured, but their growth rates are extremely slow. The slow growth as well as the unpredictable nature of symbiotic fungi in culture has limited the types of studies that can be conducted with them. Why symbiotic fungi are difficult to grow axenically is not clear. Some species may need nutrients that are not contained in the culture medium. More likely, the slow growth of these fungi is a result of their natural symbiotic adaptations. Long-lasting associations such as lichens and mycorrhizas require a balanced relationship between the symbionts because unregulated growth of the fungus would quickly kill the host. In both lichens and ectomycorrhizas, an abundance of external nutrients inhibits the symbiotic transformation of the fungus and causes a breakdown of established associations.

9 PARASITIC AND MUTUALISTIC PROTOZOANS

Protozoans live in oceans, lakes, streams, soils, and around the roots of plants. They also live in the alimentary canal, bloodstream, liver, lungs, brain, and other organs of animals. Some protozoans cause serious human illnesses, such as malaria, African sleeping sickness, and Chagas's disease (Roizman, 1996; Wills, 1996); others are pathogens of domesticated animals, such as horses, cattle, and poultry. Flagellate protozoa cause diseases in plants such as coffee, coconut palm, and oil palm (Dollet, 1984). The red tides that kill fish along coastal areas of the eastern United States are caused by large populations of toxin-producing dinoflagellates (Anderson, 1994). A newly discovered killer of fish is the dinoflagellate *Pfiesteria piscicida*. Its toxins have killed millions of fish along the North Carolina coast (Russell, 1998; Burkholder, 1999).

There are an estimated 200,000 protist species, which is more than twice the number of chordates. The populations of some protists may be extremely high. For example, there may be thousands of these organisms in only 1 cm^3 of soil and billions in the rumen of herbivores. Millions of flagellate symbionts inhabit a termite's intestine and may make up as much as one-third of the insect's weight. The White Cliffs of Dover in England consist entirely of the fossil remains of foraminiferan amoebae.

Photosynthetic protists are of immense importance in the ecology of marine organisms because they are the first link in the food chain. In this chapter, we examine symbioses that involve the nonphotosynthetic protists, the protozoans; in the chapter 10 we consider the role of photosynthetic protists in animal symbioses.

Studies in protistan biology have been concerned with how symbionts infect a host, overcome its defense mechanisms, and reproduce successfully. How are the host animal's antibody mechanisms and phagocytic activities of cells such as macrophages neutralized during infection (Goodenough, 1991)? In this chapter, we focus on the advances in our understanding of protozoan pathogenesis and the conditions under which a symbiont–host relationship becomes nonpathogenic.

9.1 KINGDOM PROTOZOA

Characteristics

Protozoa are aerobic or anaerobic organisms that lack organized tissues and embryos. They are unicellular or multicellular, heterotrophic; and reproduce both sexually and asexually. Protists may be immobile or may move by means of pseudopodia, eukaryotic flagella, or cilia.

Classification

The following is a taxonomic grouping of organisms that are discussed in chapters 9 and 10.

Kingdom Protozoa
Phylum Sarcomastigophora
Subphylum Mastigophora
Class Zoomastigophora
Order Kinetoplastida; *Phytomonas, Leishmania, Trypanosoma*
Order Diplomonadida; *Giardia*

Order Trichomonadida;
Histomonas, Trichomonas

Order Hypermastigida;
Trichonympha

Class Phytomastigophora

Order Dinoflagellida
(dinoflagellates); *Gymnodinium*

Class Opalinatea; *Opalina*

Subphylum Sarcodina

Class Rhizopoda; *Acanthamoeba,
Naegleria, Entamoeba*

Phylum Apicomplexa

Class Sporozoasida (gregarines and
coccidians); *Toxoplasma,
Plasmodium, Theileria*

Phylum Ciliophora

Class Kinetofragminophorea;
*Dasytricha, Diplodinium,
Entodinium Epidinium, Isotricha*

Class Oligohymenophorea;
Paramecium

Phylum Foraminifera (forams)

9.2 PROTOZOANS AS SYMBIONTS

Amoebae

Amoebae occur in all environments that support life. They are frequently ingested by potential hosts from drinking water or food. Many species of amoebae have adapted to a symbiotic existence in the alimentary canal of animals, where they are often referred to as "harmless commensals." Some amoebae, however, are human pathogens. *Entamoeba histolytica* causes dysentery, and species of *Acanthamoeba* and *Naegleria* can invade the cerebrospinal fluid of mammals. Amoebae are characterized by pseudopodia and trophozoites that feed by phagocytosis and multiply by binary fission. Some amoebae produce resistant cysts at certain stages of their life cycles.

Entamoeba

Entamoeba histolytica infects about 10% of the world's population, with about 100 million people suffering from invasive colitis or liver abscess. There may be 50,000–100,000 deaths per year in endemic areas, making *E. histolytica* the third major cause of death, be-

hind malaria and schistosomiasis (Ravdin, 1990). Pathogenesis of invasive amoebiasis includes (1) attachment to the intestinal mucus barrier, (2) mucus breakdown, (3) cytolysis of epithelial cells due to the parasite's enzymes and toxins, (4) penetration of trophozoites through the intestinal mucosa, and (5) the host inflammatory response (Ravdin, 1990; Montfort and Perez-Tamayo, 1994). An amoebic adherence lectin, N-acetyl-D-galactosamine (Gal/GalNAc) plays a significant role in the *E. histolytica* attachment mechanism. Several proteolytic enzymes, which are released on contact with the host tissue, have been isolated from *E. histolytica*. An enterotoxin has also been isolated that may interfere with the function of calcium channel proteins (Gitler and Mirelman, 1986).

Entamoeba has also been reported from monkeys, apes, and other mammals. Infection usually begins when a host ingests fecal-contaminated food or drink, which contains cysts of the amoeba. The cysts hatch and produce amoebae, which feed on bacteria and other food particles in the large intestine. After several cycles of growth and multiplication, some amoebae encyst; that is, they become spherical, secrete a thick wall, and store glycogen. Cysts are excreted in large numbers in the feces. An infected individual may excrete an estimated 45 million cysts in one day. If the cysts are ingested by a new host, they hatch, and each cyst liberates four trophozoites. Amoebic infection may occur without symptoms, or the amoebae may attack the epithelial lining of the large intestine and cause ulcers and dysentery. In more serious cases, amoebae may penetrate the intestinal wall and infect organs such as the liver and brain.

Understanding the pathogenicity of *Entamoeba histolytica* has been controversial. What was previously identified as *E. histolytica* is now believed to be a composite of at least two species, *E. dispar* and *E. histolytica*. *Entamoeba dispar* is the smaller of the two species and is nonpathogenic. At one time it was believed that all cases of *E. histolytica* infection required medication, and, consequently, individuals were treated with dangerous drugs. Eye disorders, allergies, migraine, rheumatoid arthritis, and liver disease were often falsely attributed to the presence of amoebae in infected persons.

Pathogenic strains of *E. histolytica* lose their virulence when they are cultured in the laboratory. Electron microscopic observations have revealed that amoebae from an infected person have a "fuzzy coat," the nature of which is not understood, that is absent in laboratory strains. Surface-active lysosomes that might be responsible for the breakdown of host cell membrane have been discovered in the amoebae. Cytoskeletal rearrangements in response to signals from external stimuli appear to characterize virulence of *E. histolytica* (Guillen, 1996). Golgi bodies and mitochondria are absent. Some scientists have observed that *E. histolytica* alone cannot produce disease and that the presence of bacterial or viral symbionts is essential for its virulence (Mirelman, 1987). Viruslike particles have been detected in the cytoplasm and nucleus of *E. histolytica*.

Entamoeba invadens is a common pathogen of turtles and snakes. Morphologically, the amoebae resemble *E. histolytica*, and scientists have increasingly used it as a model to investigate the pathogenesis of invasive amoebiasis (Eichinger, 1997).

Acanthamoeba *and* Naegleria

Species of *Acanthamoeba* and the amoeboflagellate *Naegleria* are common in stagnant lakes and ponds and may cause primary amoebic meningoencephalitis (PAM) in humans and other mammals (Ferrante, 1986). Young children swimming in such areas are most susceptible to *Naegleria* infections. Amoebae of *N. fowleri* enter the nasal passages, where they penetrate the olfactory epithelium and move along the olfactory nerve. Electron microscopy of trophozoites of *N. fowleri* has revealed unusual "feed cups" or *amoebostomes*, which are used in the phagocytosis of host cells (John et al., 1985; Marciano-Cabral, 1988).

There are several species of *Acanthamoeba* that cause meningoencephalitis, inflammation of internal organs, corneal ulcers, gastritis, and diarrhea. Bacterial contamination of contact lens cleaning solution enhances multiplication of *Acanthamoeba* that are associated with eye infection (Bottone et al., 1992). In the 1920s, when acanthamoebae were isolated from diseased tissues, they were dismissed as contaminants by scientists. Later it was shown that intranasal inoculation of mice with *Acanthamoeba* killed the mice, and the amoebae could be observed in the brain tissues. Worldwide, more than 100 cases of PAM have been identified, most of which were fatal. Pathogenicity of these amoebae has been attributed to their ability to release a phospholipolytic enzyme that demyelinates nerves.

Ciliates

Ciliates are well adapted to live in aquatic habitats. Most ciliates are free-living heterotrophs, but some live in the alimentary canals of vertebrates, insects, and annelids. Ruminants maintain a large number of ciliate species. Ciliate symbionts also inhabit the liver, blood vessels, and gonads of animals. Two different types of nuclei, micro- and macronuclei, are always present. Asexual reproduction is by binary fission, and conjugation is involved in sexual reproduction. The ciliate trophozoites feed on host fluids by pinocytosis or through cytostomes. *Balantidium* spp. are commensalistic symbionts of humans, pigs, and monkeys.

Rumen ciliates: the mutualistic protozoans

Mutualistic associations have evolved between ciliates and herbivorous mammals. The rumina of cattle and sheep contain, in each 1 ml of gut contents, an estimated half million ciliates consisting of 30–50 species, whose role is controversial. Ciliates form a large proportion of the rumen biomass and have a slow turnover. They survive by attaching firmly to the gut and thus washout is reduced. Ciliates ingest rumen bacteria, starch, and chloroplasts and have been blamed for the low efficiency of rumen fermentation (Van Soest, 1994).

Rumen ciliates are obligate anaerobes. New hosts acquire them through a mother licking her calf or from eating grass that is dropped after being regurgitated from a rumen. The environmental conditions of the rumen are complex and have not been duplicated in the laboratory. The rumen has a temperature of 39°C, is anaerobic, contains particulate matter that is resistant to the host's digestive enzymes, and is deficient in glucose and amino acids. Although the alkaline pH of the rumen favors growth of the ciliates, the posterior portion of the stomach, the omasum, has an acid environment, which is lethal to the ciliate symbionts. Ciliates may supply their host with about one-fifth of its protein and with acetic, propionic, and butyric acids, and also with a better balance of amino acids than do the bacterial symbionts of a rumen.

Two kinds of ciliates live in the rumen: entodiniomorphs and holotrichs (fig. 9.1). *Ento-*

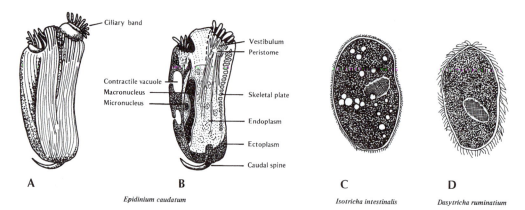

Fig.9.1   Types of mutualistic rumen ciliates. (A) and (B) External and internal morphology of a typical entodiniomorph. (C) and (D) Two examples of holotrich symbionts. Adapted from Furness and Butler. (1983).

diniomorphs have a semirigid pellicle that covers the cell and gives each species a distinctive morphology. They also have unique internal plates, which form a type of cytoskeleton. Around the mouth of each symbiont is a band of cilia, which function in a coordinated fashion. The number and location of ciliary bands are used to identify species of entodiniomorphs. In most species, the ciliary zone can be retracted and the pellicle drawn over it. *Diplodinium*, *Entodinium*, and *Epidinium* are important genera of symbiotic ciliates and have been the subject of many physiological studies. *Entodinium bursa* is the largest of the entodiniomorphs, and it consumes large quantities of plant material. Rumen ciliates produce hydrogen and volatile fatty acids. The hydrogen they produce serves as a substrate for methanogens, which live as ectosymbionts of ciliates. Entodiniomorphs digest cellulose and starch, as well as rumen bacteria. Many scientists have attempted, unsuccessfully, to grow rumen ciliates in the laboratory.

Rumen *holotrichs* resemble free-living ciliates. Their number in sheep and goats has been estimated to be 160,000–200,000/ml of rumen content. Some common holotrichs are *Dasytricha ruminatium*, *Isotricha intestinalis*, and *I. prostoma*. Holotrich ciliates, in contrast to entodiniomorphs, can incorporate large quantities of sugar and convert it to starch for storage. Holotrich starch is indistinguishable from plant starch.

Opalinata

Opalinids possess many cilialike organelles in oblique rows that cover the body surface.

Their phylogeny is a matter of dispute (Patterson, 1985).

Opalina ranarum: an amphibian symbiont

About 150 species of the genera *Cepedea*, *Opalina*, *Protoopalina*, and *Zeleriella* inhabit the large intestines of frogs and toads. The symbiosis of *Opalina ranarum* and frogs has been studied extensively in Europe (Smyth and Smyth, 1980). *Opalina obtriganoidea* is found in frogs and toads of North America. The symbionts have flat or cylindrical cells, up to 1mm long, and their reproductive physiology is controlled by gonadal hormones of the host. Opalinids do not harm the host, which may harbor thousands of these symbionts in its large intestine.

The symbionts have 2–200 nuclei. Electron microscopy of opalinids shows a complex of cortical folds, ribbons of microtubules, and pinocytotic vesicles forming at the bases of the cortical folds (Patterson and Delvinquier, 1990). A cytostome is lacking. Generally, adult individuals of *O. ranarum* multiply by binary fission in the rectum of frogs and toads. The symbiont life cycle is synchronized with that of the host. In the spring, at the onset of the frog's breeding period, the symbionts divide rapidly and produce small, precystic forms. These become transformed into cysts and are eliminated with feces from the host. Cysts ingested by tadpoles hatch in the duodenal region, and the symbionts then migrate to the large intestines. Microgametes and macrogametes are formed, and fusion of the sex cells results in zygotes, which give rise to a new generation of opalinids (fig. 9.2). Researchers have induced encystation of *O. ranarum* by injecting hypophysectomized and gonadectomized frogs with pregnancy urine, nonadotrophin, testosterone, and adrenaline (Smyth, 1994).

Flagellates

Flagellated protozoa differ widely in their morphology and physiology. Some have evolved a high degree of intracellular differentiation and possess unique structures whose functions are not fully understood. All *flagellates* have flagella and can swim, which is a distinct advantage over an amoeboid lifestyle. Some flagellates colonize the digestive, circulatory, and lymphatic systems of animals. Flagellate symbionts can be divided into three broad groups: intestinal flagellates, which inhabit the alimentary canal and genital tract of the host; hemoflagellates, which inhabit the blood, lymph, and other tissues of vertebrate hosts and the intestines of insects; and phytoflagellates, which cause plant diseases.

Intestinal flagellates

Intestinal flagellates are obligate anaerobes and do not reproduce sexually. In some species, the trophozoite stage alternates with a resting stage, the cyst. Intestinal flagellates ingest food by phagocytosis and pinocytosis, and some species have cytostomes. Little is known of the nutrition and metabolism of these flagellates, and only a few species have been cultured. Most species are nonpathogenic, but some cause disease.

Most animals acquire flagellate symbionts from contaminated food or drink. Both the cysts and trophozoites can spread the infection. The symbionts usually occupy the lower portion of the host's alimentary canal, where digestive activities are lacking (e.g., the cecum in mammals and birds and the cloacal cavity in amphibians and reptiles). Below are descriptions of some intestinal flagellates.

Trichomonads. All *trichomonads* have an anterior tuft of three to five flagella, a rigid median rod, an axostyle and an undulating membrane. They lack mitochondria but have hydrogenosomes. Asexual reproduction is by longitudinal binary fission, and it produces large numbers of trophozoites which are present in the digestive and reproductive systems of most animals. Following are some examples of trichomonad symbioses with vertebrates and humans.

Trichomonas gallinarum causes avian trichomoniasis in pigeons, chickens, turkeys,

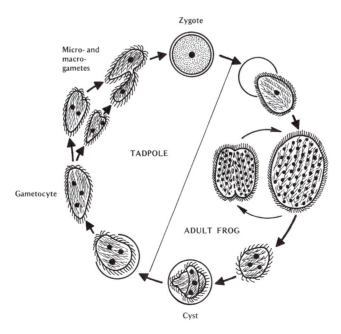

Fig. 9.2 Life cycle of *Opalina ranarum*. Adapted from Smyth (1994).

and other domesticated birds. The symbiont inhabits the upper intestinal regions of an infected bird and may cause diarrhea, weight loss, and ruffled feathers. The trichomonad does not produce cysts. Young birds become infected by their parents. Different strains of the flagellate have proved almost 100% infective in pigeons. *Trichomonas gallinarum* has been used extensively to study mechanisms of pathogenicity. *Trichomonas foetus* causes a serious genital infection in cattle and often results in abortions.

Trichomonas vaginalis occurs commonly in the genital tract of humans and is sexually transmitted. Under certain conditions, which are still not fully understood, it becomes pathogenic on the epithelial cells of the urethra and vagina and erodes the mucosal lining. The flagellate also infects rats, mice, hamsters, and guinea pigs and has been cultured *in vitro*.

Pathogenicity of trichomonad symbionts is correlated with their migration from the large intestine, where they are nonpathogenic, to other sites in the body. The pathogenicity of trichomonads is greatly reduced under axenic conditions. For example, *T. gallinae*, normally a highly virulent pathogen of pigeons, loses its pathogenicity after 4–5 months of axenic culture. Some unknown cytoplasmic factors may become diluted when the symbiont is grown in culture. In addition, cultured flagellates stimulate less antibody production in rabbits. Pathogenic strains have either small amounts of antigens or incomplete antigens and thus trigger only a limited antibody response from the host. Although great advances have been made in understanding trichomonad pathogenesis, many intriguing problems still remain (Honigberg, 1989).

Giardia. *Giardia lamblia* is a common inhabitant of the small intestine of humans, monkeys, and pigs, especially in warmer climates. Unique traits of *Giardia* include (1) it can grow in axenic culture, (2) it is one of the earliest branches of eukaryotes, and (3) it has symmetrically arranged two sets of organelles (Sogin et al., 1989; Kabnick and Peattie, 1990). The symbiont attaches to the intestinal wall of the host by means of an adhesive disc and feeds on secretions of the intestinal mucosa. Trophozoites of this symbiont multiply by binary fission. A heavy infestation of an individual can result in the production of more than 14 billion cysts per day. Some forms of *Giardia* produce the disease giardiasis, symp-

toms of which include diarrhea, vomiting, and poor vitamin B_{12} absorption. At least 10% of the human population in the United States carries a milder form of *Giardia*, which usually is harmless and asymptomatic, although the feces from an infected person may contain large amounts of mucus and fats.

Giardia has a convex dorsal and a concave ventral side that contains an adhesive disc. The cell has two nuclei, each with a prominent nucleolus. Mitochondria, smooth endoplasmic reticulum, and Golgi bodies have not been observed in the cells (Gillin et al., 1996). The symbiont has four pairs of flagella, which are positioned to help the symbiont swim as well as to make the initial contacts with the host. The middle portion of *Giardia* contains microtubules that form a unique and distinctive cytoskeleton, which is involved in both attachment and contractibility. *Giardia* attaches to the host cell surface by means of its adhesive ventral disc (fig. 9.3). The indented margin of the disc penetrates the cell and forms an interlocking system with microvilli of the host cell. *Giardia* microtubule cytoskeletons can be isolated by means of sonic treatments. *Giardia* and giardiasis have been studied and reviewed extensively (Adam, 1991; Thompson et al., 1993).

Unlike other organisms, *Giardia lamblia* and *Trichomonas vaginalis* cannot synthesize purines and pyrimidines. They obtain the nucleic acid components from their host cells by phagocytosis and lysosomal activities. Enzymes for the purine and pyrimidine pathways are similar in both of these species (Jarroll et al., 1989; Wang, 1989). *Giardia* shows surface antigenic variations and the nature of its variant-specific surface proteins is being explored (Nash, 1995).

Histomonas meleagridis. Blackhead disease of turkeys was first identified in Rhode Island in 1895 and by 1920 it was killing up to 45% of the commercial turkey flocks in the United States. In the 1920s, Ernest Tyzzer, a Harvard University parasitologist, identified *Histomonas meleagridis* as the cause of the disease and described its biology and pathology. The pathogen first infects the intestine and then spreads to the liver and causes death of the host. *Histomonas meleagridis* alone is unable to cause blackhead disease and requires the participation of intestinal bacterial species such as *Escherichia coli* and *Clostridium perfringens* (Roberts and Janovy, 1996). Tyzzer discovered that chickens and nematodes were

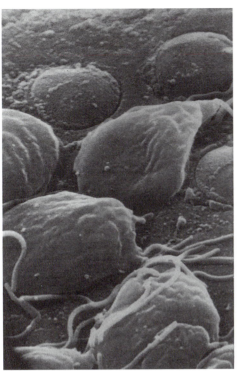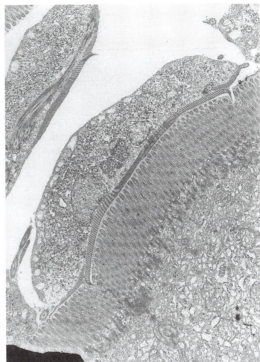

Fig. 9.3 *Giardia* ultrastructure. (Left) Scanning micrograph of *Giardia* trophozoites attached to surface of intestinal villi of rat. Note dome-shaped lesions, which were previous sites of parasite attachment. (Right) Electron micrograph of sectioned *Giardia* attached to microvillous border. Note the microtubular cytoskeleton of adhesive ventral disc (Courtesy Stanley L. Erlandsen, University of Minnesota).

carriers of the flagellate. The disease declined by 1940 primarily because of improved sanitation and management practices such as not housing chickens and turkeys together.

Histomonas meleagridis forms two types of cells. One type of cell is round, lacks flagella, and infects the liver and mucosal lining of the cecum. The other type of cell has one or two flagella and occurs in the intestine.

Birds that are susceptible to the cecal nematode *Heterakis gallinarum* have a high incidence of infection by *H. meleagridis*. Nematodes of both sexes as well as eggs of the female contain the flagellate, and when the nematodes infect poultry, the eggs hatch and produce infected larvae. Disintegration of some of the larvae by the host's digestive enzymes frees the parasite, which then invades the cecal mucosa and migrates to the liver. The symbionts cause prominent lesions on the cecal walls, a thickening of the cecal mucosa, and coagulation necrosis of liver parenchyma. Earthworms also have a role in black-

head disease of poultry. Infected nematode eggs are often eaten by earthworms, which in turn are eaten by poultry.

Histomonas meleagridis can be cultured in the presence of bacteria from the cecum of a natural host. The cultured symbiont loses some or all of its pathogenicity, but this can be restored, either fully or partially, by serial passages through poultry. Because of the unusual nature of the flagellate's life cycle and the involvement of another parasitic symbiont, a nematode, as an agent of transmission, *H. meleagridis* continues to be studied by scientists.

Flagellates of termites and roaches. Flagellate symbionts of the wood-eating roach *Cryptocercus punctulatus* and termites of the genera *Reticulitermes* and *Zootermopsis* have been the focus of much research. Lemuel R. Cleveland, a pioneer in this field, who devoted more than 40 years of his life to research on this subject. The flagellates inhabit the posterior portion of the lower termite's alimentary canal and belong to the orders Hypermas-tigida, Oxymonadida, and Trichomonadida.

An estimated 30–50% of the total weight of a wood-eating termite consists of protozoa. Some, such as *Trichonympha*, can digest cel-lulose and thus allow the host insect to sub-sist on a diet of wood. The symbionts break down cellulose anaerobically by means of the enzyme cellulase to glucose, which is then converted to acetic acid, carbon dioxide, and hydrogen in hydrogenosomes. The nature of this symbiosis is complicated because the host insects also have bacterial and fungal symbionts, which provide vitamin B complex to the hosts. These organisms occur in the al-imentary canal, in the Malpighian tubules, or within mycetocytes. Nitrogen-fixing bacteria also occur in the termite and provide nitrogen as well as growth factors and vitamins for the host and its flagellates. Other termites also have methane-producing bacteria. Termite gut shares many similarities with rumen of cattle and are often described as the world's smallest bioreactors (Brune, 1998).

There are more than 2000 species of ter-mites in the world. They occur mostly in warm climates, but 41 species are known from North America. Wood-digesting, mutu-alistic protozoa are characteristic of the lower termite families (Hodotermitidae, Kaloter-mitidae, Mastotermitidae, and Rhinotermiti-dae). Only about 10% of termites are colonial insects with a high degree of social order and a distinctive "caste system" based on morpho-logical and physiological features. Termites build complex nests consisting of galleries and chambers, and each species produces a nest of specific shape and size that may be built underground or above ground as mounds (fig. 7.11).

Termites undergo simple metamorphosis from egg to four nymphal instar stages and fi-nally to the adult stage. The first nymphal in-star acquires its flagellate symbionts by feed-ing on anal droppings of infested termites. During each molt, the entire epithelial lining of the alimentary canal is shed, resulting in the loss of intestinal flagellates. Each newly molted instar also acquires its symbionts from anal droppings. The termite digestive tract consists of three parts: foregut, midgut, and hindgut. Most cellulose digestion takes place in the paunch region of the hindgut, which is dilated and contains the bacteria and flagel-lates. The flagellates digest cellulose to sim-pler compounds that can be used by the ter-mites, and the host in turn provides an anaer-obic chamber and food for the symbionts. Termite symbionts can be eliminated by starv-ing the host or by maintaining it at high tem-peratures or high oxygen pressure. Termites that lose their symbionts do not survive on their normal wood diet and die within weeks. Survival of these symbiont-free termites can be prolonged by feeding them glucose or fun-gus-decomposed wood. Termites that reac-quire their flagellates will recover and live normally. Not all flagellates in the termite hindgut have the cellulose enzyme system. Only the cellulolytic strains are mutualistic symbionts of termites, although the remain-ing flagellates have diverse lifestyles.

Mixotricha paradoxa is a large wood-eat-ing flagellate that lives in the gut of the termite *Mastotermes darwiniensis*. At first the sym-biont was thought to be a ciliate because its outer surface is covered with structures that look like cilia. However, studies have shown that *M. paradoxa* is covered with spirochetes and rod-shaped bacteria, not with cilia. The spirochetes undulate in a coordinated man-ner and propel the symbiont through the in-testinal fluid. The four flagella of the sym-biont are believed to be used only for steering (fig. 9.4). The association is considered to be mutualistic because the spirochetes receive nutrients from the flagellate. Lynn Margulis has suggested that the flagella of eukaryotic organisms evolved from similar spirochete-like prokaryotes that were ectosymbionts (Margulis, 1991b). The flagellate symbionts of termites possess a variety of bacteria as intra-cellular symbionts. The bacteria are responsi-ble for the flagellate's ability to digest cellu-lose.

There are many similarities between anaer-obic fermentation in ruminants and cellulose degradation in termites. Both ruminants and termites ingest plant cellulose and degrade it to usable compounds by means of mutualistic symbionts.

Hemoflagellates

Intestinal flagellates of bloodsucking insects have successfully invaded the reticulo endo-thelial tissue of vertebrates. In a vertebrate host, the flagellates swim in the bloodstream or become intracellular inhabitants of macro-phages. *Hemoflagellates* assume different morphological forms during their life cycle, with their mitochondria also changing shape. Some of the flagellates can resist the host's an-tibodies. Diseases produced by hemoflagel-

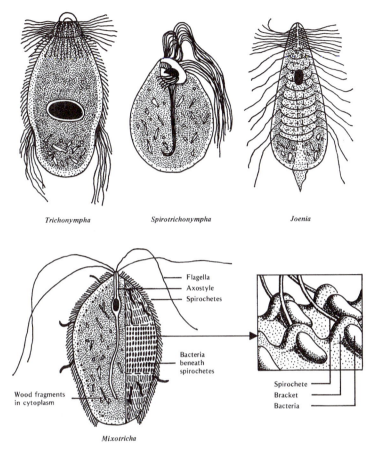

Trichonympha   *Spirotrichonympha*   *Joenia*

Flagella
Axostyle
Spirochetes

Bacteria
beneath
spirochetes

Wood fragments
in cytoplasm

Spirochete
Bracket
Bacteria

Mixotricha

FLAGELLATES OF TERMITES

Fig. 9.4   Common genera of symbiotic flagellates of termites. Adapted from Cleveland and Grimstone (1964).

lates in humans and domesticated animals have cost millions of lives and include African sleeping sickness, Chagas's disease in South America, and leishmaniasis in Asia and Africa. Hemoflagellates have been studied extensively by parasitologists, cell biologists, immunologists, and evolutionary biologists.

Trypanosoma brucei rhodesiense: *antigenic variation.* African sleeping sickness affects thousands of people throughout equatorial Africa. Both sexes of the tsetse fly, *Glossina* sp., are vectors of the trypanosome that causes the disease. The symbiont develops into the infective form in the insect's salivary glands. When the insect feeds, it injects trypanosomes into the lymph and bloodstream of the host. The trypanosome population increases rapidly and reaches a peak in about 5–7 days. The flagellates are of the slender-form type (fig. 9.5). A week after the initial infection, two classes of host antibodies appear in the blood serum: IgG antibodies, which are directed against cell antigens, and IgM antibodies, which are active against cell surface antigens. The surface antigens of trypanosomes are believed to be glycoproteins. The antibodies cause agglutination and lysis of the trypanosomes, and the disease then enters a remission phase as the number of flagellates declines sharply. During this phase, large numbers of flagellates of the short, stumpy form appear in the blood. These forms cannot develop further in the mammalian host. A few slender-form trypanosomes develop new antigenic properties and are unaffected by the host's antibodies. These flagellates then multiply until the host's im-

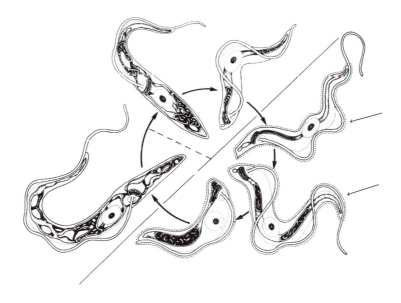

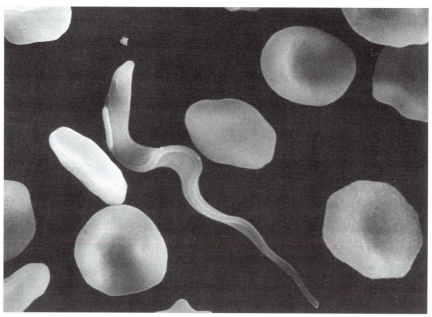

Fig. 9.5 Schematic diagram of developmental stages of *Trypanosoma brucei* in mammals and tsetse flies. (Top) Changes in cell surface and mitochondria. Note presence of surface antigens (arrows) around forms of the parasite occurring in bloodstream. From Vickerman (1971), with permission. (Bottom) Scanning electron micrograph of trypanosome among mammalian red blood cells. (Courtesy Steven Brentano, University of Iowa).

mune system manufactures new antibodies that are specific for them. The antibodies destroy most of the new trypanosomes, but again a few symbionts alter their antigens, and the process is repeated. Each new antigenic variant of the symbiont is matched by the host's production of antibodies against it, but eventually the host is overwhelmed. After the insect has a blood meal from an infected individual, the stumpy-form symbionts undergo several transformations before they migrate to the salivary glands, where they develop into the infective form. The insect may inject as many as 40,000 trypanosomes into a host during a blood meal. The developmental stages of a symbiont last from 17 to 50 days, depending on biotic and abiotic factors, and the insects remain infective for their lifetime, which is about 2–3 months.

Symbionts in the mammalian blood have a distinguishable coat on the outside of their cell membranes. This surface coat is present only in the infective form of the trypanosomes that emerge from the insects. Antigenic variation of the trypanosome is associated with changes in its surface coat. Surface coat antigens are phenotypic expressions of trypanosome genes (Deitsch et al., 1997). Up to 10% of the trypanosome genome may consist of genes that code for antigens. A single trypanosome can produce at least 100 antigenic variants, and up to 1000 genes may be involved. The surface coat of the trypanosome is composed of about 10 million glycoprotein molecules known as *variant surface glycoproteins* (VSG). Each VSG molecule is made up of about 500 amino acids, to which are linked two types of oligosaccharide side chains. One of the carbohydrate chains links the VSG molecule to the trypanosome cell membrane. Variations in the amino acid sequence of the VSG molecule results in new antigenic properties. Several VSG genes have been sequenced (Mansfield, 1995). Three mechanisms of genetic rearrangement involved in a VSG gene expression are (1) duplicating and transposing a gene to another position near the telomeric region of a chromosome, replacing the resident gene, (2) duplicating and transposing an unexpressed telomeric VSG gene to another telomeric site where it is expressed, and (3) expressing an inactive telomeric gene. There is some evidence that there may be two or more expression sites for VSG genes (Borst, 1991).

The system of antigenic variation in trypanosomes is an innovative evolutionary adaptation that allows the symbionts to survive in the host and makes the task of developing effective vaccines against the diseases they cause a difficult one.

Trypanosoma cruzi: *intracellular parasite.* To live and multiply in host cells such as macrophages, whose function is to recognize and kill foreign invaders, requires a great deal of evolutionary audacity. Yet, some flagellates have successfully conquered the macrophages, one of the most hostile intracellular

environments. *Trypanosoma cruzi* causes Chagas's disease in nearly 8% of people living in Central and South America (Schofield, 1985). The flagellate enters the host through cuts in the skin or mucus membranes. The vector of *T. cruzi* is a brightly colored bug of the insect family Reduviidae. The insect voids the trypanosomes along with its feces during a blood meal. The flagellates appear in the blood and periodically invade the reticuloendothelial tissues, especially those of the heart. Inside the tissues, the symbiont assumes a spherical form, multiplies as an intracellular parasite for about 5 days, and forms a cystlike cavity, the *pseudocyst.* Some of the flagellates break through the pseudocyst membrane and reenter the bloodstream to infect other tissues. The symbionts that remain in the pseudocyst disintegrate and cause a lesion. Reduviid bugs ingest *T. cruzi* during the blood meal from an infected individual. The trypanosomes multiply and undergo several transformations, first in the midgut and later in the hindgut. *Trypanosoma cruzi* is a versatile parasitic symbiont that can invade and replicate inside a wide variety of host cells, entering them by forming membrane-bounded vacuoles. Host cell lysosomes gather at the attachment site of the parasite and fuse with the plasma membrane of the vacuole. Destruction of the vacuolar membrane is brought about by release of a pore-forming toxin by *T. cruz* (Tc-tox). The symbionts escape from the vacuoles into the host cytoplasm, where they multiply (Burleigh and Andrews, 1995; Mauel, 1996).

Unlike *T. brucei,* which defeats the host's immune system by sequentially changing its surface antigens, *T. cruzi* expresses several glycoproteins simultaneously. Some of the principal surface antigens are neuraminidase, GP-72, and GP-85. GP-72 is a major surface glycoprotein that seems to regulate the host's complement activation. GP-85 is a lectin that binds to fibronectin receptors and helps the parasite invade the host cell. Neuraminidase of *T. cruzi* controls the infection process by a negative mechanism that may check pathogen replication and allow it to remain latent (De Souza, 1984; Pereira, 1995).

Molecular biology of Trypanosoma. Trypanosomes have an elongated, flat body with a single nucleus that has a prominent, central nucleolus. The flagellum arises posteriorly, extends along the free margin of the body to form an undulating membrane, and continues anteriorly as a free flagellum. The flagellum originates from a basal body and kinetoplast, which are enclosed in a mitochondrion. The mitochondrial DNA is in the form of a *kineto-plast*, which is a mass of circular DNA that consists of thousands of maxicircles and minicircles. Of the total cellular DNA of *Trypanosoma*, 10–25% is kinetoplast DNA (kDNA). Studies have suggested that maxicircles have genes that are similar to the mito-chondrial DNA of other eukaryotic cells. The minicircles, which make up approximately 95% of the kDNA, encode for small-guide RNAs, which control editing specificity (Stuart and Feagin, 1992). RNA editing is a type of RNA processing that regulates only the mitochondrial genes of kinetoplastid fla-gellates (Stuart, 1995; Cavalier-Smith, 1997). Mitochondrial morphology varies a great deal in hemoflagellates. The mitochondria are usu-ally in the form of a single or branched tube that runs the length of the cell, and they pos-sess characteristic cristae. In *Trypanosoma brucei*, the mitochondrial shape changes along with the transformation the symbiont undergoes when it passes from the vertebrate bloodstream to the insect's alimentary canal. In the bloodstream, the symbiont has a slen-der form and a long, tubular mitochondrion with a double membrane but only a few cristae. In the alimentary canal of the insect, the symbiont assumes the stumpy form and has a well-developed spherical mitochon-drion with a tubular network of prominent cristae. The mitochondrion of the slender form is inactive, lacking enzymes of the Kreb's cycle and electron transport chain, while that of the stumpy form is active.

Trypanosomes without functional mito-chondria depend on glycolysis and substrate-level phosphorylation for their ATP produc-tion. The parasite has acquired a novel way of increasing its level of glycolysis by compart-mentalizing various glycolytic enzymes into a membrane-bound organelle, the *glycosome*. The high concentration of enzymes in glyco-somes allow glycolysis to proceed at a rate 50 times that of a mammalian cell. Unlike mito-chondria, a glycosome does not contain nu-cleic acids and is bounded by a single phos-pholipid bilayer membrane. There may be 200–300 glycosomes per cell. Glycosomal proteins in *T. brucei* are encoded by nuclear genes (Boothroyd, 1995; Cavalier-Smith, 1997).

Leishmania: *intracellular survival in macrophages.* Several species of *Leishmania* cause diseases in humans. *Leishmania tropica* is the causal agent of oriental sore, which is endemic to Africa, the Middle East, and India; *L. mexicana* causes New World cutaneous leishmaniasis, or chiclero ulcers, a disease common in Mexico, Guyana, and other countries of Central and South America; *L. braziliensis* causes an infection called *espundia*, a disease endemic to the jungles of South America, which involves mucous membranes of the mouth, nose, and pharynx; and *L. donovani* is the causal agent of visceral leishmaniasis in most of the tropical and subtropical countries of the world.

Leishmania has two different morphologi-cal forms depending on the type of host: an oval-shaped body without an external flagel-lum when it occurs in a vertebrate host, and an elongated body with a flagellum when it inhabits the intestines of sand flies. In the mammalian host, *Leishmania* lives exclu-sively in macrophages. The establishment of the symbiosis involves recognition, intracel-lular entry, surviving the host's lytic enzymes, and growth and multiplication of the sym-biont (Chang, 1995).

As soon as flagellates from the insect host enter the mammalian bloodstream, a complex process of cell recognition takes place. Some type of chemotaxis may take place because the flagellates are strongly attracted to macrophages and, to a lesser degree, to lym-phocytes, but show no affinity with red blood cells. The symbionts possess cell surface com-ponents that fit receptors in the macrophage cell. Macrophages permit internalization of the symbiont by phagocytosis. The flagellar end of the symbiont may play a significant role in its interaction with the host cell. Entry of the symbiont into the macrophage is ac-companied by chemiluminescence and the release of hydrogen peroxide. Macrophages treated with cytochalasin B fail to phagocy-tize the symbionts, which suggests that the cytoskeletal proteins actin and myosin partic-ipate in the process of symbiont entry. *Leishmania* may develop a resistance to the lysosomal enzymes of the macrophage, or it may release substances that inactivate the en-zymes. Once inside the macrophage, the sym-biont assumes its nonflagellated form, which is better suited for intracellular existence. The symbiont grows within the membrane-bound vacuole of the macrophage and multiplies slowly. Some *Leishmania* strains have devel-oped stable associations with the host.

Leishmania interaction with macrophages

provides an excellent model for several important aspects of cell biology such as receptor-mediated endocytosis, membrane transport, and molecular signals and gene expression and also intracellular targeting and movement of internalized molecules. There is a common surface glycoprotein (gp63) found on surfaces of several pathogens, and its role in leishmanial virulence has been studied extensively (Chang, 1995).

Phytomonas: plant pathogenic flagellates

Flagellate protozoans were first observed in 1909 in latex-bearing cells of *Euphorbia*, the rubber plant. Since then the pathogen has been found to cause disease in plants that belong to the families Asclepiadaceae (milkweed), Moraceae (figs), Rubiaceae (coffee), and Euphorbiaceae (cassava). Recently, flagellate protozoa have been isolated from tomatoes (Agrios, 1997). All species of *Phytomonas* belong to the family Trypanosomatidae and require a plant and an insect to complete their life cycle. In plants, some *Phytomonas* flagellates live in the phloem sieve tubes, while others live in latex-containing cells. Phloem necrosis of coffee commonly occurs in South American coffee-growing regions and is caused by *Phytomonas leptovasorum*. The infected plant shows symptoms of wilt, which is accompanied by yellowing and dropping of leaves. The plant dies within 3–6 weeks. Vectors of this disease are several species of pentatomid insects of the genus *Lincus*. In the early 1980s more than 15,000 coconut trees were killed by *Phytomonas staheli* in Trinidad (Dollet, 1984).

Apicomplexans

Members of the phylum Apicomplexa have an apical complex that consists of a polar ring, conoid, rhoptries, micronemes, and subpellicular microtubules (fig. 9.6). The polar ring is located at the anterior end of the symbiont, just beneath the cell membrane. The *conoid* is a spirally coiled structure within the polar ring. There may be two to seven *rhoptries* that lead to the apex of the cell. *Micronemes* are small, convoluted structures that extend over most of the anterior end, and their ducts join those of the rhoptries. *Subpellicular microtubules* radiate from the apex and extend most of the length of the symbiont. A mouth is located laterally, and along its edge are two concentric rings of some yet-unknown material that may help the symbiont acquire nutrients. The apical complex participates in the penetration of the host cell by secreting proteolytic enzymes.

Apicomplexan life cycles may require two host organisms and involve several kinds of asexual and sexual stages. The following are examples of symbionts included in the phylum.

Gregarines

Gregarines are extracellular symbionts that inhabit the alimentary canal and body cavity

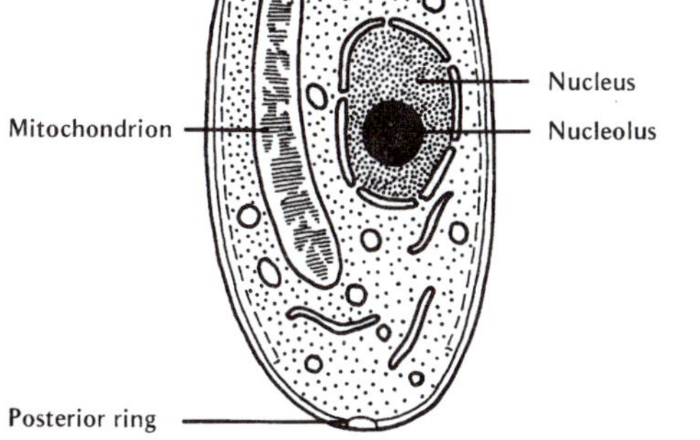

Fig. 9.6   Ultrastructural details of a typical apicomplexican trophozoite.

of invertebrates, especially arthropods, molluscs, and annelids. The life cycles of several gregarine species are synchronized with those of their hosts. Sporocysts are produced only in the sexually mature host and are liberated along with the host sex cells, so that the next generation of larvae will become infected with the symbionts. The host's hormones may play a significant role in the development of the symbiont.

Toxoplasma gondii: a cosmopolitan coccidian

The coccidian symbiont *T. gondii* occurs in many animal species throughout the world. In humans, *toxoplasmosis* is generally a mild or asymptomatic infection, but in some cases it may cause blindness and damage to the central nervous system. The symbiont invades the macrophage and circumvents its lysosomal system. Sexual stages in the life cycle begin after cats ingest cysts of the symbiont from tissues of an intermediate host such as a rodent or bird (fig. 9.7). The cysts contain hundreds of sporozoites, which infect the host epithelial cells and then transform into merozoites. Merozoites multiply rapidly and are released into the intestinal lumen when the host cells rupture. Each merozoite can infect another epithelial cell. After several such cycles, some merozoites become gametocytes. Two gametocytes fuse together to form a cyst, which is then expelled with the feces.

Asexual reproduction of *T. gondii* begins when a host other than a cat ingests cysts from the meat of an infected animal. Spores liberated from the cysts penetrate the intestinal wall and enter the circulatory system and infect the macrophages. In the macrophage, spores are called *tachyzoites*. Living tachyzoites use macrophages as an intracellular habitat, divide repeatedly by binary fission, and eventually kill the host macrophage. When the host defense mechanism is mobilized, macrophages surround large numbers of symbionts and form a wall around them and form pseudocysts that contain dormant symbionts. Pseudocysts occur in all tissues of the host but are most common in the brain, retina, and liver. Raw hamburger is believed to be an excellent source of pseudocysts. Most infections of *T. gondii* occur when undercooked or raw meat containing pseudocysts is consumed. Tachyzoites will cross placental barriers and infect the developing embryo.

Tachyzoite invasion of macrophages. Early workers had reported that when tachyzoites were liberated in a host environment, they exhibited various types of movement such as gliding or rotating in a somersault fashion. Penetration of the macrophage was thought to be achieved through a collision with the host membrane. Later evidence, however, supports the view that a symbiont's penetration is mainly the result of the host cell's phagocytic activities.

The sequence of events that results in the destruction of a pathogen in a macrophage is called the *respiratory burst* and involves a number of metabolic activities. When a macrophage is unable to degrade the pathogen contained in a vacuole phagosome, there is no respiratory burst. *Toxoplasma gondii* strains differ in their pathogenic abilities. A proteinaceous substance has been isolated from one strain of lysed *Toxoplasma* that enhances the virulence of another strain. *Toxoplasma* penetration of macrophages was greatly facilitated when lysozyme or hyaluronidase was added to the culture medium. The process of entry is initiated when the symbiont attaches to the macrophage by its anterior end and forms a small depression in the host cell membrane. Once inside the cell, the symbiont becomes enclosed in a phagosome, which is a membrane-bound vacuole (Lingelbach and Joiner, 1998). The host cell lysosomes fail to fuse with the phagosome, suggesting that *T. gondii* contains a substance that interferes with the lysosome–phagosome fusion (Dubremetz, 1998). *Toxoplasma gondii* also changes the phagosome to allow the gathering of endoplasmic reticulum and mitochondria along the vacuolar membrane. Macrophages can degrade previously killed tachyzoites, but living symbionts gain immunity to the host's destructive enzymes by altering their properties. Newly developed tachyzoites escape from the host cell by twisting through the host cell membrane.

Malarial parasites

Malaria is caused by the multiplication of the parasite *Plasmodium* in the blood and tissues of a vertebrate host. *Plasmodium* species are obligate intracellular parasites that have several morphologically different developmental stages. Although there are more than 100 species of *Plasmodium*, which affect a wide range of vertebrates, each species has a narrow host range.

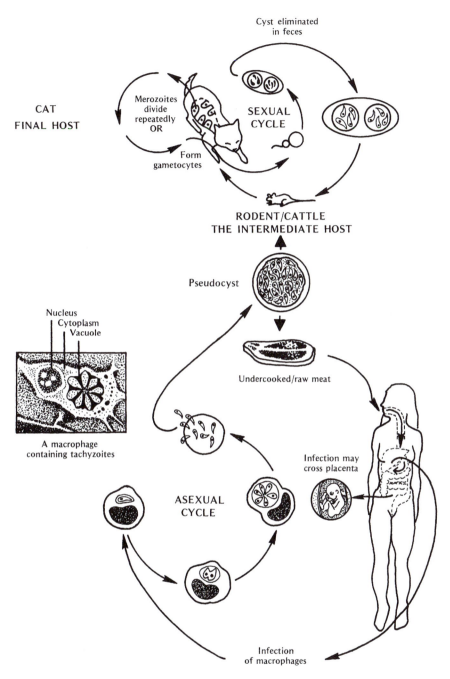

Fig. 9.7   Life cycle of *Toxoplasma gondii*. Adapted from Katz et al. (1982).

Four species of the malarial parasite infect humans: *P. falciparum*, *P. malariae*, *P. ovale*, and *P. vivax*. *Plasmodium vivax* accounts for 43% of all cases of malaria and occurs primarily in Asia and Africa. Only people with Duffy blood groups are resistant to the disease. *Plasmodium vivax* can reoccur many years after the first infection due to the presence of dormant hypanozoites in the liver of infected individuals. *Plasmodium falciparum* is the most virulent form of malaria and accounts for more than 50% of all cases. It occurs primarily in tropical and subtropical parts of the world. The malarial parasite of chimpanzees and monkeys, *P. knowlesi*, occasionally infects humans and produces mild symptoms. *Plasmodium-berghei* in rodents and *P. gallinaceum* and *P. lophurae* in birds have been used extensively as experimental models to explore the ultrastructural details of the symbiont–host interaction.

History of the disease. Malaria is one of humankind's oldest diseases and has played a significant role in history as well as in the development of modern science. Almost half the world's population lives under the threat of malaria. Each year an estimated 150 million people contract the disease (80% of them in Africa), and about 1 million die of it. Malaria remains as one of the greatest threats to childhood survival in Africa. Global warming raises the specter of bringing even more areas under the threat of malarial infections.

Discovery of the malarial parasites and understanding of the intricate details of their life cycle represent the golden age in parasitology. Ronald Ross, and Battista Giovanni Grassi, working with human malaria, showed in 1898 that the parasites developed in the mosquito and that humans became infected when bitten by mosquitoes. In 1880, Alphonse Laveran described the parasites in blood cells, but it was not until the early 1950s that the role of the liver in the disease was resolved. Both Laveran and Ross were awarded Nobel Prizes for their discoveries.

The first drug to be used against malaria was quinine, which was extracted from the bark of the chinchona tree. Peruvian Indians knew of the therapeutic value of quinine, and when the European colonists learned of it, the use of quinine became global.

The malarial symbiont is transmitted by *Anopheles* mosquitoes, and the development of the pesticide DDT in the 1940s was an important breakthrough in controlling the spread of the disease. Unfortunately, as mosquito control programs spread throughout the world, DDT-resistant strains of mosquitoes began to appear (Crampton, 1994).

During the Second World War, hundreds of thousands of American soldiers contracted malaria. The U.S. Army launched a project to combat the disease, eventually developing antimalarial drugs such as chloroquine, artemisinin, and mefloquine. A malaria eradication program was developed, based on the use of DDT and chloroquine. During the 1960s the incidence of malaria declined sharply in all parts of the world, with expectations that it would be totally eradicated. Then came the realization that *Plasmodium falciparum* and other species were producing strains that were resistant to chloroquine, while the mosquitoes were becoming resistant to DDT. The reemergence of malaria after more than 2 decades of relative success in controlling it has alarmed scientists, who are now trying to understand the molecular basis of the disease. Scientists are using techniques of immunology and molecular biology in attempts to develop an effective vaccine against the malarial symbiont (Facer and Tanner, 1997), a difficult task considering the complexity of the disease and the variability of the parasite (Good et al., 1998). Much has been learned from the malarial parasite about the molecular mechanisms, genetics, immunology, and biochemistry of symbiont–host interactions. Efforts by several collaborating laboratories are underway to sequence the complete genome of *Plasmodium falciparum*. Recent research has once again focused on the mosquito vector in an attempt to create, with DNA technology, strains of the mosquito that would be incompatible hosts for the malarial parasite (Collins and Paskewitz, 1995).

Life cycle of Plasmodium. In humans, infection begins when a female *Anopheles* mosquito penetrates the host skin and injects into the blood saliva that contains sporozoites of *Plasmodium* (fig. 9.8A). Sporozoites are covered by a single protein, the *circum-sporozoite protein* (CSP). Both the genes involved and the protein have been cloned and sequenced. Sporozoite antigens have been considered as one of the principal antigens for the antimalarial vaccine. The parasites migrate to the liver, where they multiply in the parenchyma cells. The sporozoites are transformed into merozoites, and when all the liver host cells rupture, they liberate 100,000–300,000 merozoites. In most mammals, merozoites infect red blood cells and grow and multiply intracellularly. The mature stage of the symbiont in a red blood cell is called a *schizont*. The schizont divides to form merozoites, which are released when the red blood cell ruptures and then infect other red blood cells. This asexual part of the life cycle is the *schizogony*. The clinical signs of malaria are associated with the cyclical blood stage of schizogony. In each cycle of schizogony, some of the invading merozoites differentiate into male (micro-) and female (macro-) gametocytes. The gametocytes mature in about 10 days and do not develop further in the vertebrate host.

When a female mosquito bites an infected individual, micro- and macrogametocytes are ingested with the blood. In the mosquito's stomach, each microgametocyte forms eight microgametes, which fertilize the macrogametes formed by the macrogametocyte. The resulting zygotes become transformed into motile, wormlike structures, the *ookinetes*, which penetrate the stomach wall and form sporocysts. Each sporocyst breaks open and releases sporozoites, which migrate to the salivary glands. The sexual part of the parasite life cycle is the *sporogony* (Beier, 1998).

Merzoites: intracellular symbionts of erythrocytes. Within the vertebrate host, sporozoites and merozoites have to invade specific cell types. Proteins associated with their outer surfaces have been implicated as ligands for receptors on the host cell surfaces. Liver cells possess receptors for CSP. Therefore, the CSP protein is involved in the initial interaction and attachment to the hepatocyte surfaces.

On the merozoites, a surface protein 1(MSP1), along with rhoptry-associated protein, plays a role in the invasion of the erythrocytes (Wakelin, 1996). These proteins undergo a complex series of modifications between their synthesis and successful erythrocyte invasion.

Glycophorins are major surface proteins of the human erythrocytes, and *Plasmodium* merozoites depend on these molecules for invasion of erythrocytes. The thick cell surface coat is removed during the invasion process. The coat consists of fine filaments that stand erect, like hairs, on the plasma membrane. The filament tips are either Y or T shaped and can stretch when attached to the cell. The filaments may attach to the underlying cytoskeleton of the merozoite. Two subplasma membranes make up a system of interconnected cisternae that lie under the apical complex. The polar ring of the apical complex is responsible for a particular merozoite shape, and it forms a convergent point for rhoptries and micronemes.

It is not known how merozoites are attracted to erythrocytes, but once contact is made the parasite adheres to any part of the erythrocyte surface. The invasive process begins, however, only when the symbiont's apical complex contacts the host cell surface, which has specific receptor molecules (figs. 9.8B, 9.9A,B). Merozoites can attach to a glass surface, to other merozoites, and to white blood cells. If the merozoite–erythrocyte attachment is incorrect, the merozoite detaches itself and makes another attempt. The host membrane shows a characteristic bending when the apical complex of the merozoite contacts the erythrocyte. Secretions released from rhoptries and micronemes cause the erythrocyte membrane to invaginate and form a cup, into which the merozoite advances. The surface coat of the merozoite becomes detached as the symbiont moves into the cup (fig. 9.8C). Finally, the entire merozoite becomes surrounded by the host membrane and enclosed within a vacuole. It has now been established that the apical complex of the parasite is responsible for the invasion of the erythrocyte. The rhoptries and micronemes are thought to be derived from the Golgi bodies and to possess lysosomic enzymes. After the initial contact, the movement of a merozoite into the host cavity is rapid. This is interesting because merozoites lack obvious motility mechanisms. It has been suggested that suction produced by the invagination of the erythrocyte pulls the symbiont into the vacuole (Lingelbach and Joiner, 1998). In the vacuole, microspheres near the cell surface of the merozoite protrude into the host cytoplasm. The contents of the microspheres are released in the vacuole and cause it to enlarge. The cisternae that lie underneath the plasma membrane of the merozoite, along with the microtubules in the cortex, disappear. The merozoite loses its spherical shape and becomes transformed into a trophozoite (Aikawa, 1988).

Malarial pathogenesis. *Plasmodium* infections cause severe anemia and cerebral malaria. *Plasmodium* sp. produce toxins that stimulate the release of cytokines such as tumor necrosis factor α (TNF-α) from host T-cells. It is now believed that excess release of nitric oxide by the vascular endothelium in response to TNF-α diffuses through the brain tissue and disrupts neuronal activities. Rising levels of TNF-α are associated with body temperature during a malarial paroxysm. When children are given anti–TNF-α therapy, they soon recover in intermediate steps. The development of malarial pathogenesis has been a subject of intense research (Miller et al., 1994).

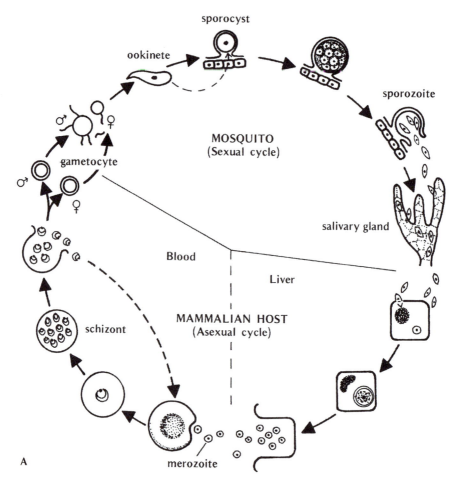

Fig. 9.8. Malarial parasitic symbiosis. (A) Developmental stages in the life cycle of *Plasmodium*. (B) Stages in the invasion of an erythrocyte by a *Plasmodium* merozoite. (Courtesy Lawrence Bannister, Guy's Hospital, Medical and Dental Schools, London.)

Theileria parva

The piroplasm *Theileria parva* causes *theileriasis* in cattle and water buffalo. The symbiont is an intracellular inhabitant of erythrocytes and leukocytes and is spread by ticks (Mehlhorn and Schein, 1984). The host–symbiont relationship is unique. The parasite causes the host cell to proliferate and undergo synchronous S-phase and nuclear division as they share a common cytoplasm (Dyer and Tait, 1987). *Theileria* becomes a permanent resident of the macrophage cytoplasm by escaping from the vacuole. The symbiont synchronizes its growth and duplication with that of the host cell. The schizont of *T. parva* segregates along with the host chromosomes

into the two daughter cells. Electron microscopic evidence suggests that the schizont of *T. parva* is closely involved with the host's mitotic apparatus. The symbiont becomes encased in microtubules during centriole replication and aster formation and becomes compressed between the microtubules during elongation of the spindle. By the time the nuclear membrane disintegrates, the schizont is fully integrated into the mitotic spindle along with the host chromosomes. During anaphase, the schizont elongates with the polar microtubules, and when the host cell constricts in the middle to produce two new cells, the schizont is also partitioned into two. Schizont-microtubule attachments have not been observed (Carrington et al., 1993).

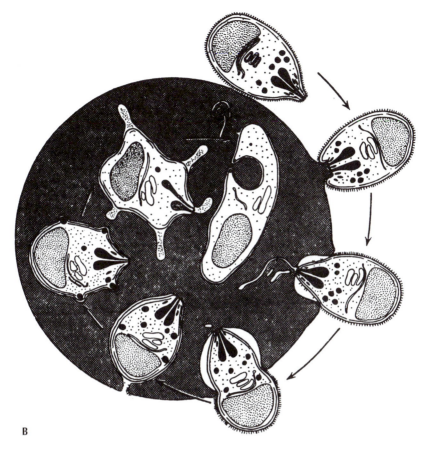

B

Fig. 9.8.   (*continued*)

35-Kilobase Apicomplexan DNA:
A Ghost of Photosynthesis Past

Malarial parasites have 14 chromosomes in addition to two extrachromosomal DNAs, 6-kb linear and 35-kb circular molecules. The 35-kb circular DNA was originally thought to be the mitochondrial DNA of *Plasmodium*. The 6-kb fragment carries the genes for cytochromes and cytochrome oxidase that are typical of mitochondrial DNA. Recent sequence evidence, however, suggests that 35-kb *Plasmodium* DNA is more like that of plastid DNA (R.J.M. Wilson et al., 1991; Palmer, 1992). Similar 35-kb circular DNAs have been identified in *Babesia*, *Eimeria*, and *Toxoplasma* and are believed to be common in other species of this group. The 35-kb DNA in apicomplexans is contained within a vestigial plastid (Feagin, 1994). In *Toxoplasma*, the plastid is sur-rounded by four plasma membranes, and during cell division of the parasite the plastid divides into two in a way similar to that of plastids in plants and algae. *Apicoplast* is the name given to the vestigial plastids of apicomplexan protists. The 35-kb apicomplexan DNA appears to be a plastid genome that has "forgotten" how to photosynthesize (Vogel, 1997). These findings suggest that apicomplexans are closely related to dinoflagellates. As discussed in the next chapter, dinoflagellates are common symbionts of invertebrates.

Apicomplexan plastids offer an excellent opportunity to develop drug therapy against diseases that affect millions of humans. We can now understand how antibiotics such as doxcycline (antimalarial), rifampicin, thiostrepton, and spramycin control toxoplasmosis and *Cryptosporidium* (McFadden and Waller, 1997).

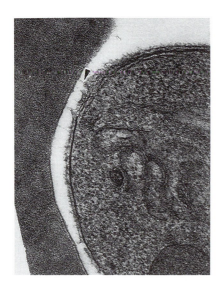
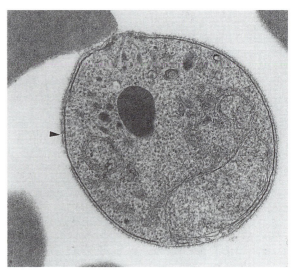

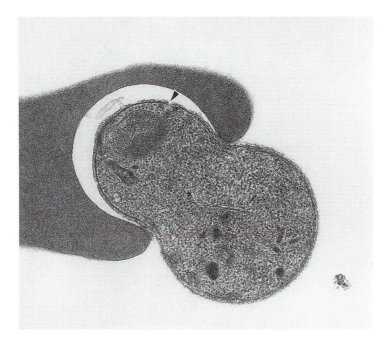

Fig. 9.9 Electron micrographs of *Plasmodium* merozoites invading red blood cells. (Top left) A merozoite showing nonspecific type of attachment to an erythrocyte. Note the presence of surface coat (arrow). (Top right) Long-range attachment between parasite and host cell involving merozoite coat filaments (arrow). (Bottom) A partially depleted rhoptry is visible at the anterior end of the parasite (arrow). Note the formation of a vacuole that surrounds the symbiont (Courtesy Lawrence Bannister, Guy's Hospital, Medical and Dental Schools, London).

9.3 SUMMARY

Protozoans cause diseases in millions of humans and domesticated animals. Many of these parasites have complex life cycles that involve two hosts and several types of spores. Species of amoebae range from harmless commensals in a host's alimentary tract to virulent pathogens. *Entamoeba histolytica* causes dysentery, and species of *Acanthamoeba* cause meningoencephalitis in humans and mammals. Opalinids are commensals in the large intestine of frogs and toads.

Two kinds of ciliates, entodiniomorphs and holotrichs, live mutualistically with herbivorous mammals. The ciliates live in the rumen of the host along with bacteria.

There are two groups of flagellated protistan symbionts: intestinal flagellates and hemoflagellates. Representative intestinal flagellates include trichomonads, *Giardia lamblia*, *Histomonas meleagridis*, and various genera that inhabit the alimentary canal of wood-eating termites and roaches. *Mixotricha paradoxa*, a wood-eating flagellate found in the gut of termites, is covered with spirochetes that propel the flagellate. Hemoflagellates include trypanosomes and *Leishmania*, which cause devastating diseases in humans and animals such as African sleeping sickness, Chagas's disease, and leishmaniasis.

Apicomplexans include gregarines, which are extracellular symbionts of invertebrates, *Toxoplasma gondii* a symbiont of many animal species, *Plasmodium*, a parasite that causes malaria, and the piroplasm *Theileria parva*, which causes theileriasis in cattle and water buffalo. Apicoplasts are vestigial plastids that have been found in apicomplexans.

Symbiotic associations may be more complex than first imagined and may involve several layers of organization. For example, scientists now believe that *Entamoeba histolytica* cannot produce the disease amoebiasis alone; rather, the presence of bacterial and viral symbionts in the cytoplasm and nucleus of the amoeba may be necessary for its virulence. Similarly, the ability of the termite symbiont *Mixotricha paradoxa* to digest cellulose may be the result of intracellular bacterial symbionts. Symbioses in the rumen of herbivorous animals and in wood-eating termites consist of complex interrelationships between bacteria, ciliates, and fungi. Similar patterns may exist in other associations.

The life cycles of the symbionts of many protistan associations are synchronized with each other, thereby ensuring continuity of the symbiosis from one generation to another.

10 PHOTOSYNTHETIC ASSOCIATIONS OF PROTOZOANS AND INVERTEBRATES

Many invertebrates, such as sea anemones, corals, flatworms, and protozoans, have formed mutualistic associations with photosynthetic microbes (Saffo, 1992b; Trench, 1993). The photobionts supply nutrients to their hosts, which allows them to colonize habitats they normally could not because of limited supplies of prey. Two groups of algae (protists), collectively called zoochlorellae and zooxanthellae, are the most common photobionts. These names do not have taxonomic status but are useful for identifying the general type of endosymbiont in a host. *Zoochlorellae* are mainly species of *Chlorella*, a freshwater, unicellular alga which is a well-known symbiont of at least 30 genera of ciliates, amoebae, hydras, sponges, and clams (D.C. Smith, 1991). About 10 species of *Chlorella* are recognized. DNA–DNA hybridization studies have revealed that many of the species are not closely related and that the *Chlorella*-type morphology evolved in several lineages of the Chlorococcales. *Zooxanthellae* are mostly dinoflagellates, which appear yellow-brown in the many marine organisms they colonize. The dinoflagellate chloroplasts have efficient light-harvesting complexes that include chlorophyll a, chlorophyll c, and large amounts of xanthophylls. Reproduction and genetics of dinoflagellates are unusual and not well understood. They have more nuclear DNA than other eukaryotes. Their chromosomes remain condensed during interphase, lack centromeres, and unwind only for DNA replication. The nuclear membrane does not break down during mitosis, as in animals and plants. Microtubules within the spindle coordinate the segregation of chromosomes. Dinoflagel-

lates are diverse in form. Their cell walls are composed of two interlocking plates; two flagella are inserted in the cell wall at about the same location. A transverse flagellum wraps around a groove in the cell, and a longitudinal flagellum is perpendicular to the cell. Dinoflagellates are surrounded by a two-membrane structure, the amphiesma.

A common dinoflagellate of marine invertebrates is *Symbiodinium microadriaticum* (fig. 10.1). Comparison of small ribosomal subunit RNA sequences show that *Symbiodinium*-like zooxanthellae represent a collection of distinct species of the genus *Symbiodinium* (Rowan and Powers, 1992; Rowan et al., 1997). *Symbiodinium microadriaticum* is greatly modified when it lives inside animal cells. The algal cell wall becomes thinner, which allows for close contact with the animal's cytoplasm. The cell loses its grooves and flagella and divides only by binary fission. The algal cells also change physiologically. In the animal, zooxanthellae excrete large amounts of glycerol, which the host uses to synthesize lipids and proteins. In some invertebrates, almost 50% of the carbon fixed by the algae is translocated as glycerol within 24 h to animal tissue. Zooxanthellae also excrete glucose, alanine, and organic acids, and perhaps other compounds in amounts too small to detect. When the algae are isolated from animals and grown in culture, they stop excreting substances. But when cultured algae are exposed to homogenates of their host, they once again excrete glycerol and other compounds. Although the nature of the stimulatory host factor is not known, it seems to be a general one because homogenates from one type of animal, such as a clam, affect zooxan-

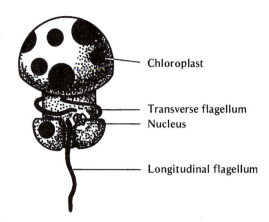

Fig. 10.1 *Symbiodinium microadriaticum*, a common
photosynthetic symbiont of marine invertebrates.

Symbiodinium

thellae from another animal, such as coral,
and vice versa. The homogenates are effective
only if they are taken from animals that contain symbiotic algae (D.C. Smith and Douglas,
1987).

In some marine animals there is a reciprocal exchange of substances between the symbionts and their hosts. The host animals supply the algae with acetate, which the algae use
to synthesize fatty acids. The fatty acids are
translocated back to the animal, which uses
them to synthesize waxes and other compounds.

The number of algal cells in each animal
cell varies. The hydroid *Myrionema am-
boinense* has 1–56 algal cells in each endodermal cell, whereas tentacle cells of the
golden brown sea anemone, *Aiptasia pallida*,
contain about 10 algal cells. Each algal cell is
housed inside a vacuole (symbiosome)
formed by the animal cell. The algae are located in translucent parts of the animal,
where they receive light for photosynthesis.
The number of photosynthetic symbionts
within host cells is controlled by nutrient
supply, factors for cell division, and pH
(Douglas and Smith, 1984; McAuley, 1985).

Although aposymbiotic invertebrates can
be reinfected with strains of *S. microadriaticum* other than their natural ones, the foreign algae grow poorly in the host. How an animal recognizes a specific algal strain is not
known, but it appears that a recognition
process takes place after the algae are phagocytized. Algal strains that are compatible with
the host are not digested because they either
avoid or resist the host's digestive enzymes.
Algal symbionts pass from one generation

of invertebrates to another in different ways.
Animals that reproduce asexually by budding
transmit some of their algae directly to their
offspring. In sexual reproduction, algae are
found in or on the eggs produced by the parent. In some invertebrates, offspring ingest
free-living algae from their surroundings.
Larvae of the jellyfish *Cassiopeia xamachana*
will not develop normally unless they become infected with algae. Certain marine
nudibranchs obtain their algae after they eat
anemones and corals, which contain symbiotic algae.

10.1 ALGAE AND MARINE PROTOZOANS AND
INVERTEBRATES

Living Sands: Foraminiferans and
Radiolarians

Foraminiferans and *radiolarians* are shelled
amoebae that commonly contain dinoflagellates as symbionts. The amoebae use carbon
compounds excreted by the algae and also digest algal cells. Foraminiferan amoebae float
or attach to objects on the ocean floor and
make thin, translucent shells that consist of
calcium carbonate. The shells have pores
through which strands of cytoplasm project
and inside of which the algae are contained.
Algae that associate with foraminiferans include unicellular green algae, diatoms, and
several unicellular red algae. One foraminiferan may associate with different algae,
sometimes simultaneously.

Foraminiferan algae are enclosed in vacuoles and lack cell walls. This makes it diffi-

cult to identify the algae to species, as the cell wall is an important taxonomic trait. If grown free from the host, the algae will produce cell walls.

The functional relationship between foraminiferans and their symbiotic algae is not fully understood. In addition to providing the host with carbon compounds, algal photosynthesis increases the foraminiferan's rate of calcification by removing excess CO_2. In return, the algae receive nitrogen, phosphorus, and vitamins from the host (Lee, 1995).

Radiolarians have siliceous shells with holes through which strands of cytoplasm project. A network of the cytoplasm surrounds the outer part of the radiolarian and contains numerous algal cells, each within a vacuole. Algal symbionts of radiolarians are yellow-brown and include dinoflagellates, prasinomonads, and a pyrmnesimonad. Each radiolarian has only one type of alga, but the number of algal cells in each host differs. Small radiolarians that are part of a colony may contain 30–50 algal cells, whereas large, solitary radiolarians may contain up to 5000 algal cells. Some hosts maintain their algal population at a constant level by digesting some of their algal cells.

Colonial radiolarians are good experimental subjects for the study of algal–protozoan relationships (Anderson, 1992). The colonies can be exposed to radioactive compounds and then separated into algal and host fractions. The central part of the radiolarian is algae-free and detaches easily from the outer network of cytoplasm, which contains the algae. The separate fractions can then be analyzed for radioactivity to determine the movement of compounds from alga to host. Studies have shown that most of the carbon compounds produced by the algae during photosynthesis pass to the host within the central capsule. The host may also obtain nutrients from prey. Compounds released from the digestion of prey may be used by the algae. The symbiotic algae assimilate waste products of the host, such as ammonia and carbon dioxide, thus enabling the radiolarian to conserve energy that it would have used to dispose of wastes. Because of their symbiotic relationship with algae, radiolarians flourish in nutrient-poor waters such as the Sargasso Sea.

In some planktonic foraminiferans and radiolarians, the dinoflagellates are moved out of their shells during the day and back into them at night. The algae are carried in and out of the shells by the cytoplasmic streaming of the host cell. This daily rhythm exposes the algae to optimum light during the day. At night the algae are brought close to the main body of the host, which facilitates nutrient transfer.

Sponges

Marine sponges contain a variety of endosymbionts, including bacteria, dinoflagellates, diatoms, and cryptomonads. The symbioses are especially common among tropical sponges. Many sponges contain endosymbiotic cyanobacteria that are unicellular or filamentous. The cyanobacteria are either intercellular, in sponge tissue directly below the outer surface, or within vacuoles inside the host cells. The sponge obtains nutrients from the digestion of bacteria or from the excretion of compounds such as glycerol and nitrogen from the bacteria (Wilkinson, 1987).

Cyanobacterial symbionts are thought to shade sponge tissue and thus protect it from the damaging effects of intense light. Sponges have not changed much since the Precambrian and may be refuges for primitive kinds of algae and cyanobacteria (Wilkinson, 1992).

Sea Anemones and Jellyfish

The sea anemone *Anthopleura xanthogrammica* contains two types of symbiotic algae: zoochlorellae and zooxanthellae. The relative proportion of each type of algal symbiont in the animal depends on the water temperature of the anemone's habitat. At high temperatures (26°C) zooxanthellae are more common, and at low temperatures (12°C) zoochlorellae are predominant. At intermediate temperatures the anemone has somewhat equal proportions of the symbionts. Curiously, the zoochlorellae excrete only small amounts of fixed carbon. Since in some anemones the zoochlorellae predominate, the algae must supply something other than carbon compounds, possibly nitrogen and phosphorus, which stimulate the anemone to grow.

Cnidarians position themselves in ways to increase the exposure of their symbionts to light. For example, when tentacles of *Anthopleura pallida* are relaxed and fully stretched, the algae in their gastrodermal cells lie in a single layer and are exposed to maximum daylight. When the tentacles contract, the gastrodermal cells shrink, and the algal cells lie on top of each other, the uppermost cells shading the lower ones. Cultured algal cells isolated from *Aiptasia* developed thicker cell walls that related to increased division rates. Algal cells in the anemone are surrounded by multiple host membranes (Palincsar et al., 1988).

Cassiopeia xamachana is a jellyfish that has been used to study how an invertebrate selects its symbiotic algae. The life cycle of *Cassiopeia* includes a sexual medusoid stage, containing algae, and an asexual polyp stage, which lacks algae. The polyps mature only after they establish an association with *S. microadriaticum* and/or bacteria (Hofmann and Brand, 1987). By exposing polyps to different strains of algae, it is possible to determine which algae are able to colonize the animal. Robert K. Trench and his co-workers (see Trench, 1993) found that although a *Cassiopeia* polyp could phagocytize different strains of algae, only one strain would survive and form the symbiosis. Thus, the host recognized its symbiont only after it was phagocytized. *Cassiopeia xamachana* does not swim freely, but rather lies upside down on the sea floor, a behavioral adaptation that allows the algae in its tentacles to receive maximum daylight for photosynthesis, and gives the animal its common name, the upside-down jellyfish.

Reef-building Corals

The symbiotic association between algae and reef-building coral animals (Scleractinia) is of great importance in tropical ecosystems and has been the subject of much study (Roberts, 1993). Coral reefs support large communities of organisms. Coral animals are similar to sea anemones, except that their polyps are small, only 10mm in diameter, and they excrete a calcium carbonate shell around their body. As the polyps die, their shells do not decay, and new polyps grow over them. After many years of this process, coral reefs are formed. Symbiotic algae live inside nutrient-rich cells of the digestive cavity of the coral polyp. In some corals more than 90% of photosynthate may be released by the symbiont to its host cell (Muscatine, 1990). Photosynthate release is one of the most studied aspects of symbiosis but still is not fully understood. It may be triggered by the action of coral digestive enzymes that increase the permeability of the algal cells.

Corals with symbiotic algae deposit calcium carbonate around their bodies much faster than animals without algae. The algae stimulate calcification through their photo-synthetic fixation of CO_2. Removal of CO_2 by the algae increases the overall reaction rates of the calcification process. High calcium ATPase activity is also linked to coral growth (Marshall, 1996). It is believed that calcium ATPase helps convert carbonate into CO_2, which is used in photosynthesis.

The algae supply the coral with oxygen as well as with carbon and nitrogen compounds. The animal obtains vitamins, trace elements, and other essential compounds from the digestion of plankton. Animal waste products, such as ammonia, are converted by the algae to amino acids, which are translocated to the animal. Such a recycling of nitrogen is an important feature in the nitrogen-poor habitats of coral.

The vast reefs that are formed in tropical waters result largely from the stimulatory effect of the zooxanthellae on the corals, which appear to be highly flexible in terms of the kinds of zooxanthellae they can take in. The high productivity of coral reefs is noteworthy, considering that they occur in shallow (less than 100 m deep) tropical waters that are low in nutrients. Pigments produced by symbiotic corals protect both the host and the algae from harmful effects of ultraviolet radiation. How symbiotic corals evolved is not known. Although all of the reef-building corals contain symbiotic algae, many other types of corals do not form associations with algae.

Coral bleaching

Coral bleaching is caused by the loss of symbiotic algae from the host or the loss of algal pigments. Bleaching may be caused by environmental stresses such as global warming, pollution, and increased ultraviolet radiation. In some cases, small reservoirs of algae survive the bleaching and can recolonize the coral (Brown and Ogden, 1993; Buddemeier and Fautin, 1993).

Understanding coral immunology is important to appreciating the nature of coral diseases. One model used to explain coral bleaching is that of the coral disease rapid tissue necrosis. A wide variety of marine bacteria live within the coral mucus (Ducklow and Mitchell, 1979), including *Vibrio alginolyticus* and *V. vulnificus*, which can release endotoxins and proteases and degrade the coral mucus (Bigger, 1988). These bacteria appear to cause rapid tissue necrosis of corals (Richardson, 1998). These and other bacterial species multiply rapidly under elevated temperatures. Phagocytosis is the primary host defense in all invertebrates. Molecules analogous to vertebrate antibodies have been found in cnidarians. In an evolutionary sense, coral bleaching may be viewed as an opportunity to create a symbiotic reshuffle that produces new associations which may be more tolerant of environmental stresses.

Convoluta roscoffensis: Green Flatworms

Convoluta roscoffensis (Turbellaria, Platyhelminthes) is a small, marine flatworm that lives in the intertidal zones of beaches in the Channel Islands of the United Kingdom and in western France. The worms are 2–4 mm long and deep green from the algae they contain; a large worm may contain up to 40,000 algal cells. During high tide, the worms are buried in the sand, but at low tide, during the daylight, they move up to the surface. During this time the algae photosynthesize until the next high tide, when the worms burrow back into the sand. When the worms are on the surface of the sand, they secrete a gelatinous substance that holds them together in colonies and protects them from drying.

The algal symbiont of *Convoluta roscoffensis* is *Tetraselmis convolutae*, a motile, unicellular green alga. The algae are ingested by digestive cells of the worm and become greatly modified. They lose their cell walls and flagella and develop fingerlike lobes that fit between the animal's muscle cells and greatly increase the surface area between the symbionts. The algal lobes are also near the host's cilia, which, like muscle cells, have high energy demands. Because of these close contacts, photosynthetic products of the alga pass directly to the most active regions of the animal. Despite their irregular shapes, all the algal cells face the same direction. The anterior end, which normally has four flagella, faces the inner part of the animal, while the lobed posterior end, which contains the chloroplast, faces the outside. The algal cells are separated from the host cytoplasm by membranes produced by the animal (Douglas, 1988).

When the worms mature, they produce eggs that hatch into larvae, which lack algae and thus are aposymbiotic. Free-living cells of *T. convolutae* are attracted to the egg cases of the worms through an unidentified chemical stimulus. The algae swim to the egg cases and attach to them. When female worms lay eggs, they also extrude algal cells and mucus that coats the eggs. Larvae that hatch from the eggs ingest the algae, which then divide until their normal population in the animal is reached.

After the worm matures and has its full complement of algae, it stops feeding and depends on the algae for nutrients for the rest of its life. Aposymbiotic larvae that do not become infected with algae will starve to death, regardless of how much food they ingest.

The physiological relationships between *C. roscoffensis* and its symbiotic algae have been determined from various studies. The algae provide alanine, glutamate, fatty acids, and sterols to the animal. In return, the algae use uric acid, a waste product of the animal, as a source of nitrogen.

Convoluta roscoffensis can be infected with algae that are related to its normal symbiont. These foreign algae undergo the same modifications as the natural symbiont, but the animal usually does not grow as vigorously with these secondary symbionts. If an animal that is infected with a secondary symbiont is exposed to cells of its natural, primary symbiont, it will discard or kill the foreign alga and replace it with its natural one.

The *Convoluta*–alga symbiosis is an early example of detailed physiological studies conducted by Keeble and Gamble during 1903–1907 that attracted public attention to the broader significance of symbiosis in nature (Keeble, 1910).

Tridacnid Clams: Climax Forests of Coral Reefs

Among the molluscs, only tridacnid clams form symbioses with algae. These large clams grow among coral reefs in shallow, tropical waters of the Indo-Pacific. Taxonomically, giant clams belong to two *Hippopus* species and six *Tridacna* species. *Tridacna gigas* can weigh up to 1000 pounds and produce shells more than 3 feet long (Lucas, 1994). Like *Cassiopeia*, the clams have undergone a behavioral adaptation, and they attach to the coral reef in such a way that their valves face upward and are open. This position allows their symbiotic algae to receive optimum illumination.

Symbiodinium lives within a tube system which develops from a diverticulum duct of the stomach that branches several times and finally terminates in the mantle. The tubules are thin and narrow and allow close contact with the hemolymph that circulates through the mantle. Presence of the tubules near the surface of the mantle allows *Symbiodinium* to capture sunlight (Norton et al., 1992). Young clams ingest free-living algae, which move from the alimentary tract to the mantle. As with other associations, the symbiotic algae supply the clams with substantial amounts of glycerol. Clams kept in the dark in nutrient-rich waters die quickly, whereas those kept in sunlight flourish.

Giant clam farming

Tridacna dersa and *Tridacna gigas* are two large and heavily harvested species in coral reef environments of Indonesia, the Philippines, Micronesia, and southern Japan. The commercial exploitation of these giant clams has caused a decline in their populations. By using farming techniques of land-based nurseries in coral reef environments, scientists are exploring the possibility of raising giant clams as a sustainable food source (Heslinga and Fitt, 1987).

The clams reach sexual maturity in about 5 years and produce large quantities of sperm and eggs. Spawning can be synchronized by using pheromones and environmental cues (Knop, 1996). The planktonic phase of a tridacnid life cycle lasts about a week, during which there are significant losses from predation. The surviving larvae attach to corals and develop into juvenile clams, which orient their mantles toward the sun. Only young clams can form symbioses with zooxanthellae. Young clams up to about 3 years old may be eaten by reef carnivores that include crabs, fish, snails, and octopuses. Adult tridacnids appear to be immune to predation (Heslinga and Watson, 1985). Giant clams have been compared to climax forests because they are sedentary, grow slowly, resist predation, and depend on photosynthesis.

The hatchery at the Micronesian Mariculture Demonstration Center in Palau produced more than 250,000 juvenile clams in 1985 for introduction on reefs of Guam, American Samoa, the Philippines, and Hawaii (Heslinga and Fitt, 1987). The successful culture of giant clams may provide another experimental system to study the molecular biology of marine symbioses (Yellowlees et al., 1993)

Sea Slugs: Animals that Express Chloroplast Genes

An unusual type of marine relationship, which strictly speaking is not a symbiosis because it involves only part of an organism, is that between sea slugs, such as *Elysia* sp., and the chloroplasts of algae. Sea slugs ingest the large cells of siphonaceous, or tubular, algae, especially *Caulerpa* and *Codium*, and the digestive cells of the animals phagocytize the algal cells. The chloroplasts resist digestion and continue to function for several weeks or even months within the host cells. The slugs contain so many chloroplasts that they appear green. A chloroplast may be enclosed within a vacuole or lie free in the cytoplasm of the animal.

Chloroplasts photosynthesize inside the animal cell and release glucose, which the slug uses for some of its nutrient needs. Functional chloroplasts are not digested by the animal, but when a chloroplast stops photosynthesizing, it is quickly digested. How the animal cell distinguishes between active and inactive chloroplasts is not clear. Chloroplasts cannot divide inside the animal because they are unable to synthesize chlorophyll without the enzymes that are present in the algal cytoplasm. A slug constantly replenishes its chloroplast supply by feeding on new algae. In *E. chlorotica* some chloroplast proteins were synthesized in the host animal cell cytoplasm and transferred to its chloroplasts, thereby extending their survival for up to 9 months (Mujer et al., 1996; Pierce et al., 1996).

The factors that cause chloroplasts to release glucose inside the slug's digestive cells are not known. Chloroplasts that are removed from the animal stop excreting glucose or only excrete small amounts. But if homogenates from symbiotic slugs are added to the isolated chloroplasts, they will once again excrete glucose. This stimulation of excretion by unknown host factors is like that in other marine symbioses.

In a similar relationship, some planktonic foraminiferans and other protozoans do not readily digest the chloroplasts of diatoms and other algae on which they prey. These chloroplasts also continue to photosynthesize while they are inside the protozoans (Laval-Peuto, 1992).

Tunicates

Some tropical marine tunicates contain an unusual photosynthetic symbiont, *Prochloron*, that has characteristics of both cyanobacteria and green algae (Lewin and Cheng, 1989) (fig. 10.2). *Prochloron* cells are green, but their cell structure is prokaryotic; that is, the cell lacks a nucleus, has a wall like that of cyanobacteria, and has organelles such as polyhedral bodies. The cells, however, lack phycobiliproteins and phycobilisomes, which are universal in cyanobacteria. Like green algae, *Prochloron* cells have chlorophylls a and b, whereas cyanobacteria have only chlorophyll a. *Prochloron* has such a unique mix of characteristics that it has been placed into a separate phylum, Prochlorophyta (Matthijs et al., 1994).

All strains of *Prochloron* are unicellular and have spherical cells. In some tunicates the symbiont cells lie within a cellulose matrix that surrounds the outer surface of the animal, whereas in other tunicates the symbionts are loosely attached to the cloacal wall. *Prochloron* cells always have an extracellular position in the animal.

The larvae of some tunicates have specialized pouches that carry *Prochloron* cells obtained from the parent. As the tunicate develops, it is infected at an early stage by the symbiont. Thus, the continuity between *Prochloron* and the tunicate is uninterrupted from one generation to the next.

Experimental studies on these associations are limited because *Prochloron* does not grow in laboratory culture. All studies, therefore, must be carried out on symbionts newly squeezed out of the tunicates or from dried or preserved specimens. Preliminary findings have revealed that the symbiont photosynthesizes in the tunicate and releases carbon compounds. Presumably, these compounds, as well as the oxygen produced during photosynthesis, are used by the animal.

Prochlorophytes have been found in symbiosis with marine invertebrates and they are also dominant members of the phytoplankton of open oceans (Chisholm et al., 1988; Mullineux, 1999). They may have been common in earlier periods, such as the Upper Cambrian, when the low concentrations of oxygen in the sea and atmosphere were more conducive for their free-living growth. Prochlorophytes in marine invertebrates may represent relict populations, isolated remnants of a once widespread group.

10.2 ALGAE AND FRESHWATER INVERTEBRATES

Many freshwater invertebrates contain symbiotic *Chlorella* (Chlorophyceae) (table 10.1). The algae are intracellular inhabitants of host cells that are capable of endocytosis. The algae are enclosed within perialgal vacuoles, which protect them from the host's digestive enzymes. Symbioses range from temporary, as in *Coleps*, to complex hereditary associations, as in *Paramecium bursaria*. In *P. bursaria*, growth rates of both the host and symbiont are coordinated. The *Chlorella* symbionts of these different invertebrates appear similar, although some algal strains excrete glucose instead of maltose.

The presence of similar strains of *Chlorella* in unrelated invertebrates suggests that these associations developed independently on different occasions. The associations probably began when cells of *Chlorella* were ingested by invertebrates and resisted not only the host's attempts to eject them but also the digestive enzymes of the host. Once the algae became permanent residents in the animal cell, relationships between the symbionts evolved. Why some species of invertebrates were more susceptible to algal infection than others is not known. For example, there are species of *Hydra* and *Paramecium* that resist colonization by algae.

Paramecium bursaria: green paramecia

Paramecium bursaria is one of the best studied freshwater symbiosis because both the paramecia and algal cells are easily maintained under laboratory conditions (Reisser, 1992b, 1993). Each algal cell is contained inside a perialgal vacuole (or symbiosome) that is formed by the host (Meier and Wiessner, 1989) (fig. 10.3). The alga somehow changes the vacuole so that it does not fuse with lysosomes, and the alga therefore avoids being digested. When an algal cell divides, each of its four daughter cells becomes enclosed in its own vacuole. After the algal cells are ingested, they move from the center of the animal cell to the periphery. Paramecia normally feed on bacteria, and when the food supply is low, those animals with algae live longer than those with none. When food is plentiful, however, both types of paramecia multiply at the same rate. Only certain strains of *Chlorella* sp. form symbioses with *P. bursaria*. All other *Chlorella*-like algal species do not trigger the formation of perialgal vacuoles and are digested in food vacuoles. *Chlorella* symbionts may use a lectinlike mechanism to become enclosed within the perialgal vacuole of the host (Reisser, 1992a).

Algae of *P. bursaria* excrete large quantities

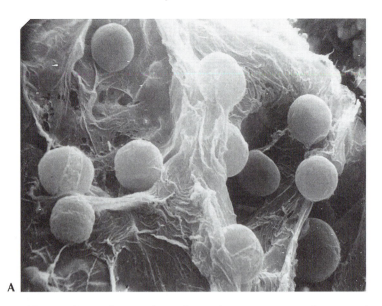

A

Fig. 10.2 *Prochloron*, a photosynthetic symbiont of tropical marine tunicate, *Didemnum midori*. (a) Scanning electron micrograph of *Prochloron* in the host tissue. (b) An electron micrograph of *Prochloron* showing cellular details of organized chloroplast and the absence of nucleus. (c) Details of the stalked thylakoids with concentric regions of stroma that contain carboxysomes, large crystalloids and dense granules. (a Courtesy of Ralph A. Lewin, Scripps Institution of Oceanography, University of California, San Diego; b and c Courtesy of Hewson Swift, University of Chicago).

of maltose and glucose and receive ammonia and glutamine as nitrogen from the host (Reisser, 1993). About 45% of the total photosynthetically fixed carbon is released by the algal symbiont. The excretion is not triggered by any host factors. The amount and type of substances that are excreted is pH dependent, and with increasing acidity, more sugars are released. At alkaline pH, amino acids such as alanine are excreted. The excretion of metabolites from symbiotic *Chlorella* is an altruistic behavior which is not seen in free-living *Chlorella*.

When fed bacteria in the dark, paramecia divide more rapidly than their symbionts, and aposymbiotic paramecia develop. Aposymbiotic strains can be reinfected by different algae, including free-living strains, but only those that have been removed from a previous symbiosis grow well in the animal. During the establishment of the symbiosis, algae in the perialgal vacuoles divide slowly until there is a population of about 600–1000 algal cells in each host cell.

Hydra–Chlorella Symbiosis

Hydra is a common inhabitant of freshwater ponds and lakes. It remains attached to living or decaying vegetation and feeds on protists and on small animals such as shrimp. Some species of *Hydra* contain the green alga *Chlorella* (fig. 10.4). The algae divide asexually, by mitosis, and each mother cell produces four daughter cells. The animal appears green because its cells are packed with algae. The animals obtain nutrients from the algae and can live in areas where food is scarce and survive periods when the external food supply is low. The animals ingest algal cells, as they do other particles of food. Algae that become symbiotic escape digestion and are retained inside vacuoles. When the animal divides, some of the algal cells are transmitted to the offspring, so that every individual in the animal population contains algal cells (Muscatine and McNeil, 1989).

Green hydras live in the same habitats as brown hydras. When the natural food supply is abundant, both types of *Hydra* multiply at about the same rate. When the external food supply is scarce, however, green hydras, because of their symbiotic algae, have an advantage over brown hydras.

Hydra viridis is a common species of green hydra. Several strains of this species have been established in laboratories throughout

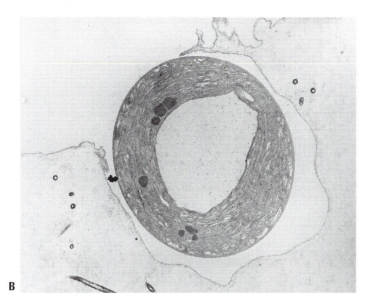

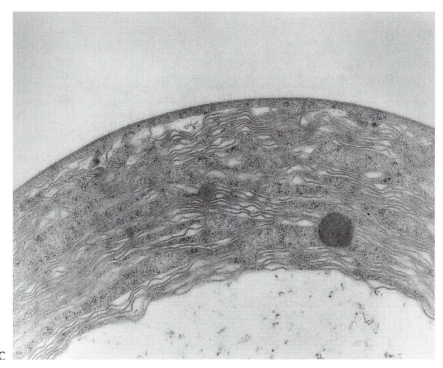

Fig. 10.2 (*continued*)

the world. These strains have been studied in terms of the relationships between the algae and their animal hosts. *Hydra* grows well in laboratory cultures. They are fed brine shrimp, kept in the light to allow the algae to photosynthesize, and maintained in clean, bacteria-free water. Under these conditions a hydra population doubles by asexual budding about every 2 days.

When *H. viridis* is exposed to high light intensity in the presence of 3-(3,4-dichlorophenyl)-1,1-dimethylurea, a chemical that in-

Table 10.1   Freshwater protozoa and invertebrates which have Chlorella and related species as symbionts

Host	Carbohydrate Released
Protozoa	
Amoeba	
Acanthocystis turfacea	Maltose
Amphitrema flavum	
Chaos zoochlorellae	
Diffugia oblonga	
Heleopara sphagni	
Hyalosphenia papilio	
Mayorella viridis	
Ciliates	
Acaryophyra sp.	
Climacostomum virens	Glucose, fructose, xylose
Coleps hirtus	Xylose
Disematostoma butschlii	
Euplotes daidaleos	Fructose, xylose
Frontonia leucas	
Frontonia vernalis	
Halteria bifurcata	
Holosticha viridis	
Malacophrys sphagni	
Ophrydium versatile	
Paramecium bursaria	Glucose, maltose, trehalose
Prorodon viridis	
Prorodon ovum	
Spirostomum viridis	
Stentor polymorphus	Maltose
Stentor niger	
Stentor roeseli	
Teutophrys trisulca	
Vorticella sp.	
Porifera	
Ephydatia fluviatilis	Glucose
Heteromeyenia sp.	
Spongilla lacustris	
Cnidaria	
Hydra virdis	Maltose
Hydra magnipapillata[a]	
Platyhelminthes	
Turbellarians	
Castrada viridis	
Dalyellia viridis	
Phaenocora typhlops	
Typhloplana viridata	
Mollusca	
Bivalves	
Anodonta sp.	
Linnaea sp.	
Unio pictorum	

Adapted from Reisser (1992b).
[a]Algal symbiont: *Symbiococcum hydrae*

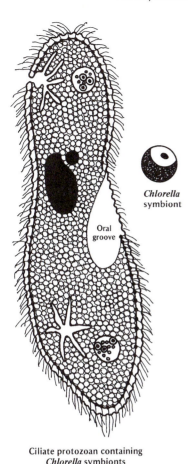

Chlorella
symbiont

Oral
groove

Ciliate protozoan containing
Chlorella symbionts

Fig. 10.3 *Paramecium bursaria* with endosymbiotic *Chlorella*.

hibits photosynthesis, the animal loses its green algae and becomes aposymbiotic or bleached. A bleached hydra will survive and grow if it has sufficient food. *Chlorella* symbionts from several strains of *H. viridis* have been grown separately from the animal. The cultured symbionts differ from the symbiotic forms in that they excrete smaller amounts of photosynthetic products. When the algae are placed back into the animal, they once again excrete compounds at the rate of symbiotic forms. If provided with suitable algae, bleached *Hydra* will ingest them and become green again (Rahat, 1992). A foreign species of *Chlorella* was shown to infect bleached *H. viridis*, but the animal did not multiply as rapidly as normal strains.

Reinfection of aposymbiotic *H. viridis* occurs through five stages. During the contact stage, the algae attach to the membrane of a hydra cell. During engulfment, the animal cell ingests the algae by means of membrane projections, or microvilli. The animal cell takes in anything that contacts it, including latex spheres and foreign algae. In the recognition stage, by means of an unknown process, the digestive cells recognize potentially symbiotic algae. Several factors may be important in the recognition process, including antigens and surface charges on the algal cells and the type of microvilli produced by the animal cell during the engulfment stage. Algae that the cell rejects are expelled from the cell; acceptable algae are retained and surrounded by vacuoles. In the migration stage, algae that are accepted by the animal are moved to the base of the digestive cell. Because the algal cells are nonmotile, their movement is controlled by the animal cell, possibly by means of microtubules. Finally, during the repopulation stage, algal cells divide asexually until the normal algal population of an animal cell is reached. Further division of the algal cells is closely linked to that of the host cell. When the host cell divides, so do the algal cells.

A single hydra contains about 150,000 al-

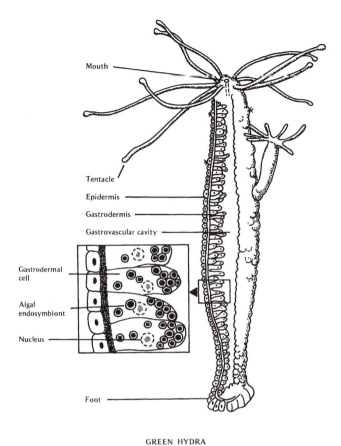

Mouth

Tentacle

Epidermis

Gastrodermis

Gastrovascular cavity

Gastrodermal
cell

Algal
endosymbiont

Nucleus

Foot

GREEN HYDRA

Fig. 10.4 Green hydra's gastrodermal cells with endosymbiotic *Chlorella*.

gal cells as well as numerous bacteria that occur in vacuoles with or without algae. The
bacteria may help the algae take up and store
phosphate. The number of algal cells in each
hydra cell varies depending on the location of
the digestive cell within the animal. In the
central region of the animal, each digestive
cell has about 20 algal cells, and in the head
and tentacles each cell has about 10 algae.

Algal and bacterial symbionts are transmitted to hydra progeny during asexual and
sexual reproduction of the animal. Cells of the
symbionts are attached to the outer surface of
the eggs and are ingested by the young hydra
when they emerge from the eggs. In some hydra strains extensions of endodermal cells
carry algae into the developing eggs.

Under normal conditions symbiotic algae
are not digested by hydra. There are two main
reasons for this. First, cell walls of the algae
contain *sporopollenin*, a compound that re-
sists digestive enzymes. Second, vacuoles

containing algae do not fuse with lysosomes,
organelles that contain digestive enzymes and
normally fuse with food particles the animal
ingests. But if a digestive cell takes in more algal cells than its normal algal population, the
extra cells are either digested or ejected.

A bilateral movement of nutrients takes
place between the symbionts of *H. viridis*. The
algae supply the animal with photosynthetic
products such as maltose (McAuley et al.,
1996). At pH 4.5, almost 60% of the carbon
fixed by the algae is excreted as maltose, but
at neutral pH very little maltose is excreted. It
is not known whether the host can change the
pH of its cytoplasm to control maltose release
by the algae. The animal rapidly hydrolyzes
maltose to glucose and forms the storage compound glycogen. Algae also provide the animal with oxygen, which they produce during
photosynthesis. The animal provides the algae with nutrients, including precursors of
proteins and nucleic acids. These compounds

and others that are breakdown products from the hydra's food enable the algae to survive if the animals are kept in darkness.

The *Hydra–Chlorella* symbiosis appears to be a finely tuned, nonpathogenic equilibrium between the way the symbionts avoid digestion and the ability of host cells to regulate algal reproduction (Muscatine and McNeil, 1989).

10.3. PARASITIC ALGAE

Pathogenic Green Algae

Cephaleuros is a common pathogen of vascular plants in tropical and subtropical regions. It is a member of the family Trentepohliaceae, which also includes the familiar epiphytic alga *Trentepohlia*. There are about a dozen species of *Cephaleuros*, the most common one being *C. virescens* (Chapman and Waters, 1992).

Cephaleuros causes red rust disease on commercial plants such as citrus fruits, coffee, and tea. The alga forms a flat, disclike orange thallus, 5–10mm in diameter, that is usually subcuticular in leaves and has hairlike filaments that break through the cuticle and have a velvety appearance; the alga also grows on fruits and stems. The damage caused by *Cephaleuros* depends on the type of host and its condition, the site of infection, the season, and the number of algal thalli on the plant. Algae that penetrate leaves often kill the host cells that lie beneath the thallus, either by shading them or by toxic substances they release. The infection does not extend beyond the algal thallus, however, and the overall damage to the plant is slight. In leaves of the coffee plant, the infected host produces thick-walled cells that form a barrier between the parasite and healthy leaf tissue. Algae may girdle young stems and kill the plants. Infection of fruits such as guava and mango reduces their attractiveness and marketability.

Cephaleuros does not penetrate the host cells, regardless of the extent of the infection. The alga obtains water and minerals from the host and a substrate on which to grow. It also changes the carbon and nitrogen metabolism of the infected leaves, but whether it receives organic compounds from the host is not known.

Cephaleuros produces motile zoospores that spread the infection. The zoospores swim along the leaf surface or are washed by rain onto other leaves, and when they come to rest, they divide and form a thallus, which penetrates the host epidermis.

An interesting aspect of *Cephaleuros* is that it may be parasitized by a fungus to form a lichen. The fungus forms haustoria inside the living algal cells and eventually kills them. This type of host–parasite relationship is an example of hyperparasitism. The most common lichen formed by a fungus–*Cephaleuros* association is the genus *Strigula*, which occurs on leaf surfaces throughout the tropics. Lichenized and nonlichenized *Cephaleuros* may occupy the same habitats, but the lichenized form does not harm the leaves and grows more slowly than the unlichenized form.

Several parasitic green algae, not as common as *Cephaleuros*, cause limited damage to their hosts. *Chlorochytrium* is an endophyte of aquatic plants, such as duckweed, and also of several mosses. *Rhodochytrium* occurs in various angiosperms, such as milkweed and ragweed. *Rhodochytrium* lacks chlorophyll, and when first discovered it was mistaken for a primitive fungus.

Parasitic Red Algae

The phylum Rhodophyta consists of about 6000 species of mostly marine organisms with complex life cycles. More than 100 species, within 50 genera, of red algae parasitize other red algae. About 80% of the parasites occur on closely related species (i.e., ones in the same family or tribe) whereas others parasitize unrelated hosts. Parasitic red algae occur only in the subclass Florideophyceae, especially in the orders Ceramiales and Cryptonemiales, and not in the more primitive subclass Bangiophyceae.

Parasitic red algae are small (0.2mm to 4cm long), colorless or with reduced pigments, and host specific. They attack only one or two genera, to which they are closely related, and they penetrate the host tissue by means of rhizoids. Some parasites consist only of a rhizoidal system and do not form a vegetative thallus external to the host. Some parasites complete their life cycle 3–5 weeks after their spores germinate on the host; other parasites take about 18 weeks to complete their life cycle.

Red algal parasites generally do not cause disease and thus are not considered pathogens. Their parasitic nature, however, has been confirmed from radioisotope studies, which have demonstrated movement of organic compounds from the host to the parasite. For example, in one study, a segment of the red alga *Gracilaria*, with its attached parasite *Holmsella*, was placed in sea water that contained ^{14}C-labeled bicarbonate and allowed to photosynthesize for several hours. The segment was removed from the labeled solution, washed, and either separated immediately into host and parasite plants or allowed to photosynthesize for several more hours in an unlabeled solution, before host and parasite were separated and analyzed for radioactive compounds. The results of this study clearly showed that the amount of radioactivity in the parasite increased with time, indicating a movement of compounds. The compound that moved from host to parasite was the alcohol glycoside floridoside. When this compound reached the parasite, it was converted to mannitol and floridean starch. Similar results were obtained with the alga *Polysiphonia lanosa* and its parasite *Choreocolax polysiphoniae* (Goff, 1983). In this case, sodium mannoglycerate, another alcohol glycoside, was translocated to the parasite and converted to an unidentified sugar. Thalli of *Choreocolax* that were detached from their host fixed CO_2 more rapidly than thalli that remained on the host. This finding suggested that the host inhibited photosynthesis by the parasite.

Morphological relationships between parasitic red algae and their hosts vary considerably. Some parasites attach to the host by only one or two short rhizoids, whereas others develop extensive rhizoids in the host tissue. Parasites that penetrate deeply usually damage the host cells (e.g., they become plasmolyzed), but such damage remains localized. Many parasites are linked with host cells by means of pit connections. When pit connections develop, nuclei of the parasite are transferred to the host cell. The red alga *Leachiella pacifica* inserted some of its nuclei into the cells of its host, *Polysiphonia confusa*. The foreign nuclei caused the host cell to enlarge to 20 times its original size and to develop a thick cell wall. The number of host nuclei, mitochondria, and chloroplasts increased, as did the DNA content of the host nuclei; products of photosynthesis also accumulated in the cell (Goff et al., 1996).

How parasitic red algae evolved is a matter of speculation (Goff et al., 1996). One hypothesis is that they arose when mutated spores of a red alga produced small gametophytes that were unable to live independently but had the ability to penetrate cells of the parent plant. There are present-day red algae whose tetraspores germinate and give rise to dwarf gametophytes that attach to the parent plant. The gametophytes form rhizoids and develop pit connections with the parent cells. A second hypothesis is that the algae were originally epiphytes that became parasites. The first hypothesis seems more plausible because of the close taxonomic relationships that exist between parasitic red algae and their hosts.

Some scientists believe that ascomycetous fungi may have evolved from parasitic red algae that lost their plastids. Ascomycetes and red algae have many common morphological and cytological features.

10.4 SUMMARY

Symbiotic algae live in the cells of many marine invertebrates. The algae excrete glycerol and other compounds that the host uses to produce lipids and proteins; waste products of the host are used by the algae. *Anthopleura xanthogrammica*, a sea anemone, has two types of algae, zooxanthellae and zoochlorellae; the former is more prevalent when the anemone grows in warm waters and the latter is more common in anemones found in cold waters. The life cycle of the upside-down jellyfish, *Cassiopeia xamachana*, cannot be completed until its asexual polyp stage becomes infected with algae. Symbiotic algae greatly enhance the activities of reef-building corals, tridacnid clams, foraminiferans, and radiolarians. A single marine flatworm, *Convoluta roscoffensis*, may contain up to 40,000 unicellular green algal cells. The algae live inside the digestive cells of the worm and are greatly modified. A mature worm is completely dependent on the algae for food. Larvae of the worm are aposymbiotic and must be infected with algae in order to develop further. Giant clams are being cultivated as a source of food. Marine sponges contain symbiotic cyanobacteria. Sea slugs obtain glucose from chloroplasts of algae they ingest. *Prochloron* occurs in tropical marine invertebrates and has characteristics of both green algae and cyanobacteria. *Hydra viridis* and *Paramecium bursaria* contain symbiotic algae that belong to the genus *Chlorella*. Infection of *H. viridis* involves different stages of contact, engulfment, recognition, migration, and repopulation. Several green algae are parasitic on vascular plants, and many small red algae parasitize other red algae.

Symbioses between algae and invertebrates involve common features such as integration of the symbionts, regulation of the algal population, specificity, bilateral nutrient exchange, and recognition mechanisms. Integration of a symbiont into a host generally results in reduction of parts, thin walls, loss of sexuality, and slow growth. Algal populations are regulated by the host by several mechanisms, including digestion or ejection of surplus algae and inhibition of symbiont cell division. Anemones and corals eject surplus algae, and in other symbioses algal division is restricted. Studies of recognition systems between symbiont and host in different associations have not revealed a common pattern, a finding that is consistent with the different lines of evolutionary development among symbioses.

11 ANIMAL PARASITISM

Flukes, Tapeworms, Nematodes, and Parasitoids

Scientists estimate that up to 50% of all animal species are parasitic symbionts. Some phyla such as the Platyhelminthes, Nematoda, and Arthropoda contain large numbers of parasitic species. Hosts and parasites have evolved together and under natural conditions many have become mutually tolerant. Host organisms can live independently, but, in many cases, the parasite's association with its host is obligatory (Price, 1980; Jenzen, 1988; Thompson, 1994).

Animal parasites affect the health of humans and domesticated animals throughout the world. In most warm climates parasitic infections from flukes, nematodes, and arthropods greatly diminish the quality of life. Scientists estimate that three out of four people in the world are infected with parasites, and many people carry multiple infections. The prevalence of parasitic diseases is extremely high among people who are poor and illiterate (Evans and Jamison, 1994). Because of public health concerns, the science of parasitology continues to attract public support to study host–parasite relationships and to develop effective strategies to control and cure parasitic infections. In this chapter, we examine symbioses that involve flukes, tapeworms, nematodes, and arthropods. We describe the morphological adaptations of parasites, how the host defends itself against parasitic invasion, how parasites evade a host's defenses, and coevolution of parasites and hosts.

Helminths are widely distributed parasites of plants and vertebrates and are surpassed only by insects in the variety and diversity of hosts that they parasitize. There are more than 100,000 species of helminthic parasites, but only a relatively few species are of economic and public health concern. Infections caused by helminths such as *Schistosoma*, hookworms, and filarial nematodes are the major causes of sickness in humans inhabiting tropical regions (Warren, 1988). Several species of nematodes are parasites of vegetables, fruits, and other crop plants. Some free-living nematodes, such as *Caenorhabditis elegans*, are being used to study metazoan genetics and developmental biology because of the ease with which they can be cultured and analyzed biochemically.

Helminths have complex life cycles. They live for a long time within the host animals, and they often possess a remarkable ability to evade the host's defense mechanisms. The prevalence of helminthic infections in some areas is high; however, only a few either very young or old hosts develop disease. Helminths do not multiply in humans, and therefore the severity of the disease depends on the extent of the original infection. Some helminths may accumulate after repeated infections of a host.

11.1. CLASSIFICATION

We have included in this chapter a broad classification of helminthic symbionts:

Phylum Platyhelminthes (flatworms)

Class Trematoda (flukes): *Clonorchis, Diplozoon, Fasciola, Paragonimus, Polystoma,* and *Schistosoma.*

Class Cestoda (tapeworms): *Diphyllobothrium, Echinococcus, Hymenolepis, Lingula,*

Schistocephalus, Spirometra, Taenia,
and *Taeniarhynchus.*

Phylum Nematoda (roundworms)

Human and vertebrate parasites:
Ancylostoma, Ascaris, Brugia, Diro-
filaria, Dracunculus, Haemonchus,
Litomosoides, Necator, Onchocerca,
Trichinella, and *Wuchereria.*

Insect parasites: *Deladenus, Mermis,*
Neoaplectana, Romanomermis, and
Sphaerularia.

Plant parasites: *Anguina, Bursapha-*
lenchus, Globodera, Heterodera, and
Meloidogyne.

11.2. TREMATODES: THE FLUKES

Some Fluke Symbioses

Adult *flukes* are obligate endoparasites of vertebrates and include the liver fluke, the lung
fluke, and the blood fluke of humans. After
sexual reproduction, the female fluke produces eggs that exit through the genital pore
into the host environment and then are
passed out of the host with the feces or urine.
There are several larval stages, which multiply asexually in snails serving as the intermediate host. A larval stage (cercaria) with a
characteristic tail emerges from a snail and either penetrates a vertebrate host immediately
or encysts and attaches to grass blades, which
may be ingested by a herbivorous vertebrate
(fig. 11.1).

Polystoma integerrimum

Polystoma integerrimum inhabits the urinary bladder of frogs throughout Europe and
North America. Up to 50% of the frogs in the
United States carry the fluke. The life cycle
of the fluke is unusual in that its maturation
is synchronized with the sexual maturation

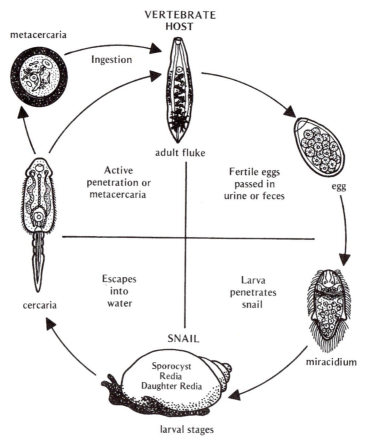

Fig. 11.1 A typical fluke life cycle.

of its host. When the frog spawns, the flukes become sexually mature and release their eggs. The synchronized cycles enable the parasite to produce larvae at a time when tadpoles are abundant. The fluke larvae attach to the tadpole gills and feed on mucus and other materials. The mature larvae migrate from the gills to the bladder. Experimental evidence shows that fluke maturation is regulated by host hormones. Pituitary extracts injected into an immature frog induce sexual maturation of the parasite. The exact nature of the hormonal control of fluke development is not known. Sexual maturity of the protozoan *Opalina* is also synchronized with reproduction of its frog host (Smyth, 1994).

Diplozoon paradoxum

Diplozoon paradoxum is a parasite of freshwater fish and has a unique life cycle that involves permanent fusions of pairs of flukes. Such fusions raise interesting questions about the biochemical, genetic, and immunological compatibility of the fused individuals. The attachments begin when the larvae pair and fuse with their posterior suckers. The reproductive openings of each worm exit on the opposite end of their partner in order to permit cross-fertilization. Larvae that do not form pairs die before becoming sexually mature. In *D. nipponicum*, the reproductive cycle is synchronous with that of the fish host (Smyth, 1994).

Fasciola hepatica

Fasciola hepatica, the sheep liver fluke, commonly inhabits the bile duct, liver, or gall bladder of cattle, horses, pigs, and other farm animals. The parasite's life cycle, ultrastructure, nutrition, and biochemistry have been studied extensively. Upon hatching from metacercariae, young flukes penetrate the intestinal wall, move through the body cavity, and enter the liver. During their migration, the young flukes feed on muscle, intestinal, and liver parenchyma cells. When the parasites reach the bile duct, they feed on epithelial cells, blood, and liver tissue. The fluke also absorbs amino acids, glucose, and fatty acids through its tegument. Epithelial cells of the bile duct become hyperplastic during the fluke's infection. The fluke somehow stimulates production of the host tissue on which it feeds (Roberts and Janovy, 1996).

Clonorchis sinensis

The Chinese liver fluke, *Clonorchis sinensis*, is an important parasite of humans and other fish-eating mammals in Far Eastern countries. The fluke makes its way to the liver and bile duct directly from the alimentary canal and feeds on epithelial and blood cells. In endemic areas, the prevalence of fluke infection may range from 14–80%, and the number of flukes in an infected individual varies from 20 to 200. Adult flukes can live for more than 8 years in humans. Fish become infected with larvae that are liberated from a snail host. People acquire these flukes by eating uncooked or smoked fish whose muscle contains metacercariae of the parasite. Fish farming in East Asia is a major source of fluke infections in people. In other areas, dogs and cats serve as reservoir hosts of *C. sinensis* (Komiya, 1966).

Paragonimus westermani

Species of *Paragonimus* inhabit different organs of vertebrate hosts and are excellent examples of zoonosis. *Paragonimus westermani*, the lung fluke, infects humans as well as cats, dogs, and rats. In humans the flukes become encapsulated in the lung tissue and produce eggs, which pass upward into the trachea to the mouth and then down the intestinal tract to exit with the feces. This fluke is extremely prevalent in the people of China, the Philippines, Thailand, and other Asian countries. Humans become infected with the fluke by eating raw crabs and crayfish. Crab juice is frequently used to prepare food in Korea and the Philippines.

Schistosoma

Next to malaria, schistosomiasis is the most important parasitic disease in the world, affecting more than 200 million people in more than 75 countries (Webbe, 1981). *Schistosomes* are blood flukes, and they reside in the mesenteric blood vessels of humans. Adult schistosomes have a unique morphology and physiology. Male and female flukes are elongated and wormlike, and the female fluke is held permanently in the ventral groove of the male fluke (fig. 11.2). The presence of eggs of blood flukes in various host tissues triggers an immunological response, causing the affected person to show disease symptoms that include enlargement of the liver and spleen,

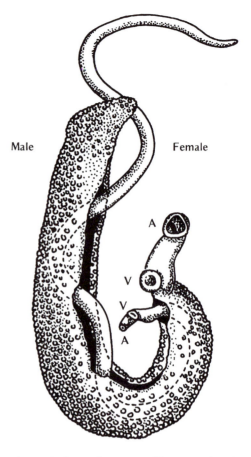

Fig. 11.2 An adult female of *Schistosoma mansoni* in permanent copulatory union within the ventral groove of the male fluke.

A = anterior sucker   V = ventral sucker

bladder calcification, deformity of the ureter, and kidney disorders. Schistosomiasis is common among poor people, especially farmers who live and work under unsanitary conditions, and is increasingly being recognized as an anthropogenic disease. The development of irrigation projects increases the range of the snail species that are the intermediate hosts of *Schistosoma*.

Three important blood flukes that infect humans are *Schistosoma mansoni*, *S. japonicum*, and *S. haematobium*. *Schistosoma mansoni* causes intestinal bilharziasis in South America, Central America, and the Middle East and has been studied extensively by scientists because of the ease with which it grows in laboratory mice, hamsters, and rats. *Schistosoma japonicum* affects millions of people in Asia and is also prevalent among rats, pigs, dogs, cattle, and horses. Urinary schistosomiasis is caused by *S. haematobium* and occurs in about 40 million people in Africa and the Middle East. It is a disease of young adults between the ages of 10 and 30 years, and rarely occurs in older persons. Scientists in the Sudan have noted a correlation between bladder cancer and urinary schistosomiasis. Research in the development of an effective vaccine in the control of schistosomiasis is making significant progress (Cherfas, 1991; Bergquist and Colley, 1998).

Fluke–Host Relationships

Studies of the biology of host–parasite relationships of flukes have focused on structure and function of the body wall (tegument), nature of the host's immune response to the fluke, and evasion of the host's immune mechanism by the parasite. On penetration, the cercaria's tegument undergoes profound changes in structure and physiology.

Until the early 1960s, it was believed that

all flatworms were covered with a cuticle, a tough, inert structure. Ultrastructure of the fluke body covering, however, has revealed it to be a unique and biologically active structure. The term *tegument* is now applied to the outer covering of flatworms instead of cuticle, which is characteristic of roundworms.

The tegument is composed of (1) an outer region, the *distal cytoplasm*, which consists of cytoplasmic extensions of tegumental cells forming a syncytium, and (2) an inner *proximal cytoplasm*, which consists of tegumentary cells (fig. 11.3A). The tegument is bound externally and internally by a plasma membrane. The tegumentary matrix contains many mitochondria, ribosomes, vacuoles, endoplasmic reticulum, and Golgi bodies and is capable of nutrient uptake, osmoregulation, and excretion and can support various sensory structures. The tegument contains secretary products believed to have an immunological role in protecting the parasite from host digestive enzymes. The tegument is also covered with a carbohydrate–protein complex, the *glycocalyx*, which is continuously produced by

the tegument (McLaren and Hockley, 1976).

The free surface of the tegument is folded into ridges or fingerlike projections. In blood flukes the tegument has an elaborate network of channels that open to the outside. Although flukes have a well-developed digestive system, the tegumentary uptake of nutrients is significant. In vitro studies have shown that small molecules are absorbed through the fluke tegument, and large nutrient molecules are taken in by pinocytosis (Wakelin, 1996).

The tegument of the adult *Schistosoma* is unusual because it does not have a glycocalyx. Scanning electron microscopic observations have shown that the tegument of the male schistosome has an irregular, spiny surface, whereas that of the female is mostly smooth with spines concentrated at the tail end. When larvae of *Schistosoma mansoni* emerge from the intermediate snail host, they are surrounded by a glycocalyx (fig. 11.4). Immediately after penetration of a vertebrate host, the larvae become transformed into the schistosomulum. The larval surface membrane and associated glycocalyx are replaced by mem-

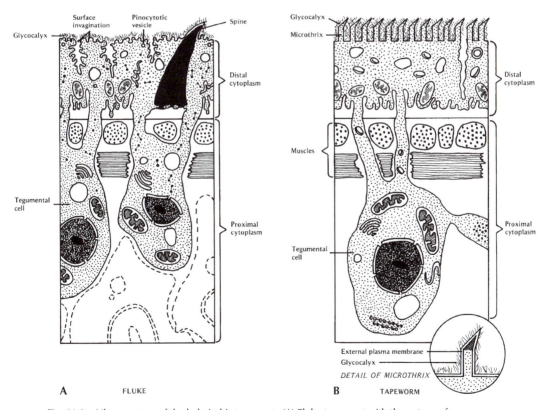

Fig. 11.3   Ultrastructure of the helminthic tegument. (A) Fluke tegument with the outer surface covered with a glycocalyx. (B) Tapeworm tegument with characteristic microtriches and glycocalyx.

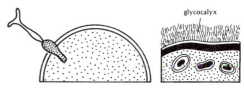

Cercaria penetrates skin of non-immune host

Schistosomulum acquires immunological disguise from host

Fig. 11.4   Tegumental changes in *Schistosoma mansoni*.
(A) Schematic representation of outer surface changes in
cercaria following penetration of host skin. Note relative
amounts of glycocalyx on cercaria, schistosomulum, and
adult. Adult worms have a characteristic multilayered
membrane system. (B) Electron micrograph of the fibrillar
nature of the glycocalyx of a schistosome. (C) Electron micrograph of an adult schistosome and the three membrane outer covering (B and C: Dianne J. McLaren,
National Institute for Medical Research, London).   **A**

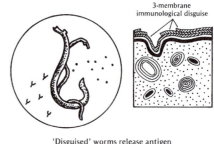

'Disguised' worms release antigen

branes derived from the host and membranous
inclusions that are formed near the surface of
the larval tegument within minutes of skin
penetration (McLaren, 1989) (fig. 11.4).

The fluke migrates from the skin to the
lungs and then to the liver, and in the process
loses its tegumental spines. The lung stage of
the fluke is particularly resistant to the host's
hormonal and cellular immune mechanisms.
The schistosomulum changes or masks its
original surface antigens and thus does not
trigger the host's immune responses. In 1964
Raymond Damian, a pioneer in parasitic immunology, proposed the concept of molecular
mimicry to describe how the parasitic symbiont has surface antigens similar to those of
its host (Damian, 1964). These antigens disguise the parasite against the host's immune
response. The question of whether the host-
like antigens are synthesized by the fluke or
obtained from host cells has been a subject of
intense research. Evidence has confirmed that
schistosomula can acquire A, B, and H blood-
group antigens from the host's red blood cells,
along with glycoproteins of the major histocompatibility complex. These antigens and

others required by the symbiont seem to play
a central role in the phenomenon of molecular mimicry (Damian, 1987; 1989).

Although the first parasites successfully
evade the host's immune responses, they induce the host to reject all further parasites of
the same kind. This phenomenon is called
concomitant immunity (Hagan and Wilkins,
1993).

Adults of *Fasciola*, like *Schistosoma*, can
live in the host's bile duct for long periods of
time, and mechanisms of molecular mimicry
fail to explain how the liver fluke evades the
host's antibodies. Scientists have suggested
that *Fasciola* periodically sloughs off its glycocalyx along with associated host antibodies
and replaces it with a new coat (Coles, 1984).

11.3 CESTODES: THE TAPEWORMS

Tapeworms represent the ultimate in biological adaptation to live in another organism. All
tapeworms are obligate symbionts of vertebrates and arthropods. The adult tapeworm
body consists of a head, the *scolex*, which

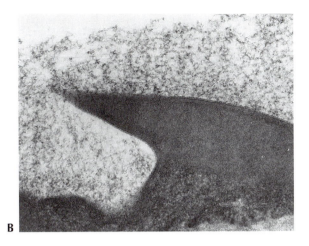

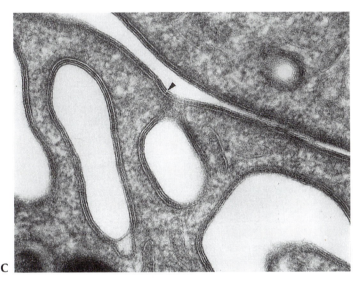

Fig. 11.4   (*continued*)

may possess anchoring devices such as suckers and hooks, a neck or zone of proliferation, and a body made up of segments called *proglottids* that may number from 3 to 3000. Sexually mature tapeworms live in the small intestines of vertebrates; their larval stages develop in the visceral organs of an alternate host, which may be a vertebrate or an arthropod. The development of larval stages in the muscles and nervous tissue of the vertebrate host produces serious disease. Some scientists view the adult tapeworms in the alimentary canal as endocommensals living in a nutrient-rich environment. Adult tapeworms are well known for their rapid growth and production of large numbers of eggs.

The extent of the disease in the host depends on the number of larvae involved, their survival, and the host's immune response. Infected individuals rapidly become resistant to further infection, and this immunity is passive and can be transferred to healthy hosts. Scientists have noted the similarity between the antigens of various tapeworm species. Scientists hope to collect immunogens for developing vaccines from a nonpathogenic tapeworm such as *Taenia hydatigena*, which is easily maintained and propagated in the laboratory.

There is increasing evidence that the establishment and survival of adult tapeworms in the host intestine is immunologically controlled. The crowding effect, often noted in tapeworm infections, means that the size of individual parasites is inversely related to the number of parasites in the host. Solitary tapeworms are commonly found in humans, for reasons that are not clear. It is possible that an established tapeworm secretes substances that inhibit the development of other tapeworms. Alternatively, the host may tolerate only a certain amount of parasitic load before its immunological mechanism becomes operative. Scientists believe that the competition for carbohydrates and nucleotides plays an important role in producing the crowding effect (Zavars and Roberts, 1985).

The Tapeworm Tegument

Tapeworms lack a digestive system and absorb all their nutrients through their tegument, which is remarkably similar to that of the flukes. A unique feature of the tapeworm tegument are minute projections, called *microtriches*, that cover the outer surface (fig. 11.3B). Microtriches are points of attachment for the tapeworm in the host intestines, and they significantly increase the absorptive surface of the tegument. The glycocalyx of the tapeworm tegument enhances the host's amylase activities while inhibiting the host's production of trypsin, chymotrypsin, and pancreatic lipase. The glycocalyx protects the tapeworm from the host's digestive enzymes. Both the larvae and adult tapeworms live in an environment rich in amino acids, fatty acids, glycerols, acetates, and nucleotides. These molecules are absorbed by diffusion and active transport, which take place at specific sites on the tegument (Threadgold, 1984).

Some Tapeworm Symbioses

Diphyllobothrium latum: *the fish tapeworm*

The adult form of the fish tapeworm is a common inhabitant of the alimentary canal of fish-eating mammals, birds, and fish. In temperate climates, people who eat fish often carry *D. latum* in their small intestine. The fish tapeworm is well known for its ability to absorb vitamin B_{12}, thereby causing the host to be deficient in a vitamin that is essential for the development of red blood cells. As a rule, frogs, fish, or snakes are the intermediate hosts of this tapeworm. Inhabitants of south-

east Pacific Islands often use crushed frogs to prepare food and in this way acquire the tapeworm larvae. Humans infected with adult tapeworms pass eggs in the feces. Ciliated larvae (coracidia) emerge from the eggs and infect animals such as water fleas, in which the larvae become transformed into procercoid larvae. Fish acquire the tapeworm by eating water fleas. The tapeworm larva penetrates the alimentary canal of the fish and develops into the final larval stage, the plerocercoid, in the fish musculature.

Diphyllobothrium (Spirometra) mansonoides: *the beneficial parasite*

The adult *Spirometra mansonoides* occurs in the alimentary canals of dogs and cats. From eggs passed in host feces, larvae hatch and develop in water fleas. Frogs become infected with the larval stage by eating water fleas. Mice that eat frogs become infected with the final larval stage of the tapeworm. Scientists have observed that infected mice showed increased growth and efficiency of nutrient uptake. J.F. Mueller (1974) considers *D. mansonoides* to be a beneficial parasite because the larger size of the infected mice makes them sluggish and therefore an easy catch for the final host, a cat. The plerocercoid larvae of *D. mansonoides* produce plerocercoid growth factor (PGF), which is similar to human growth factor (HGF). The gene for PGF in the tapeworm came from a human gene for HGF that became sequestered in the symbiont during the course of its evolution (Phares, 1987).

Taenia solium: *the pork tapeworm*

Humans become infected with pork tapeworm when they eat undercooked pork which has a speckled appearance because of the presence of larvae. *Taenia solium* is a dangerous tapeworm because of autoinfection with cysticercus larvae. The larvae may occur in every organ and tissue of the body, but they are most commonly found in subcutaneous tissues, eyes, brain, and muscles. The incidence of infection is high in parts of Europe, the Middle East, East Africa, and Mexico.

Taeniarhynchus saginatus: *the beef tapeworm*

Humans acquire the beef tapeworm by eating raw hamburger or other uncooked beef. Adult tapeworms occur in the human small intestine, where they may grow to lengths of >20 m. Humans pass the eggs of *T. saginatus* with feces, which are ingested by cattle. The larvae hatch in the digestive tract of the cattle, penetrate the intestinal wall, enter the circulatory system, and finally become lodged in the muscles and other tissues as cysticeri. The flesh riddled with tapeworm larvae is called "measly" beef.

Hymenolepis diminuta: the rat tapeworm

Hymenolepis diminuta has been a favorite experimental subject to investigate the nutrition, biochemistry, immunology, and developmental biology of tapeworms. The adult tapeworm commonly occurs in the alimentary canals of rats, mice, and hamsters. Insects such as beetles, fleas, cockroaches, and flies are the intermediate hosts. Rat tapeworms can live for an indefinite period and show no signs of aging. Adults of *H. diminuta* have been kept alive for more than 14 years by using serial surgical transplantations. The host shows no symptoms. The tapeworm has a diurnal migration in the intestine, moving anteriorly in the morning and returning to the posterior portion of the small intestine in the evening. The migration is correlated with the feeding behavior of the host (Arai, 1980).

Schistocephalus solidus: a case of parasitized host behavior

Schistocephalus solidus inhabits the alimentary canal of fish-eating birds and mammals. The birds pass tapeworm eggs, which hatch and produce swimming larvae that are ingested by water fleas. When a fish such as the three-spined stickleback eats infected water fleas, the final larval (plerocercoid) stage of the tapeworm develops in the fish's body cavity. The stickleback–*Schistocephalus* association has been widely used as a model to study the effects of parasitic symbiosis on host behavior. The plerocercoid larvae metabolize energy more efficiently than the host (Milinski, 1990). A tapeworm that closely resembles *S. solidus* is *Ligula intestinalis*, whose plerocercoid larvae are >1 m long. *Ligula intestinalis*-infected fish show *parasitic castration*, which is characterized by suppressed growth of the gonads (Smyth and McManus, 1989).

11.4 NEMATODES: THE ROUNDWORMS

Roundworms are second only to insects as the most abundant animals on earth. Most nematodes are free-living. They occur in freshwater, marine, and soil habitats and feed on microorganisms and decaying organic matter. Many nematodes are adapted for a parasitic lifestyle in plants, fungi, and animals. Nematode parasites of plants have syringelike mouthparts, which they insert into plant tissue to absorb nutrients. Some nematodes are ectoparasites on plant roots; others are endoparasites and cause abnormalities of the host tissue. Scientists estimate that every kind of animal is inhabited by at least one parasitic nematode. Nematodes parasitize organs such as eyes, tongue, liver, and lungs, often causing destruction of the host tissues.

Roundworms are elongated, cylindrical helminths ranging in length from microscopic to several inches. Sexes are separate, and in most species mature females lay resistant eggs. The nematode life cycle consists of four larval stages and the mature adult. Each larval stage is transformed into the next by a molting process, during which the old cuticle is shed and a new one is formed. Sexually mature roundworms do not molt. The study of nematodes has provided some significant insights into modern biology. For example, the process of meiosis was first elucidated in *Ascaris megalocephala* by Edouard van Benden in 1883, and later in 1899, Theodor Boveri, using the same roundworm, described the role of egg and sperm nuclei in fertilization (Roberts and Janovy, 1996). Modern developmental genetics is based on studies of *Caenorhabditis elegans*. Nematodes also show the phenomenon of eutely or *nuclear constancy*, which is characterized by the presence of all cells of the adult in the fully developed embryo. This gives biologists the best documented case of germinal lineage in the animal kingdom (Crofton, 1966). A major landmark in modern DNA biology has been the complete genomic sequencing of *Caenorhabditis elegans*.

Parasites of Humans and Vertebrates

Intestinal roundworms

Ascaris: *large roundworms. Ascaris lumbricoides* is one of the largest intestinal nematodes of humans and is prevalent throughout the warmer climates. More than 1 billion people are infected with this nematode but most are asymptomatic carriers (Crompton, 1988). The nematode may cause intestinal obstructions and interfere with host nutrition. Adult worms live in the small intestine, and their eggs are passed with host feces to the outside. When eggs are ingested with contaminated food or drink, the larvae hatch in the small intestine, penetrate the gut lining, and are carried by the bloodstream to the liver, heart, and lungs. From the lungs the nematodes travel to the mouth and then back into the alimentary canal. It is not known why the nematode adopts such a circular path in its development in a host. During the larval migration, the parasite molts, feeds on host tissue, and grows. The young adult roundworms in the small intestine increase in size and become sexually mature in about 2 months. A mature female roundworm produces about 200,000 eggs each day. Most infected humans carry only one to a few nematodes, but some may have hundreds of worms. Individuals are susceptible to ascarid infections because of genetic predisposition and social, behavioral, and environmental factors either alone or in combination (Holland et al., 1992). The adults of *A. lumbricoides* can remain in a host for up to 1 year. Because of its size and simple one-host life cycle, *A. lumbricoides* has been studied for its morphology, physiology, biochemistry, and host–parasite relationship (Nadler, 1987).

Haemonchus contortus: sheep stomach worm. *Haemonchus contortus* is a bloodsucking nematode that resides in the abomasum of sheep, goats, cattle, and other ruminants and causes severe anemia in heavy infections. Female nematodes pass their eggs with host feces onto soil, where under favorable conditions the eggs hatch and the larvae undergo three developmental stages. The last larval stage does not feed and has to be ingested by a grazing host animal before it can develop further. In the host's stomach, the parasite resumes its growth, undergoes a final molt, and develops into an adult. A single sheep may carry thousands of adult worms, which attach firmly to the stomach villi and feed on blood.

A phenomenon called *self-cure* takes place in the host sheep. When an infected host receives a challenge infection, a large number of previously acquired adult worms are expelled, and the host becomes resistant to reinfection. The worm expulsion has many characteristics of an immediate hypersensi-

tive response to antigen release by the incoming larvae (Lloyd and Soulsby, 1987). In other words, parasites acquired early in life make the host resistant to heavy infections by the same parasite. From an evolutionary perspective, according to the welcome mat hypothesis, both the host and the parasite avoid mutual extinction.

Research on the mechanism of the self-cure phenomenon has shown that newly ingested larvae produce substances that trigger a localized allergic reaction in the host. Host animals that have expelled adults of *H. contortus* have high levels of histamines in their blood. The host animal becomes sensitized against the parasitic antigens after repeated exposure to *H. contortus*.

The self-cure phenomenon has been studied extensively in rats infected with *Nippostrongylus brasiliensis* because both organisms can be maintained under laboratory conditions (Kassai, 1982). Infected rats release eggs of the parasite in the feces by day 6 after infection and show peak egg production on day 10. Within the following week, egg production drops to almost zero, and there is a massive expulsion of adult worms from the host intestine. A small number of nematodes survive the expulsion phase and remain in the host intestine for a long time. The host animal becomes resistant to further reinfections. It is believed that the host's high levels of IgE, along with the mast cell activities, trigger an immediate hypersensitive response and are responsible for the worm expulsion from the intestines (Rothwell, 1989). An increase in oxygen radicals has also been linked with worm expulsion, suggesting that the expulsion phenomenon may involve many components of the host's immune system. The nematodes release larger amounts of acetylcholinesterase, which interferes with the host neuromuscular transmission preventing peristalsis and thus helping the worms survive. The hosts produce specific antibodies against acetylcholinesterase activity of the nematodes. The worms that survive expulsion respond by producing larger amounts of acetylcholinesterases. During the entire process, the host's intestinal cells become highly inflamed and damaged, and it is unable to feed (Sanderson and Ogilvie, 1971).

The phenomenon of *lactational rise* (or spring rise) has been observed in *Haemonchus* and other gastrointestinal nematodes of herbivores. For example, the number of eggs in a lamb's feces greatly increases 4–8 weeks after the young are born. Several weeks later, egg production falls to a low level. Studies have confirmed that the host's cellular immunity becomes depressed during pregnancy and lactation. From an evolutionary perspective, the lactational rise may be viewed as an adaptation for the survival of the nematode because it facilitates transmission of the parasite to the offspring (Smyth, 1994).

Ancylostoma and Necator: hookworms. An estimated 1 billion people who live in the warm climates of the world are infected by hookworms (Crompton, 1988). The two most important hookworms are *Ancylostoma duodenale*, the oriental hookworm of China, Japan, Asia, North Africa, the Caribbean Islands, and South America, and *Necator americanus*, the American hookworm which is prevalent primarily in South and Central America but also occurs in Africa and Asia.

The mature female hookworm resides in the small intestine of its host and lays about 10,000 eggs each day, which pass out of the host with the feces. The eggs hatch in warm, moist soil and develop into infective larvae. The larvae penetrate the human skin and enter the blood circulation, traveling to the lungs, trachea, and down the esophagus to the small intestine, where they become sexually mature. The nematodes damage the host intestine as they feed on mucus, blood, and associated tissues. Infected persons with poor nutrition often show symptoms of anemia and iron deficiency. Epidemiologists have found that during the first decade of their lives humans steadily acquire hookworms. The level of infection remains stable from the second to the fifth decade but increases again from the sixth decade. During much of the adult host's life the static parasitic population is believed to result from acquired immunity (Shad and Warren, 1990; Gilles and Ball, 1991).

Filarial nematodes

Filarial nematodes are obligate parasites with complex life cycles involving humans and other vertebrate and arthropod vectors. The adult female worms give birth to larvae, the microfilariae, that invade new tissues and develop to the third larval stage. Larvae in this stage migrate to the skin or peripheral blood circulation. A bloodsucking insect becomes infected with these larvae during a blood meal. *Wuchereria bancrofti* and *Brugia malayi* occur in the human lymphatic system

and cause the disease elephantiasis. These nematodes affect about 200 million people throughout the tropics (Sasa, 1976). In Africa, the same mosquito that transmits malaria is also the vector of *W. bancrofti*. The filarial worm *Onchocerca volvulus* causes skin tumors and blindness in some 20% of the people of Africa and the Middle East. The nematode is spread to humans by bloodsucking black flies (Evered and Clark, 1987).

The heartworm, *Dirofilaria immitis*, is parasitic in the heart and pulmonary artery of dogs and other mammals. The worm is transmitted by several species of mosquitoes. The disease is produced from lesions caused by dead or dying worms.

The Guinea worm, *Dracunculus medinensis* (which, strictly speaking, is not a filarid) has been known since the early recorded history of the Middle East, India, and Africa. The adult female nematode may be >12 m long and takes more than 1 year to reach the skin of the ankles, where it causes a blister and emerges to lay eggs. The eggs are ingested by water fleas, such as copepods, where they develop into the infective larval stage. Ancient physicians learned the importance of extracting the entire worm by slowly winding it on a stick. In the Bible, there is a reference to God sending "fiery serpents" among the people of Israel if they disobeyed his commandments. Many scholars believe that the "fiery serpents" were Guinea worms. Humans become infected with Guinea worms by drinking water that contains copepods. A number of countries where dracunculiasis is endemic are making significant efforts to eliminate this disease (Hopkins, 1990).

A typical filarial life cycle begins when humans acquire the parasite from the bite of an infected bloodsucking insect. The parasites show a high degree of host specificity as well as tissue preference. Once in the bloodstream or lymphatic vessel, the nematode larvae become sexually mature. Adult worms of *Wuchereria* and *Onchocerca* live in the lymphatic vessels for many years; Guinea worms live for little more than a year. The mature female gives birth to microfilariae, which are the larval stages still ensheathed in the egg membrane. Microfilariae of various nematodes have a 24 h periodicity. The number of microfilariae of *Wuchereria bancrofti* reaches a peak in the peripheral blood circulation late at night. The microfilariae are acquired by night-feeding mosquitoes along with their blood meal. In the insect host, the larvae develop in the thoracic muscles, undergo two molts, and then migrate to the mouth parts. Laboratory studies on *Litomosoides carinii*, a filarial parasite of the cotton rat, have provided fundamental insights into the host–parasite relationship of filarial worms. This nematode infection of the cotton rat is prevalent throughout the open grasslands of North and South America.

The development of disease from a filarial parasite varies a great deal among individuals, for reasons that are not clear. Some humans develop acute filarial disease with recurrent episodes of fever, yet the microfilariae are not always detected. Other individuals are asymptomatic but produce microfiliae, thus contributing to the spread of the infection. Some individuals are immune to filarial infections.

Scientists generally agree that symptoms of filarial disease are the result of host immune response. Obstructive lesions are produced by a delayed hypersensitivity reaction to dead or dying nematodes. Elephantiasis results from the host's immune reaction to adult worms, and the skin and eye disease of onchocerciasis is a reaction to the microfilariae. Acute disease symptoms develop after treatment with antifilarial drugs, which suggests that the sudden release of dead parasites disturbs the equilibrium between the host and parasite. A nonpathogenic relationship results when the host's immune response to the parasitic antigens is suppressed (Kazura et al., 1993).

It is ironic that treatment to eliminate nematode parasites from a host precipitates a disease condition that does not occur in stabilized infections. The number of filarial parasites in the lymphatic system remains constant for many years, indicating that the host somehow manages to control the intensity of infection (Maizels and Lawrence, 1991).

Trichinella spiralis

Trichinella spiralis, the largest known intracellular symbiont of vertebrates, has been described as "the worm that would be a virus" (Despommier, 1990, p. 163). The adults of this parasite reside in the intestinal epithelium, while the larvae occur in parasite-induced nurse cells associated with skeletal muscles (Despommier, 1993). The nematode has been studied extensively by physicians, public health officials, experimental biologists, and ecologists. The severity of the disease trichinosis depends on the intensity of infection in

a host. Symptoms include intestinal discomfort, diarrhea, and nausea from the presence of adult nematodes in the small intestine, and edema, fever, fatigue, and muscle pain caused by the migration of larvae to muscle tissue. *Trichinella spiralis* has a wide host range among carnivorous vertebrates. Human trichinosis is associated with the consumption of undercooked pork containing parasitic larvae. The nematode has a short life cycle, does not need an intermediate host, and is easily maintained in laboratory animals.

Trichinella infection begins with the ingestion of meat containing the infective first-stage larvae (fig. 11.5). After the meat is digested, the larvae become liberated and enter their intracellular habitat, a row of columnar epithelial cells in the small intestine. A larva may occupy more than 100 epithelial cells without damaging the host cells. The columnar cell membranes fuse with each other and form a syncytium. It is not known how the parasite enters the columnar cells; the larvae do not possess any special organs for cell penetration. After about 30 h the larvae undergo four molts and become adult worms. The speed of the molting is unusual and might be related to the host immune response. The adult nematode increases in size and occupies an average of 400 columnar cells. During the molting process, the symbiont undergoes extensive morphological changes in its cuticle, hypodermal gland cells, and alimentary canal. Hypodermal glands produce secretions that are believed to possess unique antigenic properties. Little is known about how the adult worms obtain their nutrients. Stichosomes are unique discoid cells that occur in the anterior half of the infective larval stage. These cells produce secretory granules that are strongly antigenic. Some scientists speculate that the host immune response is directed against the secretory granules.

Sexually mature worms copulate and the females give birth to 200–1600 larvae, depending on the host species. The newborn larvae migrate to the lymph, and some move into the general circulation, infecting organs such as the liver and lungs. Recent studies have shown that striated muscle cells are the preferred intracellular habitat for newborn larvae, which are attracted to the muscles by chemical and electrical stimuli. A larva penetrates the muscle cell by a mechanical process and then redirects the host cell metabolism to its own survival. The infected muscle cells lose their myofilaments, undergo

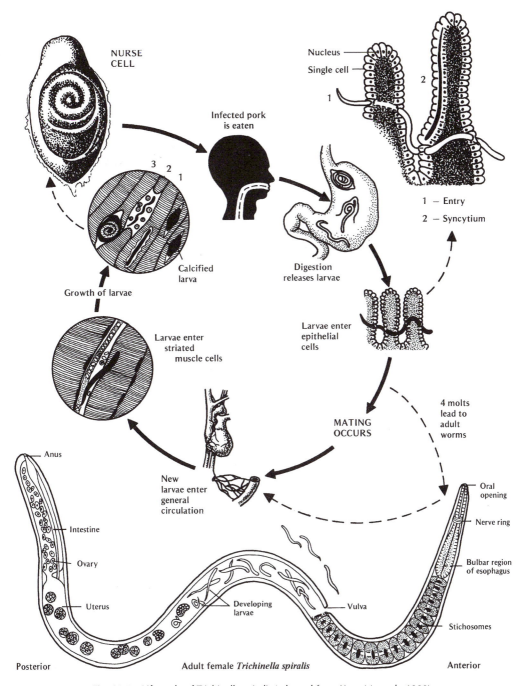

Fig. 11.5   Life cycle of *Trichinella spiralis* (adapted from Katz, M. et al., 1982).

nuclear hypertrophy and mitochondrial breakdown, and increase their collagen synthesis. A small network of blood vessels forms around each worm–nurse cell complex (Despommier, 1993). The antigen present in stichocytes of the larvae is believed to be responsible for altered gene expression of the host cell. During the next 3 weeks, the *Trichinella* larva develops stichosomes and a thick cuticle and grows to 270 times its original size. The muscle cell containing the coiled larva is transformed into a nurse cell. Muscle cells function within a narrow range of physical and chemical conditions; few parasites have successfully invaded the intracellular muscle environment. *Trichinella spiralis* survives in skeletal muscle cells by transforming the structure and biochemistry of the host cell. The nurse cells lose their ability to contract like normal muscle cells. Calcification of the nurse cell is a complex host response, often developing several years after the initial infection, and results in the worm becoming encysted. For reasons not yet understood, the muscles of the eye, tongue, jaws, diaphragm, and the ribs are most susceptible to *Trichinella* infection.

Trichinella spiralis evokes a powerful host immune response. As a new generation of larvae becomes established in the muscles, adult worms residing in the small intestines are expelled from the host (Appleton and McGregor, 1984). Animals previously exposed to *T. spiralis* resist further infection, and a host mother transfers the immunity to her offspring. In an immune host, *T. spiralis* grows slowly and does not reproduce. Different larval and adult stages of *T. spiralis* in the same host produce different antigens. The host immune system responds by developing antibodies against the three stages of the parasite's life cycle: adult, infective larvae, and newborn larvae. Multiple antigen–antibody interactions facilitate the migration of newborn larvae from the intestine to the muscle. *Trichinella* antigens have been used to develop a vaccine to control trichinosis in pigs.

Nematode–Insect Symbioses

Insects are the dominant form of life on earth, and nematodes have successfully evolved symbioses with many of them. There are an estimated 3000 nematode–insect associations. Nematodes of insects have intricate life cycles that are synchronized with those of their hosts. The symbionts cause changes in the insect's morphology, physiology, and behavior and reduce the reproductive potential of the host. *Heterorhabditis* and *Steinernema* are nematodes that parasitize insects and transmit bacteria that kill their hosts. There has been renewed interest in nematode–insect symbioses because of the potential for controlling insect pests by means of pathogenic symbionts (Poinar, 1979; Kaya and Gaugler, 1993).

Deladenus

The complex life cycle of *Deladenus* involves two hosts, a wood wasp and the fungus *Amylostereum*. The fungal part of the life cycle occurs in dying or dead pine trees, where the fungus grows inside galleries dug by the wasp larvae. Nematode larvae feed on the fungal hyphae and grow and molt until they become sexually mature adults. Old fungus colonies somehow trigger the parasitic behavior of the nematodes. A fertilized female nematode penetrates a wasp larva and enters its body cavity, where it increases in size and molts. The old nematode cuticle ruptures, and the hypodermal cells become transformed into absorptive cells with microvilli. When the wasp larvae pupate, the female nematode gives birth to live larvae, which migrate to the reproductive organs of the newly developed female wood wasp. The wasp disperses the nematodes by depositing them on a suitable tree along with its own eggs.

Mermis nigrescens

Mermithid nematodes are well-known obligate parasites of insects, spiders, and other invertebrates. *Mermis nigrescens* is a common parasite of grasshoppers and katydids. The parasite usually kills its host. The nematode larvae are found in all stages of the insect, obtaining nutrients from the host body cavity. The female nematode is >10 cm long and lives in the soil. After a rainfall, the nematode climbs onto vegetation and lays eggs on the foliage. The eggs remain viable for up to 1 year. Grasshoppers and other susceptible hosts ingest the eggs along with the foliage. The larvae hatch in the grasshopper's alimentary canal and reach the body cavity by penetrating the gut wall. The parasite increases in size from 3mm to >10cm. In the last larval stage, the parasite leaves the host body and burrows in the soil. During the escape process, the insect host is killed. The larvae remain dormant until the following spring, when they molt and become adults. A mature female produces more than 14,000 eggs and can overwinter and resume egg laying the next spring (Nickle, 1984).

Romanomermis

The possibility of developing a biological control for mosquitoes has heightened interest in the mermithid nematode *Romanomermis*, which kills mosquito larvae. Sexually mature adults of the nematode live for up to 6 months in the shallow waters of lakes, ponds, and streams (fig. 11.6). The female lays many eggs that hatch in about 7 days. The larvae are strong swimmers and seek out mosquito larvae at the water surface. The nematode penetrates the cuticle of the larvae and develops in the host for 5–10 days. The parasite is not carried through the pupal or adult stages of the mosquito because either the infected larvae usually die or the immature nematodes leave the host. The nematode larvae return to the bottom mud, where they develop into adults and the life cycle is repeated. Once introduced into an area, the nematode becomes a permanent resident of the ecosystem. Scientists have developed methods to mass produce the eggs of *Romanomermis* for large-scale applications (Petersen and Cupello, 1981).

Attempts are being made to control black flies in Africa by using parasitic nematodes that kill the insect larvae. Black flies are vectors of the filarial nematode *Onchocerca volvulus*, which causes blindness in humans.

Neoaplectana

The nematode *Neoaplectana* carries a bacterial symbiont in a specialized intestinal pouch. Nematodes either are ingested by an insect or enter the host's body through natural openings. Once inside the host alimentary canal, the nematode releases its bacterial symbionts, which move into the body cavity and multiply, causing death of the host (fig. 11.7). Nematodes develop rapidly in the dead host and become sexually mature. The females produce larvae that emerge from the host carcass carrying the bacteria and burrow into the soil, where they can survive for 3 years or more. Scientists have attempted to use this parasitic nematode as a way to control insect pests (Maggenti, 1981).

Nematode–Plant Symbioses: Plant Nematology

Most nematodes that attack plants are obligate parasites and include root-knot nematodes and cyst nematodes, which cause destructive infections in crop plants. The nematodes have a hollow, needlelike structure, the *stylet*, which they insert into plant tissue, like a syringe, to draw out nutrients. Most plant nematodes feed on roots but some invade parts of a plant above ground. There are more than 2000 described species of plant parasitic nematodes, but only a few species are pests. Scientists are beginning to realize that serious nematode infestations are the result of unsound agricultural practices such as monocropping and reliance on genetically uniform plants.

Life cycles of plant nematodes are simple, consisting of adult and juvenile stages associated with the same host. Cell proliferation, giant cell formation, suppression of cell division, and cell wall breakdown are some host responses to nematode parasitism. *Arabidopsis thaliana*, a small crucifer, is a host to several plant parasitic nematodes, including *Heterodera* and *Meloidogyne* species. Molecular biology approaches have opened new perspectives to understanding host gene expression and accompanying molecular changes in the plant–nematode interaction dynamics (Niebel et al., 1994). The infected plants are stunted and have yellow leaves, and in the case of crop plants, yields are significantly reduced. Most plant nematodes are ectoparasites, but a few species are endoparasites (Atkinson, 1995; Zacheo and Bleve-Zacheo, 1995).

Since the end of the Second World War, soil fumigation has been the preferred method for control of nematode pests. Unfortunately, when fumigation is discontinued or suspended, the nematode populations increase to destructive levels. Because many of the common soil fumigants are now considered to be harmful to humans and other organisms, scientists are searching for alternative methods of nematode control, such as the use of biological control methods, resistant varieties of plants, crop rotation, and agricultural practices that maintain the stability of the soil-microbe equilibrium.

Meloidogyne: the root-knot nematode

The root-knot nematode infects more than 2000 cultivated plants throughout the world, contributing to an estimated annual loss of $77 billion. Plant nematologists have conducted extensive studies on the biology, physiology, and host–parasite interactions of the *Meloidogyne*–tomato symbiosis. Juvenile nematodes penetrate the root tips of a host plant and move between the cells to the

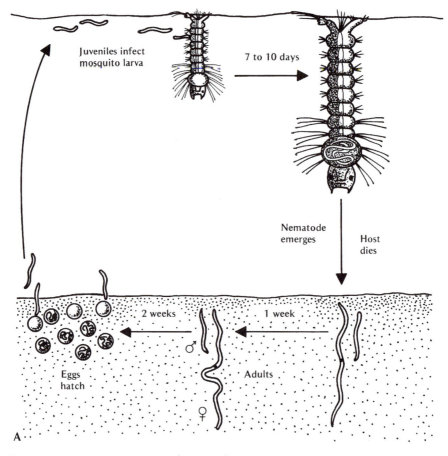

Fig. 11.6 *Romanomermis* parasitism of mosquito larvae. (A) Stages in the life cycle of
Romanomermis culcivorax. Adapted from Nickle (1984). (B) Nematodes collect at the bottom of
tank after emerging from mosquito larvae. (C) Nematodes in the thorax of mosquito larvae (Parts B
and C courtesy James Peterson, USDA, Lincoln, Nebraska.)

vicinity of the vascular tissues. Plant cells
surrounding the nematode undergo hypertrophy and hyperplasia and form a gall. Most
galls contain giant cells that are formed in response to the nematode's feeding. The juveniles in the gall molt rapidly and become
adults. The female nematode becomes enlarged and spherical and lays eggs, which are
extruded in a protective gelatinous matrix
(fig. 11.8A). Under favorable conditions, the
juveniles hatch from the eggs and infect other
roots. Plants parasitized by root-knot nematodes are less resistant to infections from soilborne fungi and bacteria and more susceptible to environmental stresses such as temperature change and drought. Yields of
heavily parasitized plants are greatly reduced.

Heterodera and Globodera: the cyst nematodes

The cyst nematode *Heterodera* is a major parasite of agricultural plants in temperate regions. A unique feature of these parasites is
that the female body containing the eggs is
transformed into a resistant cyst. The infection begins when juveniles penetrate root tips
of the host and migrate through the cortex.
The juvenile begins the feeding process by
orienting its head toward the vascular tissue.
Giant cells are formed, but galls do not develop. After egg production the female dies
and becomes a resistant cyst, which remains
in the soil for several years. The eggs in the
cyst hatch when a specific stimulus is received from a host plant. Cyst nematodes

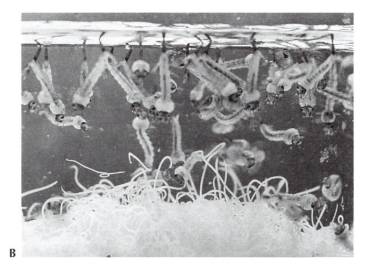

Fig. 11.6   (*continued*)

cause extensive losses in crops such as sugar beets, soybeans, potatoes, and cereals. Scientists have studied the hatching factors in the hope of developing an effective cyst nematode control (Niebel et al., 1994).

Anguina tritici: *the wheat gall nematode*

In 1743, Turbevill Needham, an English naturalist, claimed support for the spontaneous generation of life when he crushed wheat galls in water and observed that the contents of the galls became alive. The galls crushed by Needham contained large numbers of juveniles of *Anguina tritici* in a state of anhydro-

sis. Wheat galls are unknowingly sown along with wheat seed. Under warm, moist conditions, the galls break open and liberate juveniles, which infect nearby wheat seedlings. The juveniles first feed as ectoparasites and later penetrate the plant tissues as the wheat flowers begin to develop. After the nematodes mate, the female migrates to the ovary of the flower and lays thousands of eggs, which hatch to produce juveniles. The ovary is transformed into a resistant gall. Galls are also produced on the leaves and stems. Nematode-infected plants are usually stunted or wrinkled and have twisted leaves. *Anguina tritici* is prevalent throughout Asia, the Middle East,

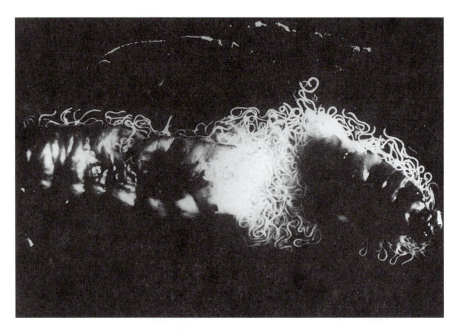

Fig. 11.7   Wax moth larva parasitized by *Neoaplectana carpocapsae* (Courtesy W.R. Nickle, USDA, Beltsville, Maryland).

and Brazil. Because the seed galls float in water, they are easily separated from wheat seeds. With improved agricultural practices, the incidence of *A. tritici* is declining in many areas (Maggenti, 1981).

Bursaphalenchus: *beetle–pine mutualistic symbiosis*

Bursaphalenchus xylophilus is a nematode that lives in weakened or dead trees. The nematode is spread by a beetle that may carry up to 15,000 juvenile nematodes to a new location. The nematodes feed on wood tissue and are suspected of killing pine trees previously weakened by environmental stresses. Their life cycle is intertwined with that of the bark beetle. The beetle normally lays eggs under the bark. Upon hatching, the larvae dig tunnels in the wood tissue, where they pupate and overwinter. As the adult beetles emerge from the pupae, the nematode juveniles develop and gather near pupal cases. The nematodes hang from the roof of the tunnel gallery by their tails. As the adult beetle emerges from the pupal case, its vigorous movements cause the nematode juveniles to attach to it. The juveniles migrate swiftly to undersurfaces of the wings or to the air passages of the host. The

infected beetles emerge and find other pine trees to feed on. During the feeding process, nematode juveniles leave the beetle to invade the new substrate. The beetle lays eggs in the areas softened by nematode activity. The ecological relationship between the plant-parasitic nematode and the bark beetle is thought to be mutualistic.

11.5 PARASITOIDS: NEGLECTED SYMBIOSES

Parasitoids are restricted to the orders Hymenoptera (ants, bees, and wasps) and Diptera (flies) and are defined as insects that live freely as adults but as larvae they feed in or on the body of an another arthropod. There may be 1.6–2 million species of parasitoid insects, many remain undescribed (Godfrey, 1994). Parasitoids are common in all terrestrial ecosystems and are categorized by the host they parasitize and where their offspring develop. Environmental and scientific interest in these insect symbioses has been heightened by their exploitation as biological control agents. An adult parasitoid must find a suitable host and lay eggs on or in it by using its ovipositor. The parasitoid *Cotesia marginventris* is attracted to plants damaged by its host and uses chemicals in the host's saliva and feces as orientation cues. Wasp parasitoids, whose females control the sex ratio, must decide whether to lay male or female eggs. Parasitoid development is greatly influenced by the quality of its host in addition to coping with defense mechanisms with counteradaptations. Adult parasitoids also must find a mate and avoid predation. Evolutionary aspects of the physiological interactions between parasitoid and host provide new opportunities to study the biological basis of biodiversity (Godfrey, 1994). The coevolution of host synchronization and diapause strategies of parasitoids is receiving attention from evolutionary biologists because of the emergence patterns for males and females that maximize their chances of finding mates and hosts. Evidence from comparative morphology and DNA segment analysis shows that parasitoids evolved from fungus-feeding ancestors (Whitfield, 1998).

Parasitoid–polydnavirus mutualism is also drawing the attention of scientists because in some parasitoids virus particles are injected in the host along with the wasp's eggs, and either venom or virus or both sup-

press the host immune response (see fig. 2.4). The polydnaviral genome consists of several pieces of circular DNA, and the nucleocapsid is surrounded by a double-membrane envelope. The virus multiplies in the wasp ovary cells and has no adverse effect on it. But in a lepidopteran caterpillar, the virus suppresses its cellular immune response and disrupts the normal endocrine functions that control insect development, and when the caterpillar dies, the new generation of wasp that emerges carries the virus (see chapter 2). Some studies have shown that viral DNA is required for a successful parasitism of caterpillars by parasitoid wasps. The polydnavirus and wasp genomes have become so well integrated that the virus depends on the wasp for its spread and reproduction (Whitfield, 1990).

Figs, Wasps, and Nematodes

Figs and wasps are one of the best studied species-specific examples of mutualism and coevolution. There are more than 1000 species of figs that grow throughout the tropics. Allen Herre (1993) explored the complex relationship of fig trees with wasps that polli-

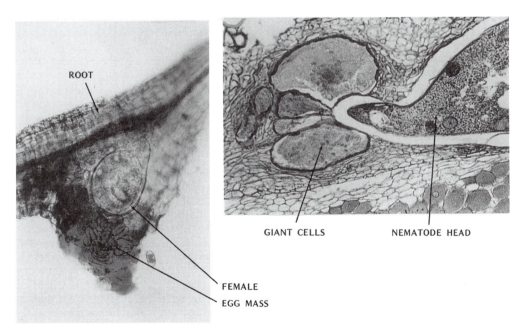

Fig. 11.8 Plant root parasitism by the root-knot nematode *Meloidogyne incognita*. (Left) A root gall has formed around the female nematode. Note protruding egg mass. (Right) Cross-section of an infected root, showing giant cells near the head of a feeding nematode (Courtesy J.D. Eisenback, Virginia Polytechnic Institute).

nate them and nematodes that infect the wasps. All fig species are pollinated by chalcidoid wasps, which develop as mutualistic symbionts in galls within the figs. (We discuss the complexity of fig pollination in chapter 12.) In fig wasps, the nematodes are species specific and often act as a virulent pathogen in some wasp species, while in others they have no effects. Herre was one of the first to provide experimental evidence on the evolution of virulence. He predicted that if a disease spreads easily among hosts as a function of higher contact rates, then the infection will be virulent and destructive (Herre, 1993). This idea is supported by the earlier observation of Paul Ewald (1993) that emergence of virulence can be a more successful evolutionary strategy than the traditional view that evolution favors the emergence of nonvirulence in various host–pathogen interactions. For an estimated 40 million years, figs, wasps and nematodes have coevolved into relationships that allow all three species to coexist.

Nonpollinating Fig Wasps

There are many nonpollinating wasp species that visit fig trees and interfere with the established mutualism between the host plant and its pollinator. Many wasp species stimulate gall formation by altering the ovules on which their larvae feed. Some wasps use ovules already initiated by other wasp larvae. It is believed that fig-pollinating wasps evolved from parasitoid ancestors. The parasitoid wasps have identical life histories to the pollinating wasps, but they carry no pollen and have long ovipositors. The larvae of pollinating wasps are parasitized by other species of fig wasps that are true parasitoids. These species are potential competitors of pollinating wasps. Some species have evolved large mandibles and fight among themselves for females of their species within the fig synconium. Males of other wasp species resemble females. In general, wasp communities associated with the fig trees are specialists and show a high degree of host specificity. Fig-wasp mutualism is sustained by a number of characteristics of each partner, such as direct adaptations to mutualism, byproducts of selection on another trait, and preadaptations (Anstett et al., 1997; Godfrey, 1994).

11.6 SUMMARY

Helminths are metazoans with complex life cycles. Humans in many parts of the world are infected with helminthic parasites, which include flukes, tapeworms, and roundworms such as hookworms and filarial nematodes. Flukes are common in vertebrates, where they infect different organs such as the liver and urinary bladder as well as the blood. In some cases, the life cycle of a fluke is synchronized with that of its host. Schistosomes are blood flukes that cause the diseases schistosomiasis and bilharziasis. The outer body wall of a fluke or tapeworm, the tegument, is a complex structure that protects the parasite from the digestive enzymes of the host and also absorbs nutrients. Some parasites disguise themselves by producing surface antigens similar to those of their hosts, a phenomenon called molecular mimicry.

Tapeworms are highly specialized parasites of vertebrates and arthropods. In the tapeworm life cycle, water fleas and various types of insects are common intermediate hosts. Nematodes, or roundworms, are parasites of plants, animals, and fungi. Some animals become resistant to reinfection after their initial infection by a parasite. This phenomenon is called self-cure. Hookworms infect almost a billion people in regions with warm climates. Filarial nematodes have complex life cycles that involve humans and other animal vectors. Filarial diseases such as elephantiasis are the result of the host's immune responses to the parasites. The nematode *Trichinella spiralis* causes the disease trichinosis in humans and other vertebrates and results in multiple antigen–antibody interactions between host and parasite. Insects have many nematode parasites, some of which are being studied as possible biological controls of pests. Plants also are infected by different roundworms.

Parasitoids are common in all ecosytems. Figs and wasps have coevolved complex relationships. Nonpollinating wasps also visit fig trees. The use of parasites to control other parasites has had much appeal among scientists. Such biological control methods, when successful, are much better alternatives than the indiscriminate spraying of pesticides. In this respect, it is important that the life cycles of parasites be thoroughly understood in order to realize the full potential of a biological control agent.

12 FLOWERING PLANT SYMBIOSES

Angiosperms, or flowering plants, make up more than 80% of all living plants. Although their origin remains shrouded in what Darwin called the "abominable mystery," the explosive diversity of flowering plants during the past 110 million years is an evolutionary success story. The fossil record confirms that many flowering plant families existed 100 million years ago. Flowering plants have shown great flexibility in the morphology of their roots, stems, leaves, flowers, fruits, and seeds. As a defense against herbivorous animals, many plant species have evolved prickles and spines, hairy leaves, and a wide range of secondary metabolites (Jones and Firn, 1991). The *growth–differentiation balance* (GDB) *hypothesis* deals with plant resource allocation patterns. Primary metabolism supports the growth of new tissues, while secondary metabolism supports the production of specialized tissues, organs, and chemical compounds. According to the GDB hypothesis, growth and differentiation require energy that comes from photosynthesis (Herms and Mattson, 1992; Lerdau et al., 1994). Thus, in a coevolutionary arms race, a cost–benefit equation is central to the allocation of a host's limited energy resources. One reason for the success of flowering plants may be their ability to form mutualistic symbioses, such as those with bacteria that fix nitrogen, fungi (mycorrhizae), and insects, birds, and mammals, which pollinate plants and disperse seeds.

12.1 PLANT–ANIMAL INTERACTIONS

Plants and Grazing Herbivores

Herbivores depend on plants for food and have been a major influence in the evolution of leaf size and shape (Brown and Lawton, 1991). In response, some plants have evolved structural and chemical defense mechanisms that discourage herbivory. Other plants encourage herbivory and regenerate quickly. Plants that are under grazing pressure often store food in roots and rhizomes and can regenerate when growing conditions are favorable.

Grasslands constitute about 25% of the earth's surface and vary widely in geography and climate. All grasslands are dominated by a low, perennial ground cover formed by species of the family Poaceae and some dicotyledonous plants. Grasses and dicots have evolved different strategies against herbivores. Grasses store food in roots or stems, and their leaves have heavily lignified midribs. Herbivores grazing on grass must contend with siliceous leaf tissues, to which they have responded by increased consumption and low retention time. Elephants, horses, and pandas follow this strategy. In general, mammalian herbivores of forests and woodlands have short-crowned teeth and feed on leaves, fruits, and seeds of trees and shrubs. Grazing mammals, however, have continually growing, elongated teeth to compensate for the abrasive action of grass leaves (MacFadden, 1997).

Grazing ruminants extract energy molecules by slowly degrading plant cell walls through anaerobic microbial fermentation. Their rumen detoxifies secondary metabolites and retains plant food longer in the digestive system. One way herbivores lessen a plant's secondary metabolite defense is by consuming different plant species. Tropical and temperate forest herbivores favor this strategy. Some animals, such as the koala bear, which feeds exclusively on leaves of eucalyptus trees, have specialized biochemical pathways in their liver cells and can detoxify plant poisons (Van Soest, 1994).

Cultivated crop plants and domesticated animals are the result of associations with humans during the past 10,000 years. Most crop plants have been selected for their increased nutritional contents and absence of antiherbivory features.

Carnivorous Plants

Three principal publications, Darwin's *Insectivorous Plants* (1875), Lloyd's *The Carnivo-* *rous Plants* (1942), and Juniper et al.'s *The Carnivorous Plants* (1989), provide historical aspects and insights into the functioning and evolution of these curious plants. About 500 species of carnivorous plants grow in wet or waterlogged, acidic soils (Zamora et al., 1998). In response to nutrient-poor habitats, carnivory is believed to have evolved in at least six independent evolutionary lineages (Thompson, 1981) (table 12.1). *Carnivorous syndrome* is a conceptual description of plants that (1) attract, (2) retain, (3) trap, (4) kill, (5) digest, and (6) absorb useful substances (Lloyd, 1942; Juniper et al., 1989). These plants eat mainly insects, but spiders, woodlice, slugs, earthworms, tadpoles, and fish are also prey. Carnivorous plants obtain nitrogen, phosphorus, sulfur, potassium, calcium, and magnesium from insect carcasses (Adamec, 1997).

In recent studies, hydrolytic enzymes have been discovered in the digestive juices that are produced by carnivorous plants (Juniper et al., 1989). Some of these digestive enzymes may be produced by bacteria living in the modified leaf traps (fig. 12.1A). Carnivorous

Table 12.1 Major carnivorous plants

Order/genus	Common name	Distribution
Sarraceniales		
Sarracenia	Pitcher plant or trumpet leaf	Eastern North America
Darlingtonia	Cobra plant	Northern California
Heliamphora	Marsh pitcher	Venezuela, Brazil
Nepenthales		
Nepenthes	Dutchman's pipe	Sri Lanka, Madagascar
Dionaea	Venus fly trap	North and South Carolina, USA
Drosophyllum	Portuguese sundew	Portugal, Spain, Morocco
Drosera	Sundews	Worldwide
Violales		
Triphyophyllum		West Africa
Saxifragales		
Byblis		Australia
Cephalotus	Australian pitcher plant	Southwest Australia
Scrophulariales		
Pinguicula		Worldwide
Utricularia	Bladderwort	Worldwide
Genlisea		Brazil, tropical Australia, Guyana
Bromeliales		
Brocchinia		Venezuela, Guyana
Catopsis		Southern Florida to Brazil

plants can now be grown axenically, so future studies may determine the nature of these digestive enzymes and the role of other microbial symbionts in the release of animal nutrients.

Most carnivorous plants have a weak root system and form clonal colonies by stolons or rhizomes. They capture prey by using one of four trapping strategies: adhesive traps (e.g., *Byblis, Drosophyllum, Drosera, Pinguicula, Triphyophyllum*), snap traps (e.g., *Dionaea*), pitcher traps (e.g., *Catopsis, Cephalotus, Darlingtonia, Heliamphora, Nepenthes, Sarracenia*), or suction traps (e.g., *Utricularia*) (fig. 12.1b).

The carnivorous traps, because of their highly developed visual and olfactory signals, are considered examples of "aggressive mimicry" to attract insects (Pasteur, 1982). Pitcher plants, with their flowerlike appendages, attract insects that are inexperienced pollinators (Williamson, 1982; Joel, 1988). Nectaries are common to all pitcher plants and, unlike floral nectaries, are designed to deceive insects. By providing nectar in certain habitats, pitcher plants support insect communities. In such a mutualism, the pitcher plants can grow in nutrient-poor habitats by obtaining nutrients from prey; in return they provide food to insects that escape capture (Joel, 1988).

Ant–Plant Mutualism

Ant–plant symbioses are significant in that they are formed between a highly developed metazoan, the zoobiont, and a flowering plant, the photobiont. The importance of ant–plant mutualism was recognized from the pioneering research of D.H. Janzen (Beattie, 1985; Janzen, 1985).

Ant–plant mutualism is based on four major themes in which ants (1) defend the host plant against herbivory, (2) disperse seeds, (3) provide plants with nitrogen, phosphorus, and potassium, and (4) pollinate plants. In return, plants provide ants with nutrients and nest sites. Plant structures used as nesting sites are called *domatia* and consist mostly of hollow stems and twigs. Domatia are formed by tissue-boring insects and their larvae. Symbioses between *Acacia* plants and ants of the genus *Pseudomyrmex* have been well studied (Janzen, 1967). Ants chew out a hole at the tip of one thorn in each pair of thorns and live in the hollow cavity. In addition to providing nest sites for the ants, the host plant secretes nectar from large nutritive organs called *Beltian bodies*. *Acacia* nectar is a rich mixture of proteins, carbohydrates, and lipids, and ant workers feed it to the larvae. *Acacia* ants continuously patrol the host plants and their immediate surroundings, repel insect or mammalian herbivores and destroy any invading vines by cutting through their petioles (Beattie, 1989).

In southeast Asia, the fast-growing *Macaranga* tree has evolved associations with ants of the genus *Crematogaster*. The host plant provides nesting space and food for the ants, which in turn protect the host against herbivores by removing eggs of other insects and removing any competing plants that are near the host (Fiala et al., 1991).

Phloem-feeding aphids are tended by ants, which harvest their energy-rich excretions (honeydew) in a symbiosis described as conditional mutualism. Ants protect aphids from predators. About 25% of aphid species have developed associations with ants, and host plants play a central role in that they allow aphids to attract ants. How mutualistic interactions are sustained remains to be determined (Cushman and Addicott, 1991a; Gaume et al., 1998).

In tropical forests, epiphytic plants offer conspicuous nesting sites for arboreal ants. The ants live in swollen petioles, leafy pouches, stipules, and hollowed-out stems. Each ant garden may be inhabited by two or more ant species. Seeds of the host plant are collected by the ants and deposited in their nest walls. Domatia of epiphytic ant gardens not only house ant colonies but also absorb nutrients from ant waste. Ant garden mutualistic symbioses are often exploited by ant species that parasitize resident ant colonies (Letourneau, 1991).

Many ant species, such as harvester ants, store seeds in their underground nests. Ant seed dispersal is commonly found in plant species inhabiting nutrient-poor soils. Ants are attracted to a food body (*elaiosome*) which is attached to the seeds. When the food body is eaten, the seed remains intact and subsequently germinates (Beattie, 1985). Harvest ant symbioses are diverse in Australia, where the host plants include herbs, shrubs, and trees (Andersen, 1991).

Though ant pollination has been reported several times, there are only two well-documented cases (Peakall et al., 1991). Ants that visit flowers are more likely to take nectar than to transfer pollen. Well-designed experiments are exploring ant pollination in differ-

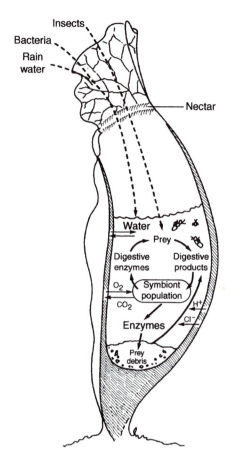

Fig. 12.1 Carnivorous plant symbiosis. (A) Schematic representation of the microenvironment of a generalized pitcher plant showing exchange of nutrients. (B) Four types of capturing strategies used by carnivorous plants: (1) pitcher trap (*Nepenthes*), (2) snap trap (*Dionaea*), (3) suction traps (*Utricularia*), and (4) adhesive traps (*Drosera*). Note the structure of the digestive glands. Adapted from Juniper et al. (1989).

a Microenvironment of *Sarracenia* (Pitcher Plant)

ent ecological habitats, and it may be more common than thought.

12.2 PLANTS AND POLLINATORS

Coevolution of Insects and Plants

Coevolution of insects and plants over the past 200 million years has resulted in relationships that have influenced the development of both groups of organisms. An example of such a relationship is pollination in flowering plants. A primary reason that flowering plants are the dominant group of plants is their exploitation of insects as pollinators. Insects transport pollen from one plant to another, while plants provide insects with a food source such as nectar or pollen. Nearly 70% of flowering plants rely on insects for pollination (Kearns and Inouye, 1997), and

30% of our food comes from bee-pollinated crops. Steep declines in populations of pollinators, particularly bees, is a cause for concern (Norstog, 1987; Barth, 1991; Pellmyr, 1992; Proctor et al., 1996; Kearns et al., 1998).

Pollination is the transfer of pollen from an anther to a stigma or similar receptive surface. Evolutionary adaptations of flowers promote cross-pollination, which results in greater hybridization and, subsequently, greater variation on which natural selection can operate. The mechanisms that plants have evolved to prevent self-pollination and to facilitate cross-pollination are many and varied. They include different maturation times for anther and stigma of the same flower, morphological adaptations of the flower, such as different style lengths in flowers of individual plants, and a self-incompatibility that is genetically determined.

Flowers of the orchid family, one of the largest families of flowering plants, are highly

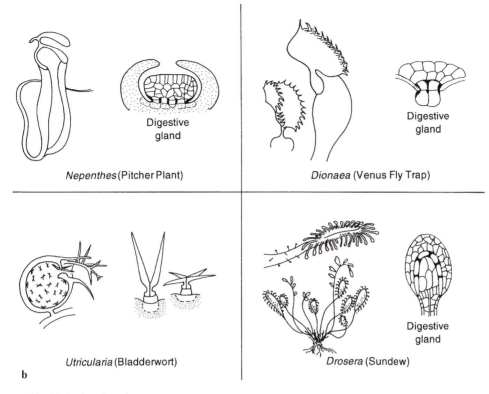

Nepenthes (Pitcher Plant)   *Dionaea* (Venus Fly Trap)

Digestive gland   Digestive gland

Utricularia (Bladderwort)   *Drosera* (Sundew)

Digestive gland

b

Fig. 12.1   (*continued*)

adapted to promote cross-pollination. Such adaptations have resulted in thousands of orchid hybrids.

Some insect–plant relationships are so highly evolved that neither symbiont can reproduce independently of the other. Examples are the yucca and the pronuba moth and the fig and its attendant wasp. Many plants have flowers whose morphological features are adaptations to specific types of animal pollinators. Animals have also adapted to specific types of flowers.

Many plants and animals have coevolved biochemically. Protective chemical compounds produced by plants to ward off predators have resulted in some insects being resistant to these compounds or even using them for their own defense against other animals. For example, larvae of the monarch butterfly feed on milkweed plants and ingest a latex from the plant that is poisonous to other insects but tolerated by the larvae. Adult monarch butterflies retain this poison and, as a consequence, are not eaten by birds (Buchmann and Nabhan, 1996).

Floral Features that Attract Pollinators

The primary reason insects visit flowers is that flowers provide a source of food (pollen and nectar). Pollen is high in proteins, fats, and sugars, and probably was the original food source for many insects. Most flowers produce sufficient pollen to compensate for that eaten by insects.

When insects visit a flower, they generally become covered with pollen grains. Because they cannot clean all the grains from their bodies, they transport the pollen to the next flower they visit and pollinate it. The anthers of most plants release pollen through longitudinal slits, but in some plants, such as blueberries and tomatoes, pollen is released through terminal pores. In these flowers, bees grasp the stamens, and by moving their bodies they cause the anthers to vibrate and release pollen. This phenomenon is called *buzz pollination*. Pollen released from anthers may be attracted to an insect because of electrostatic conditions around the insect and flower. Bees develop positive charges when flying to a flower, which has a negative charge, and when a bee lands on a flower the charges are dispelled.

Nectar is a sugar solution consisting of sucrose and its breakdown products, glucose and fructose. Nectar is secreted from small strands of phloem that terminate in specialized nectar glands, the *nectaries*, located at the base of the pistil, or in protected areas such as tubes or spurs. The sugar concentration of the nectar of different plants as well as the amount of nectar produced in a flower varies and is usually related to the energy needs of the pollinators. For example, flowers pollinated by birds produce more nectar than flowers pollinated by insects. Rare plants and those with specialized pollinators have flowers that produce nectar with a high sugar concentration. For example, some gentians have closed flowers and are pollinated by large bumblebees. The insect forces open the petals in order to enter the flower. Because of the work involved in opening the flower and the rarity of the plant, the number of gentian flowers that a bee can visit is limited. Thus, in this insect–plant relationship, flowers with a concentrated nectar are favored over those with a less concentrated nectar.

Flowers that bloom at night produce nectar only at night, and day-blooming flowers produce nectar only in the morning. In addition to nectar and pollen, flowers may have oil-secreting glands. The oil is gathered by pollinating insects and mixed with the pollen they collect.

Many insects use flowers as a place to lay eggs or to hide from predators. The classic relationship between the *Yucca* plant and the yucca moth (*Tegeticula yuccasella*) and the fig plant and its wasp illustrates the intimate nature of some insect–plant symbioses. In both relationships, the symbionts have developed traits that are a result of their coevolution. The flowers are pollinated by the insects, which lay their eggs in the flowers. Larvae hatch from the eggs and use some of the developing seeds for food.

The life cycles of the *Yucca* plant (some 30 species of *Yucca* grow in North America) and moth are synchronized in several respects. For example, when *Yucca* flowers blossom, the female moths are ready to lay their eggs, and when seeds of the flower mature the insect larvae hatch from the eggs. The female moth visits the white flowers of *Yucca* at night, and by means of specialized mouth parts gathers the sticky pollen from the an-

ther, rolls the pollen into a ball, and carries it to another flower. In the second flower the moth lays its eggs in the ovary, among the ovules or immature seeds. Then, in what seems to be almost purposeful behavior, the moth carries the ball of pollen it took from the first flower and deposits it on the stigma of the second flower, thus causing pollination and ensuring that the ovules develop into seeds. The larvae eat only about one-third of the seeds in the ovary. When the larvae are mature, they bore out of the ovary and drop to the ground, where they pupate until the next flowering season, when the adult moths emerge again (Powell, 1992).

Pollination of the fig plant is one of the most sophisticated examples of plant–insect symbioses known and is an excellent example of coevolution (Anstett et al., 1997) (fig. 12.2). The symbiosis is an obligate mutualism between a gall wasp, *Ceratosolen arabicus*, and the sycamore fig, *Ficus sycamorus*. Fig trees have been propagated from stem cuttings for centuries throughout the Mediterranean region, but in East Africa they grow wild, reproducing sexually and producing fruits throughout the year.

The fig fruit is actually an inflorescence called a *syconium*, with its margins folded to a pear-shaped structure that has a small opening, the ostiolum, which is lined with hairlike scales. The inner fleshy surface of the inflorescence contains numerous small flowers; those close to the ostiolum are male, whereas those deep within the interior are female. When fertilized, each female flower produces a hard, grainy seed.

The female gall wasp deposits an egg into the ovary of a female fig flower with its ovipositor and induces the ovary to form a tumorlike gall, which encloses the insect egg. As female wasps enter the fig fruit through the ostiolum, they frequently lose their wings, antennae, and other body parts because of the dense scales around the opening. Wasps that pass through the ostiolum are attracted to the female flowers, of which there are two types, those with long styles (1.5 mm) and those with short styles (0.8 mm). Long-styled flowers are sessile, and short-styled flowers are borne on stalks. The wasp ovipositor is about 0.8 mm long and can reach the ovary of a short-styled flower, into which it deposits an egg. On a long-styled flower, however, the female wasp inserts its ovipositor through the style, but because it cannot reach the ovary, it quickly withdraws the ovipositor. Nonethe-

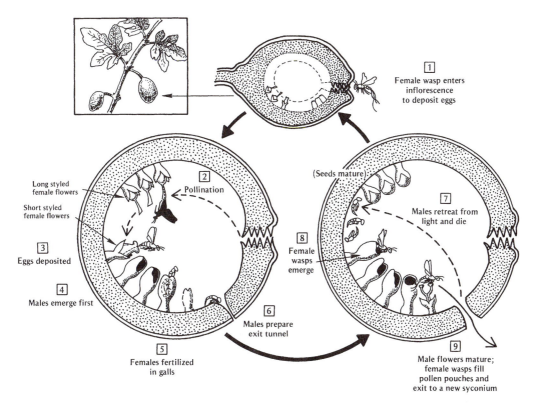

Fig. 12.2   Diagrammatic representation of the mutualistic symbiosis between the sycamore fig and the gall wasp. Adapted from Meeuse and Morris (1984).

less, the wasp still places pollen collected from other flowers onto the stigma. The female wasps die shortly after they have deposited their eggs. Larvae emerge from the eggs a few days after they have been laid and feed on the gall tissue. Larvae then transform into pupae from which adult wasps with strong mandibles develop and eat their way through the gall. Male wasps develop first and search for galls containing female wasps. The males puncture the galls and fertilize the females by inserting the tips of their abdomens into the gall. After they are fertilized, the female wasps emerge from the galls. The wasps then gather near the ostiolum and begin to tunnel through the syconium. Because of their sensitivity to light, the male wasps withdraw into the interior of the fig fruit, where they eventually die.

As the tunneling activities near completion, the male flowers mature and produce pollen. Before the female wasps leave the syconium they fill specialized pouches on their thoraxes with pollen from anthers of male flowers. On emerging from a fig fruit the

females search out new syconia, and the entire cycle is repeated. The symbiosis between the fig plant and wasp is obligatory and mutualistic. Neither symbiont can reproduce sexually in the absence of the other (Addicott et al., 1990).

Insects sometimes are attracted to flowers for reasons other than food and shelter. Orchids of the genus *Ophyrs* produce flowers that so closely resemble female bees or wasps in morphology and fragrance that the male insects attempt to mate with the flowers and in doing so transfer pollen from one flower to another. Similarly, flowers of the orchid *Oncidium* resemble enemies of some insects, and when the male insects attempt to fight the flowers they bring about pollination. These forms of floral deception also occur in plants other than orchids.

Pollinators are also attracted to flowers because of their odor, color, shape, and nectar guides. Color is an important attractant over long distances (distances >1 m), whereas odor is more effective at shorter distances. The various features of a flower are generally closely related to the characteristics and be-havior of its pollinators. Flowers can be grouped according to the types of insects that pollinate them.

Types of Flowers

Insect-pollinated flowers

Bees are the most common insect pollinators, and they are also important pollinators of crop plants such as fruit trees. Bee-pollinated flowers are generally brightly colored and often have a sweet fragrance. The classic studies of the Austrian scientist Karl von Frisch in the 1920s showed that bees can discriminate between colors and scents (Frisch, 1950). Von Frisch found that bees can distinguish the colors blue and yellow, a finding that correlates well with the fact that most bee-pollinated flowers are blue or yellow. Bees also detect ultraviolet light, and many flowers have pigments near their nectar glands that absorb ultraviolet light and create patterns directing pollinators to the stored nectar. Bees cannot discern red flowers, except from their ultraviolet markings. The pleasant odors of bee-pollinated flowers are usually the result of essential oils, such as peppermint and lavender oils.

Nectar guides are color markings or a pattern of ridges on a flower that signal to an insect where nectar is stored. In violets and orchids, the nectar guides are streaks of color on the lower petals. Bee-pollinated flowers usually have modified petals that act as platforms to allow insects to alight on the flower. Bees tend to concentrate their visits on one type of flower at a time.

Beetle-pollinated flowers are usually white or dull in color and smell like fruit or carrion. Beetles have poor vision but a keen sense of smell. They eat different parts of a flower, and the flowers they normally pollinate have their ovules embedded in protective tissues away from the regions where the beetles forage.

Moth- and butterfly-pollinated flowers usually produce their nectar at the base of long petal tubes or spurs. The nectar is out of the reach of many insects but available to the long tongues of butterflies and moths (Nilsson, 1998). Some butterflies pollinate red flowers in addition to blue and yellow flowers. Because moths feed only at night, the flowers they pollinate are white or yellow, and their fragrance develops strongly at night.

Bird- and bat-pollinated flowers

Bird-pollinated flowers are large, usually red or yellow, and odorless. Large flowers are necessary to hold the amounts of nectar birds require. Birds do not have a good sense of smell, but their vision is well developed and they can see brightly colored flowers. Hummingbirds are the most common bird pollinators in North and South America, and nectar is their primary energy source. The nectar is produced in floral tubes away from the reach of smaller insects. Hummingbirds do not land on a flower but hover before it when they are drinking nectar. Mainly tropical plants such as hibiscus and bird of paradise have flowers that are pollinated by birds.

Bat-pollinated flowers such as those of some tropical cacti open only at night. They have dull colors and fruity odors and are large enough to withstand the bat's foraging. As the bats drink the nectar and eat some of the floral parts, pollen collects on their fur and is thereby transferred to other flowers.

Floral Changes after Pollination

Flowers change after pollination. The changes are signals to animals that food is no longer available in these flowers, and, as a consequence, the flowers are not visited by pollinators. Several hypotheses have been proposed to explain the adaptive significance of post-pollination floral changes. One hypothesis is that after a certain number of insect visits to a flower, further visits may remove pollen already deposited on the stigma or damage the reproductive parts of the flower. A second hypothesis states that it is more efficient, in terms of energy expenditure of the pollinators and seed set of the flowers, for the pollinators to limit their visits to flowers that have not reached their maximum number of pollinator visits. Pollen carried by an insect to a flower that had been maximally pollinated would be wasted and, similarly, pollen picked up from such a flower would most likely not be functional.

Floral changes include changes in color and orientation of the flower, the loss of odor and nectar production, the collapse of the flower, and dropping of petals. Color changes cause the flower to become less conspicuous to its pollinators. For example, a white flower that blooms at night and attracts moths becomes darker after it is pollinated and no longer contrasts well with the darkness.

Similarly, the colors of nectar guides become duller, and fragrant flowers lose their odor and stop producing nectar. Moth-pollinated flowers change their position after pollination and thus make it impossible for the moths to fit their proboscises into the usual channels in the flower.

12.3 PARASITIC PLANTS

In 1969, publication of Job Kuijt's *The Biology of Parasitic Flowering Plants* was a major landmark in the study of parasitism among flowering plants and inspired a generation of scientists to explore this phenomenon. A recent publication edited by Malcolm Press and Jonathan Graves, *Parasitic Plants* (1995), describes some of the advances made since Kuijt's work.

Plants that parasitize other plants have been recognized for many years. The parasitic habit has developed independently in at least 11 unrelated families of dicotyledons. In some families, all the plants are parasitic; in others only one genus is parasitic. How and why this habit arose in plants, particularly in families whose other members are autotrophic, is not clear (Musselman and Press, 1995). Parasitic plants are not found in the ferns, gymnosperms, or monocotyledons. A classification of some common parasitic plants is given in table 12.2.

There are about 3000 species of parasitic plants, ranging from trees to small, herbaceous plants and some are important weeds in Mediterranean and subtropical Asia and Africa. Holoparasites obtain all of their nutrients from other plants, whereas hemiparasites still retain photosynthetic leaves and can manufacture food. About 40% of plant parasites attack the shoots of host plants, while 60% are root parasites (Norton and Carpenter, 1998).

Plant Parasite Haustorium

A *haustorium* is a complex system that not only transports but also digests nutrients that it extracts from the host. The parasite and host contain different sugars and amino acids, and haustorial cells are responsible for converting nutrients from the host into a form suitable for the parasite. The origin of haustoria in parasitic plants is obscure, and they may have evolved from roots. Haustoria have highly modified xylem cells and lack phloem. Haustoria are initiated when the roots of a parasitic plant intertwine with a host. The cortical portions of the host and parasite fuse,

Table 12.2 Some common parasitic flowering plants

Family	Genus	Common name
Cuscutaceae (dodders)	*Cuscuta*	Devil's thread
	Grammica	
Hydnoraceae	*Hydnora*	
Loranthaceae (mistletoes)	*Arceuthobium*	Dwarf mistletoe
	Nuytsia	
	Phrygilanthus	
Orobanchaceae (broomrapes)	*Conopholis*	
	Epifagus	
	Orobanche	
Rafflesiaceae	*Rafflesia arnoldii*	
Santalaceae (sandalwoods)	*Phacellaria*	
	Santalum album	
	Thesium	
Scrophulariaceae (figworts)	*Alectra*	
	Lathraea	
	Pedicularis	
	Striga	Witchweed
	Tozzia	
Viscaceae (mistletoes)	*Dendrophthora*	
	Viscum album	European Christmas mistletoe
	Phoradendron	American Christmas mistletoe

but their vascular systems remain separate. The next step in the process involves the penetration of parasitic root tissue into the host and the formation of a single vascular system. Penetration of host tissue is facilitated by mechanical force and enzymatic action of parasitic roots.

Mistletoes

The largest and best known group of parasitic plants are the mistletoes. There are about 800 species, most of which occur in the tropics and subtropics. All mistletoes in North America belong to the family Viscaceae and belong to two important genera, *Phoradendron* (American mistletoe) and *Arceuthobium* (dwarf mistletoe) (fig. 12.3). *Viscum album*, the traditional European mistletoe, has a wide range of hosts and is an important pest in orchards and forests. A few mistletoes are specific to only one host; for example, the dwarf mistletoe *Arceuthobium douglasi* attacks only Douglas fir trees. Mistletoes of the genus *Arceuthobium* show parallel phylogenetic

patterns of cospeciation with that of the host, and speciation occurs by host switching. Mistletoes can even parasitize themselves (Press and Graves, 1995).

Most mistletoes parasitize tree branches, but some, such as the giant mistletoe, *Nuylsia floribunda*, have a terrestrial habit. This mistletoe forms a tree that grows as high as 10 m and parasitizes roots of nearby grasses and other plants. It is common in West Australia and is called the Christmas tree because of the bright flowers it produces in December. The Brazilian mistletoe, *Phrygilanthus acutifolius*, is a vine that parasitizes several different trees simultaneously by means of a large network of underground roots.

Mistletoes that grow on trees and shrubs have roots that grow along the host's branches and develop haustoria. The roots are not influenced by gravity and grow up and down the branches. In some mistletoes, new shoots develop from these roots and allow the parasite to compete for light with the developing canopy of the host plant.

Many mistletoes are dispersed by birds,

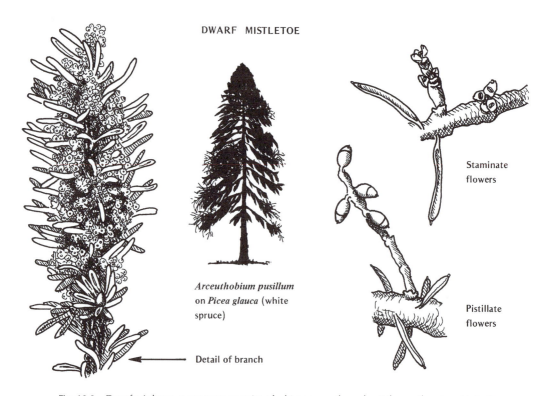

DWARF MISTLETOE

Arceuthobium pusillum
on *Picea glauca* (white
spruce)

Detail of branch

Staminate
flowers

Pistillate
flowers

Fig. 12.3   Dwarf mistletoe, a common parasite of white spruce throughout the northeastern United States and Canada. Note small staminate and pistillate flowers Adapted from Hawksworth and Wiens (1972).

which eat the fruits and void the seeds onto tree branches. The fruit of a mistletoe is a one-seeded berry. Some mistletoes have fruits that explode and disperse their seeds, such as the dwarf mistletoe, whose seeds travel up to 15 m and stick to any surface they contact. The dwarf mistletoe is a serious pest of forest trees in the western United States because of the ease with which it spreads from tree to tree.

When a mistletoe seed germinates, the emerging root forms a disclike structure that presses against the surface of the host. A column of cells arises from inside the disc, penetrates the host tissue by means of mechanical and enzymatic action, and forms a haustorium. The tissues of the haustorium fuse with the host's vascular tissue and form a passageway through which water and minerals move. A few mistletoes fuse with the host phloem tissue and in this way obtain photosynthetic products from the host. Actual penetration of a host branch by a haustorium appears to be a rapid process that follows a series of slow, preparatory stages. If a haustorium fails to penetrate the branch, the parasite forms another disclike structure from which a new haustorium develops. The parasite may make numerous attempts to penetrate the host.

In many mistletoes the haustorium stimulates the host tissue to divide, and a structure called a *placenta* forms along the haustorial surface. The shape of the placenta varies according to the species of mistletoe. Some Mexican species produce an elaborately furrowed placenta that is commonly called woodrose and is prized for its beauty. Mistletoe evolution shows the importance of host specialization in speciation either through host switching or cospeciation (Norton and Carpenter, 1998).

Sandalwoods

Sandalwoods are similar to mistletoes in that they are hemiparasitic, mostly tropical plants, and may be trees, shrubs, or herbs. Most sandalwoods are root parasites, but some attack tree branches. The sandalwood tree, *Santalum album*, has a fragrant wood that is used in cabinetry and wood carving. Its oil is used to manufacture perfumes and soaps. *Phacellaria* is a sandalwood that parasitizes mistletoes and other sandalwoods. The parasite grows inside the host tissues and produces small external, leafy branches that emerge from swollen nodes of the host.

Some sandalwood species are destructive parasites of cultivated crop plants. *Thesium humile* on wheat and barley, *T. australe* and *T. resedoides* on sugarcane, and *Osyris alba* on almonds, pears, and grapes are well-known destructive plants.

Dodders

Dodders are holoparasites that consist only of twining stems and scalelike leaves. They belong to the genera *Cuscuta* and *Grammica* and are members of the Cuscutaceae family. Dodders occur throughout the world and attack a variety of hosts, both wild and cultivated. A dodder seed is devoid of cotyledons, and the embryo is coiled. The seeds are passed through the digestive tracts of grazing animals and can remain dormant for 5 years. When a dodder seed germinates, it gives rise to a seedling that must find a suitable host within a few days or die. The seedling first grows upright and then in a spiral fashion until it contacts a suitable host. After contact, the dodder forms haustoria that grow toward the vascular cells of the host. When xylem cells of the host are reached, the haustorial filaments differentiate into tracheids or vessels. After the haustorial connections to the host have been established, the small roots of the dodder die and the parasite loses all connection to the earth. In *Cuscuta*, haustoria are unusual in that they possess a group of transfer cells that form fingerlike branches to clasp the host sieve cells. This results in a 20-fold increase of surface area to obtain host nutrients. The haustorium contains a high concentration of abscisic acid, which may have a role in sucrose uptake (Riopel and Timko, 1995). The tips of the dodder are phototrophic so the parasite always grows upward along the host plant. The *Cuscuta* stem shows nastic movements that allows the parasite to "forage" the host and searches out shoots with highest nutritional contents. Host secondary metabolites may be responsible for the location and development of haustoria (Kelly, 1992). *Cuscuta*'s survival strategy is based on continual attachments to new hosts so that overexploitation of the same individual is minimal. *Cuscuta* is a metabolic sink which can attract more than 80% of the host's photosynthate and most of its nitrogen, causing devastating losses from the host tissue. Late in the growing season the dodder produces flowers and sets seeds to repeat the cycle.

The *Cuscuta* symbiosis provides an opportunity to study the molecular evolutionary steps associated with the loss of photosynthesis. *Cuscuta europea* and *C. reflexa* have provided insights into the biochemistry of photosynthesis.

Broomrapes and Figworts

Broomrapes and parasitic figworts are mostly herbaceous root parasites of temperate regions. Broomrapes are holoparasites and belong to the family Orobanchaceae; figworts are hemiparasites and are in the family Scrophulariaceae. In broomrapes and in some advanced parasitic figworts, such as *Alectra*, *Latraea*, *Striga*, and *Tozzia*, the seeds germinate only in the presence of exudates from the roots of specific host plants. The exudates cause the parasite to grow toward the host and also to develop thicker roots with more root hairs. *Striga* (witchweed) species are obligate parasites of cereal crops, such as sorghum and millet, in Africa and India. Seeds of these parasites remain viable in the soil for many years and germinate only when near a suitable host. Compounds called *sorgoleons* that stimulate germination of *Striga asiatica* seeds have been isolated from sorghum roots. Sorgoleons may be part of a host's chemical defense system to inhibit the growth of competing plants. Once contact with the host is made, initiation and development of haustoria are triggered by secondary metabolites like phenolics, kinetins, and quinones which are present in host cell walls (Press & Graves, 1995).

Tozzia (big wort) lives underground much of the time, takes several years to mature, and then produces a green flowering shoot above ground. After flowering and releasing seeds, the plant dies. Common broomrapes include beechdrops (*Epifagus virginiana*), found only on roots of the American beech, squawroot (*Conopholis americana*), a parasite of the roots of red oaks, and cancer root (*Orobanche uniflora*), a root parasite of a variety of plants.

Hydnora

Hydnora is a genus of subterranean, parasitic plants and a member of a small, unique family, the Hydnoraceae, all of whose members are root parasites. The plants generally grow in remote areas of South Africa and in East African countries. *Hydnora* produces thick, coarse roots that are bright red inside and grow through the soil and small, thin roots that parasitize the host roots. The coarse roots produce white, fleshy flowers that have a foul odor and are pollinated by beetles. *Hydnora* fruits are large and fleshy and mature underground. They have a fruity odor and are eaten by humans and other animals. The dried and powdered roots of *Hydnora abyssinica*, called tartous, are used by traditional African healers to treat diarrhea and other sicknesses. Seeds of the parasite germinate only in the presence of exudates from the host roots.

Rafflesia

The most exotic and spectacular parasitic plant is *Rafflesia arnoldii*, a native of the jungles of Sumatra. The plant is a holoparasite on the roots of plants and produces the largest known flower (about 1 m in diameter) in the plant kingdom. The parasite lacks stems and leaves and consists only of thin, filamentous haustoria that are so fine they have been confused with fungal hyphae. The flower of *Rafflesia* is fleshy, purplish brown with white patches, and has highly fused sexual parts (fig. 12.4). The inner parts of the petals are raised to form a circular rim or diaphragm. The flower smells like carrion and is pollinated by flies. *Rafflesia* belongs to a small family, Rafflesiaceae, which consists of about 8 genera and 30 species of tropical plants. The fruits are large and fleshy with small seeds approximately 1 mm in diameter. How the seeds are dispersed and the conditions under which they germinate are not known. The family parasitizes only tropical members of the Vitaceae, or grape, family, which form woody vines.

12.4 SUMMARY

One reason angiosperms have been so successful is their ability to form mutualistic symbioses with bacteria, fungi, and animals. Plants have evolved various defenses that discourage herbivory. Carnivorous plants capture prey by using different trapping strategies. Ants help plants to disperse seeds, pollinate flowers, supply nutrients, and defend against herbivores. Insects and flowers have coevolved for millions of years, and pollination is the result of such a coevolution. Adaptations of flowers have been toward changes that promote cross-pollination, as this leads to greater hybridization and variation among the offspring. A few plant–insect

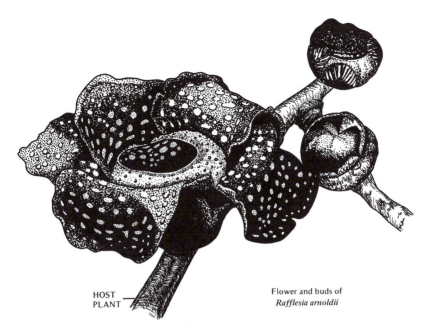

HOST
PLANT

Flower and buds of
Rafflesia arnoldii

Fig. 12.4   *Rafflesia arnoldii*, a parasitic plant, produces the largest known flower in the plant king-dom. Adapted from Kuijt (1969).

relationships such as that between the *Yucca* plant and a moth and the fig plant and a wasp are so highly evolved that neither symbiont can complete its life cycle in the absence of the other one. Some plants and animals have undergone biochemical coevolution.

Flowers attract insects with pollen and nectar as sources of food. Buzz pollination is a means by which bees cause anthers to release their pollen. The sugar concentration of different plant nectars varies and is related to the energetic needs of the pollinators. Birds require more nectar than insects. Insects are also attracted to flowers because of their color, shape, odor, and nectar guides. Some orchids produce flowers that resemble female bees or enemies of the pollinator, attracting male bees, which try to mate or fight with the flowers and in the process bring about pollination.

Flowers have features that relate to the behavior of their pollinators. Bee-pollinated flowers have bright colors, a sweet smell, and nectar guides, whereas flowers pollinated by beetles are white and have a fruity or carrion smell. Bird-pollinated flowers are large and odorless, those pollinated by bats open at night and have dull colors and a fruity fragrance. After pollination a flower undergoes changes that serve as signals to pollinators that food is no longer available at the flower.

Parasitic plants occur in many different families of angiosperms, especially in tropical and subtropical regions, and range in size from trees to small herbs. Mistletoes are the largest group of parasites, with about 800 species. Most mistletoes have a wide host range, parasitizing tree branches by means of haustoria that fuse with the vascular tissue of the host. A plant parasite haustorium is a complex system that extracts and digests nutrients from the host. Sandalwoods are mostly root parasites, and dodders consist mainly of thin stems that wind around the host. Broomrapes and figworts are root parasites that grow in temperate regions. Their seeds germinate only in the presence of exudates from the host plant. *Hydnora* is a root parasite that grows and flowers underground. *Rafflesia arnoldii* parasitizes the roots of other plants and produces the largest flower among the angiosperms.

The parasitic habit is found in all kingdoms including plants. It may seem surprising that organisms that can manufacture their own food through photosynthesis should revert to parasitism and take nutrients and water from other organisms. Why this habit has evolved among separate lines of plants is unknown. In the first chapter we indicated that over a long period of time some parasitic associations have evolved into mutualistic ones. Whether this will occur among plant parasites in the future is not clear. At present, there is no evidence to suggest that plant parasites might confer a benefit on their hosts. Because the parasitism is between members of the same taxonomic group, angiosperms, the defenses of the host plant may not be sufficiently strong to slow or stop the growth of the parasite.

The relationships between flowers and their pollinators are of great interest to ecologists and evolutionary biologists, who are using pollination systems to test various hypotheses related to general ecological and evolutionary principles. The process of co-evolution and the adaptive response of plants and animals to each other are seen clearly among a myriad of examples of flowering plants and their pollinators. There appears to be a continuum of relationships from plants whose flowers are accessible to many different insect pollinators to the extreme examples of the yucca and fig plants that depend on only one type of insect for pollination. Pollination biology has changed dramatically during the past 15 years, from a purely descriptive subject to one in which computer and mathematical models are being used to simulate gametic competition in flowers and various aspects of plant–pollinator relationships.

13 BEHAVIORAL AND SOCIAL SYMBIOSES

Darwin's book *The Expression of the Emotions in Man and Animals* (1872) was a turning point in the study of behavior. For the first time, behavior was considered from an experimental view, and it was recognized that adaptive behavior of an organism fosters its survival and reproduction. Today, behavioral scientists study four aspects of behavior: causation, development, evolutionary history, and function. Observation and description are central to behavioral studies, but the problems involved in carrying out such studies are great. For example, observations cannot be made continuously during the lifetime of an organism under study. Further, behavior is complex, and a scientist can study only one aspect of it at a time. Therefore, personal biases of the scientist become an important consideration in behavioral analysis. Causation is studied by examining the factors that influence behavior over relatively short periods of time in the life of an organism. These factors include environmental as well as sensory and physiological stimuli. Regulation of behavior in an interorganismic association involves internal as well as external factors. Changes in hormone levels or brain abnormalities may result in behavioral modifications of the host. Similarly, a symbiont's behavior may be influenced by developmental signals from the host.

The development of behavior results from the interplay between the genetic and environmental components of an organism's life cycle. Many behavioral patterns deteriorate or are modified because of symbiotic associations. Scientists also study the role of natural selection in the evolution of behavior within a species. Advances in sociobiology have brought about new insights into the evolution of social behavior. Genetic fitness and natural selection are two central tenets of sociobiology. Genetic fitness is measured as a contribution of a particular genotype to the next generation in a population. Parasitic symbioses, infectious diseases, pathogenicity, and host resistance are important factors that increase the genetic fitness of the associating species. Natural selection always maximizes fitness by increasing the chance for survival and reproduction of a species (Moore and Gotelli, 1996).

The functional aspect of behavior relates to an organism's adaptiveness to its environment. The processes of natural selection mold behavior to a particular environment. Examples of specific behavioral patterns include migration patterns among birds, fishes, and mammals; food selection among specialized feeders; and symbiotic associations between different species. Based on symbiotic interactions, behavioral ecology today is gathering a new perspective that includes population regulations (Anderson and May, 1978; Price, 1980), maintenance of genetic polymorphism (Clarke, 1979), structure of the ecological communities (Dobson and Hudson, 1986), phenotypic determination as in bird coloration (Hamilton and Zuk, 1982), and social organization (Hausfater and Watson, 1976).

Although behavioral patterns can be recognized in most symbioses, including those involving bacterial and fungal symbionts, the best known patterns are those of animals. In this chapter, we consider some examples of cleaning symbioses of marine organisms, symbioses in which parasites modify host behavior, and symbioses of social birds and insects.

13.1 BEHAVIORAL SYMBIOSES

Cleaning Symbioses

Many marine fishes are cleaned regularly of ectoparasites and diseased or damaged tissues by specialized fish or shrimp called *cleaners.* The cleaners provide a valuable service by keeping fish free of parasites and disease, and, in turn, they acquire food and protection from predators. The behavioral patterns associated with these symbioses are remarkable and include posing by host fishes to expose to the cleaners that part of their body that needs attention. Posing is a means of communication between the host and the cleaner. Host poses include opening mouths and gills to allow the cleaner to enter, or assuming an unnatural vertical position (fig. 13.1).

Cleaning symbioses have been observed in many parts of the world but are most common in tropical waters. The fish involved in this symbiosis first establish cleaning stations near prominent parts of the ocean floor, along the margins of kelp beds, or even in ship wreckage. Some stations remain constant for several years, whereas others are used only for a short time. Tropical cleaners have bright colors in patterns that contrast sharply with their background. The colors, along with ritualistic displays ("dances") put on by the fish, attract host fish to the cleaning stations. Tropical cleaners work alone or in pairs. In contrast, cold-water cleaners have dull colors and live in schools. Many species of fish are known to be cleaners, especially in their juvenile stages. These fish generally live along the coast or in coral reefs. Deep ocean fishes, such as tuna, may also have cleaners, but this has not been firmly established. Even marine turtles, manta rays, and sharks visit cleaning stations.

Parasites that are removed from the host include small crustaceans such as copepods and isopods, as well as mats of bacteria and fungi that grow from infected host tissue. Parasitic crustaceans are usually transparent or the same color as the host, which makes it difficult for the cleaner to remove them all. The food cleaners obtain from their hosts is only part of their diet; they also eat small animals from the surrounding water or from the ocean floor.

Although many different types of fish have evolved the cleaning habit, some of their adaptations are similar. Many cleaners, for example, have long, pointed snouts that enable them to probe into small crevices of the host's skin or gills and tweezerlike teeth that are used to remove parasites embedded in the skin.

The small wrasse or senorita, *Oxyjulis californica*, is a cleaner common off the southern California coast. It cleans a wide variety of fishes including the ocean sunfish (*Mola mola*), black sea bass (*Stereolepis gigas*), and blacksmith (*Chromis punctipinnis*). In the Caribbean a genus of goby (*Gobiosoma*) includes species that are cleaners even as

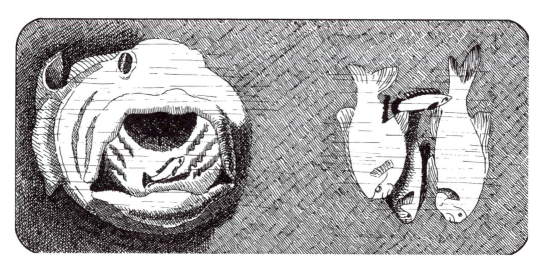

Fig. 13.1   Cleaning symbioses of marine fishes involve removal of ectoparasities by small fish called cleaners. Adapted from Limbaugh (1961).

adults and live off what they receive from their hosts. Many fish may crowd around a single cleaner and assume various poses as they wait their turn to be cleaned. Several other wrasses such as the bluehead wrasse (*Thalassoma bifasciatum*) and the Mexican rainbow wrasse (*Thalassoma lucasanum*) are also cleaners.

The more highly coevolved the symbiosis between a cleaner and host, the less likely that the cleaner will be eaten by the host. Senoritas are rarely eaten even though they enter the mouths and gills of some large hosts. Other cleaner fish, and especially shrimp, are not as fortunate and may be eaten by their hosts. The tiny neon goby (*Elaca-ti-nus oceanops*) lives among coral reefs and cleans a variety of fishes including large groupers, parrot fishes, grunts, angels, and morays. The goby swims in and out of the mouths and gills of its hosts with impunity. Several gobies may group together to clean a large fish. The large barracuda, *Sphyraena barracuda*, which grows to a length of 6 feet, is cleaned by a blue and yellow wrasse (*Bodianus rufus*) that is less than 1 inch long. Remoras are sucker fish that spend most of their lives attached to sharks and turtles and not only share in their meals but also clean them of parasites.

Several species of cleaner shrimp are common in tropical waters. The Pederson cleaning shrimp (*Periclimenes pedersoni*) is 4 cm long and has antennae that are even longer. The shrimp generally lives with an anemone, which protects the shrimp and serves as its cleaning station. When a fish approaches the anemone, the shrimp waves its antennae and body back and forth until the fish gets close enough for the shrimp to climb on it. The shrimp then crawls over the fish's body and removes parasites. In some cases, cleaner shrimps and cleaner fishes occupy the same station.

The boxer shrimp (*Stenopus hispidus*) lives in tropical waters throughout the world. It grows up to 8 cm long, has a white body with red stripes, and long, white antennae. The shrimp lives among gray corals and is easily visible to fish seeking its services.

Experiments designed to test the importance of cleaner fish and shrimp to marine fishes have shown that cleaners control the spread of host parasites and infections. In a field experiment conducted in the Bahamas, Conrad Limbaugh (1961) removed all the cleaning organisms from two small, well-populated reefs. Limbaugh noticed that within a few weeks the number of fishes in the reefs had declined sharply and the fishes that remained were infected by fungi and other parasites. A marked difference was seen between the health of fishes of reefs without cleaners and the health of those with cleaners. Other experiments have not been as conclusive, and some investigators believe that the cooperative behavior of the hosts could be a response to the tactile stimulation given by the cleaners (Losey, 1987; Poulin and Grutter, 1996).

Some fishes change color while they are being cleaned. For example, the black surgeonfish (*Acanthurus achilles*) turns bright blue, and other fishes assume a bronze color. The color changes may help to increase the contrast to ectoparasites, thereby making parasites more visible to the cleaners.

Often when fishes are being cleaned, they become alarmed and signal their alarm to their cleaners, who then quickly retreat. Cleaners working in a host's mouth during an alarm are ejected by the host or given a signal that causes them to leave.

Some fishes in the Indo-Pacific region where cleaners may have first evolved mimic the color patterns of cleaners. Such a disguise protects a fish from predators or allows it to approach prey without alarming them. Because the saber tooth blenny (*Aspidontus taeniatus*) closely resembles the common cleaner *Labroides dimidiatus*, it can approach other fish, particularly juvenile ones, which are less wary, and bite off pieces of their fins. The blenny also removes parasites from larger fishes. The blenny may be in an intermediate stage of evolution, between a predator and a cleaner.

Much of the information on cleaning symbioses comes from observations made by divers. For this reason, studies have concentrated along coastal areas and around tropical islands and reefs. Observations of deep-sea fish populations and those of colder waters such as in the Arctic and Antarctic are limited.

Cleaning symbioses also occur with land animals. The red rock crab, *Grapsus grapsus*, is a cleaner of the marine iguana *Amblyrhynchus cristatus*, which lives on the Galapagos Islands. Various types of birds, such as oxpeckers, remove parasites and infected tissue from large animals such as water buffalo, crocodiles, cattle, and zebras (Breitwisch, 1992).

Anemone–Clownfish Symbiosis

Fishes of the genera *Amphiprion*, *Dascyllus*, and *Premnas*, commonly called clownfish, form mutualistic associations with giant sea anemones that live in coral reefs throughout the Pacific Ocean. The associations are obligatory for the fish, but facultative for the anemones. The anemones eat prey they have paralyzed by means of poisonous nematocysts discharged from specialized cells in their tentacles. The clownfishes are immune to the stinging nematocysts and can nestle within the tentacles and contact them frequently without harm. The question of how clownfish develop their immunity to the anemone poison has intrigued scientists (Fautin, 1991; Fautin and Allen, 1994).

Some clownfish go through a period of acclimation before they become protected from the anemone. Studies designed to determine how acclimation is achieved appear to show that the mucous coating around the fish is changed by association with the anemone and the fish is no longer recognized as prey by the anemone.

The change in the mucous coat was first thought to result from internal secretions from the fish, but scientists now believe it is a result of the addition of mucus from the anemone (Mariscal, 1971). When the fish acquires this mucus, the anemone no longer recognizes it as foreign and thus does not discharge its nematocysts when its tentacles are contacted by the fish. To obtain mucus from the anemone, the fish undergoes an acclimation behavior. The first stage of this behavior is recognition of the specific type of symbiotic anemone by the fish. The clownfish approaches the anemone cautiously, swims around its column and oral disc, and comes as close to the tentacles as possible without touching them. After frequent passes over the oral disc, the fish makes the initial contacts with the anemone by means of its tail and fins. At first, the tentacles adhere to these structures and the fish pulls away and frees itself. More extensive body contacts then follow until the fish becomes fully acclimated and can bury itself in the anemone's tentacles. The acclimation time varies but may take up to 1 h. If a clownfish becomes separated from its host for longer than one hour, it must again perform the acclimation behavior. The immunity a fish acquires from one anemone does not protect it against another anemone should the fish decide to move on.

Studies by Richard Mariscal have demonstrated the importance of the protective cover around the clownfish (Mariscal, 1971). When Mariscal presented pieces of a grouper and an acclimated clownfish, skin-side down, to a giant anemone, the tentacles of the anemone adhered strongly to the grouper skin but not to that of the clownfish. When both pieces were presented flesh-side down, the tentacles of the anemone adhered to both pieces, which were then eaten.

Once a clownfish acclimates to its anemone, it stays there as long as the food supply is adequate. Clownfish are brightly colored and marked, and they attract larger fish to the anemone. These fish, if they venture too close, are stung by the tentacles and eaten by the anemone. The clownfish share in the meal and also remove fragments of the prey and wastes from the anemone. The fish may also bite off and eat pieces of the anemone's tentacles that contain symbiotic algae and eat the crabs and shrimp that also live among the anemone's tentacles. Thus, in addition to protection from predators, the clownfish obtain food.

A similar relationship exists between the Portuguese man-of-war (*Physalia physalia*) and the horse mackerel (*Trachurus trachurus*). The fish live between the tentacles of this cnidarian and somehow avoid being stung by its nematocysts. The bright blue and silver colors of the fish, as well as its small size, attract prey for the man-of-war.

Host–Parasite Interactions and Behavior Modifications

Many parasites, such as protozoans, helminths, and nematodes, have complex life cycles. They spend their early life in one animal species, the intermediate host, and reach sexual maturity in another animal species, the definitive or final host. A cost–benefit equation is key to understanding the evolution of behavior. From a parasite's perspective, its behavioral evolution depends largely on enhancement of its transmission. The evolution of host behavior is influenced by efforts of the host to avoid contact and to live with the infection. Therefore, ecological and behavioral models for habitat choice and patch settlement of host organism may be determined by the parasitic symbiont (Curio, 1988; Barnard and Behnke, 1990). For example, male fleas are more likely to be found on juvenile male ground squirrels, which are the part of the host population that leaves its home range to colonize new areas. By selecting juvenile male squirrels, male fleas can move to new resource patches and also avoid inbreeding (Hanley et al., 1996).

Host behavior determines parasitic infections by factors such as host exposure, host susceptibility, and parasitic virulence. A parasite may support features of another species that differs from its host. The larvae of the swallowtail butterfly, *Danaus chrysippus*, taste bad to insectivorous birds because the larvae feed on toxic plants. A parasitoid lays its eggs only on butterfly larvae that are not eaten by birds, and thus an advantage gained from the plant toxin is neutralized by the parasitoid (Gibson and Mani, 1984).

A cleaning fish's grooming behavior on its host fish decreases the number of ectoparasites it carries. Fever in an infected host is also an adaptive feature, as shown by desert iguanas, which raise and lower their body temperature to slow the growth of parasites (Kluger, 1979).

Parasitic symbionts have a variety of physiological and ecological mechanisms to ensure their survival and reproduction. Some parasites modify the behavior of their intermediate host in such a way that the host becomes more vulnerable to predation by the definitive host species (Giles, 1983; Moore, 1984a). The life cycles of symbionts and their hosts may show parallel evolutionary development (Beckage, 1997).

It is argued that parasites evolve toward reduced virulence against the host in order to prolong survival of both host and parasite (Holmes, 1983). A parasitic symbiont with its larval stages in an intermediate host must allow its survival until it passes to the next host in the parasite's life cycle. For example, larvae of the tapeworm *Schistocephalus solidus* inhabit three-spined sticklebacks, whereas the adult tapeworms occur in the intestines of fish-eating birds. The tapeworm larvae interfere with stickleback reproduction, slow its growth, increase its oxygen consumption, and hamper its movement by distending the host belly. Parasitized fish swim close to the water surface, making it easier for birds to eat them. This form of behavior is called *adaptive manipulation*. Another example is host castration and prolonged life of fluke parasitized snails, which allow longer periods for the parasite to reproduce (Minchella et al., 1985). Following are further examples of parasites that alter their host behavior:

1. Larvae of the nematode *Tetrameres americana* invade the muscles of grasshoppers, causing the host to become sluggish and therefore easily caught by chickens in which the nematodes live as adult worms.

2. Larvae of the tapeworm *Taenia multiceps* parasitize the central nervous system of sheep and cattle. The infected animal staggers in circles and becomes separated from the herd. Carnivores such as wolves and wild dogs constantly search for such weakened prey. The tapeworm becomes sexually mature in the digestive tract of the second host species.

3. The fluke *Leucochloridium macrostomum* occurs in the rectum of European birds such as crows, sparrows, jays, and nightingales. The birds acquire the flukes in a highly unusual manner. Fluke eggs are ingested by snails, where they develop into the larval, sporocyst stage. The parasite then produces a second sporocyst generation, which migrates to the head and tentacles of the snail. These structures then enlarge, and the tentacles become green, brown, or orange with prominent banding patterns. The sporocysts pulsate, and this movement attracts birds, which peck on the sporocysts and ingest them. Inside the host birds the larval cysts rupture and liberate the adult parasites. This is an example of aggressive mimicry. Sporocysts acquired by a parent bird are transferred to nestlings. In a similar manner, pulsating sporocysts of the fluke *Ptychogonimus megastoma* are ingested by shore crabs, and the fluke larvae become encysted in the crab tissue. Fish acquire the parasite by ingesting the infected crabs.

4. The lancelet fluke *Dicrocelium dendriticum* inhabits the gallbladder of sheep, deer, rabbits, and other grazing mammals. Eggs of the fluke are eaten by land snails, in which the cercarial larval stages develop. The mature larvae exit the snail in a slime ball, which is eaten by ants. The "brainworm" larvae of the fluke forms cysts in the brain of an infected ant

and radically transforms its behavior. The affected ant climbs to the tip of a blade of grass and hangs on tightly. This abnormal behavior increases the chances of the ant being transferred to a grazing host (Holmes and Bethel, 1972). One cercaria of many ingested by an ant becomes the brainworm, and it sacrifices future growth and reproduction, thereby leaving other cercariae in the ant to assume the evolutionary burden of the species. This is an example of a parasitic altruism, where a cercaria sacrifices its own life to alter an ant's behavior but gains many times over in the next generation (Smith Trail, 1980).

5. Acanthocephalan worms are endo-parasites that inhabit the alimentary tracts of vertebrates, where they attain sexual maturity. The intermediate host of these parasites is an arthropod, whose behavior is transformed after infection. The altered behavior significantly increases the chances of the parasite being acquired by the final host through preferential predation of the infected intermediate host. There are several well-known examples of this phenomenon.

(a) *Polymorphus paradoxus* is an endoparasite that inhabits the alimentary canals of beavers, mallards, and muskrats. Aquatic crustaceans such as amphipods are the intermediate hosts of the parasite. Healthy amphipods normally move away from light and rarely appear at the surface of a lake or pond. When disturbed, they dive and burrow into the mud at the bottom. Parasitized amphipods, however, are attracted to light and are commonly observed clinging to vegetation on the surface of the water. This altered behavior increases the likelihood of the amphipods being eaten by a final host species (Moore, 1984a; 1995). Similar anomalous behavior of amphipods infected with *Polymorphus marilis* allows ducks to catch the crustaceans.

(b) Songbirds preferentially feed on pill bugs or sow bugs (isopods) that have become infected with the acanthocephalan *Plagiorhynchus cylindraceus*. Songbirds do not feed on uninfected pill bugs, but infected bugs may make up more than 40% of a bird's diet. The behavior of the parasitized pill bugs is altered so that they seek out exposed areas, where they are quickly eaten by birds.

(c) Cockroaches parasitized with *Moniliformis moniliformis* are attracted to light and become hyperactive. The roaches are less likely to remain hidden and therefore more likely to be caught by rats, the final host for the parasite.

13.2 SOCIAL SYMBIOSES

Among insects and birds, symbioses occur not only between individuals of different species but also between individuals and societies and between different societies (Abrahamson, 1989; Price et al., 1991; Choe and Crespi, 1997).

In insect societies, each individual has only a limited role in relation to the society as a whole. Because of this limitation, other species of insects can easily intrude into the society, and this leads to symbiotic interactions. A similar situation exists in birds whose young mature slowly and do not form close bonds with their parents. These parents are easily deceived when eggs are laid in their nests by other birds and will incubate the foreign eggs and care for the young along with their own. Among other birds, such as ducks, and among mammals, social symbiosis is rare because the parents form close recognition bonds with their young at an early stage and repel all apparent intruders.

Edward Wilson of Harvard University is the leading authority on social symbioses. His studies of insect societies and their complex internal and external interactions have greatly increased our understanding of the subject. Wilson groups social symbioses among animals into three major categories: social parasitism, social commensalism, and social mutualism (Holldobler and Wilson, 1990; Crozier and Pamilo, 1996). We consider only these three major groups.

Social Parasitism

Parasitic ants, wasps, and other insects

More than 200 species of ants have evolved different types of social parasitism with other ants, ranging from the loosely organized to the highly integrated. The simplest form of social parasitism is one in which a colony of ants builds a nest next to a colony of a different species and steals food from and preys on the neighboring workers. Some ants may even occupy the nest of another species and maintain a separate colony within the nest (Bourke and Franks, 1995).

A highly evolved form of social parasitism among ants is shown in colonies containing different ant species living together. For example, in several subfamilies a fertile queen may enter a colony of ants of a different species and by some unknown mechanism be accepted by the workers. The host queen is killed, either by its own workers or by the intruder queen. As the host workers die of old age or injury, there are no offspring to replace them, and the nest becomes fully colonized by offspring of the new queen. Wilson calls this example *temporary social parasitism* (Holldobler and Wilson, 1990).

Another type of social parasitism among ants involves slavery. Some ant species raid the nests of different ant species, steal pupae, and return with them to their own nest, where the young ants are reared as slaves. The slave ants do all of the work in the nest such as foraging, building, and rearing the brood (Topoff, 1990, 1994).

One of the most highly integrated examples of social parasitism involves the parasitic ant *Teleutomyrmex schneideri*, which is found in the Swiss and French Alps. *Teleutomyrmex schneideri* parasitizes a close relative, *Tetramorium caespitum*. The ant parasite lives only in the nests of its host and lacks a worker caste. The parasite queen is much smaller than the host queen and spends most of its time clinging to the back of the host queen or workers. The parasite ants are fed by the host workers. Both queens lay eggs; those of the host produce only workers and not sexual forms, while those of the parasite produce only sexual forms. When the parasite ants mate, the new fertile queens either remain in the same nest or fly away and invade other nests of the host ant. It is thought that *Teleutomyrmex* represents an end point in the various evolutionary pathways of parasitic ants. The morphological adaptations of the ant and loss of the worker caste are traits of other highly evolved parasitic ant species.

Parasitic wasps behave in a manner similar to parasitic ants. The queen wasp penetrates the nests of other wasps, kills the host queen, and assumes her role. Some wasps are permanent residents in the nests of other wasps and, like ants, have evolved to a point where they do not have a worker caste and depend on the host workers to care for them. A similar type of behavior has been reported for parasitic bumblebees.

The origin of parasitic ants is of great interest to scientists. In 1909, Carlo Emery developed a hypothesis, known as *Emery's rule*, that states that parasitic ants and slave-making ants evolved from closely related species that now have become their hosts. Wilson suggests that parasitic ant species and their host species developed from a common parent species through geographical isolation and genetic differences. When the range of the two new species later overlapped, one became a parasite on the other. Wilson believes that Emery's rule applies also to parasitic wasps and bumblebees.

Termites do not exhibit the same type of social parasitism as other social insects because termite queens form new nests after fertilization rather than returning to their old nests or attempting to invade established nests.

Some beetles are accepted into an ant or termite colony because they produce substances that attract the host insects. The substances are similar to pheromones, which the host normally produces, and have a calming effect on the hosts after they lick these substances. Wilson calls these compounds *appeasement substances* because they help the beetle become adopted into an ant or termite population. After the beetle obtains entry into a nest, it feeds on the host larvae, and by stroking worker ants with its antennae can stimulate them to regurgitate food. Many of the beetles that live with ants and termites have evolved morphological features, such as a swollen abdomen, that mimic features of their hosts. Loss of eyes and wings are also common changes the beetles have undergone.

Brood parasitism in birds

Social parasitism is also present among birds, especially cuckoos and cowbirds. About 100 species of cuckoos are *obligate brood para-*

sites, which means that they depend on other birds to hatch their eggs and rear their young (Payne, 1998). The cuckoos lay their eggs among those of the host birds, and when the eggs hatch, the young cuckoos destroy the host's eggs and kill any young of the host (fig. 13.2). This association is so highly evolved that the eggs and young of the cuckoo resemble those of the host bird in color and size. One of the best studied social parasites among birds is the European cuckoo (*Cuculus canorus*). This species consists of different populations, each having a different host bird, whose eggs are mimicked. How the cuckoo populations maintain their specific relationships with their host birds is not clear, particularly since the different cuckoo populations interbreed. Cuckoos have evolved other adaptations in addition to egg mimicry for their parasitic habit. Their eggs have thick shells that resist damage when the eggs are dropped into host nests, and the birds have extendible cloacae, which enables them to deposit their eggs into small nests (Lyon, 1993). Another remarkable adaptation is the ability of some cuckoos to fly like hawks and in so doing to distract the host bird from its nest so that the cuckoo may deposit its eggs undisturbed (Lotem and Rothstein, 1995). Relationships between cuckoos and their hosts afford excellent examples of coevolution (Davies and Brooke, 1991; Rothstein and Robinson, 1998).

Cowbirds parasitize more than 200 bird species and have caused population decline in many of these species (Rothstein and Robinson, 1994). A coevolutionary arms race produced some interesting behavioral outcomes: avian brood parasites have evolved special mechanisms to obtain parental care from their host species, while the hosts have evolved mechanisms or traits to avoid and reject parasites. For example, the host may remove its own eggs along with those of the parasite. The presence of brood parasites can be advantageous if they remove ectoparasites from the host nestlings.

In response to wide-ranging parasitic symbionts, many hosts have adopted behavioral countermeasures. Hamadryas baboons and Hadza tribesmen in Tanzania avoid schistosome-infested waters and collect water by digging holes in dry river beds (Holmes, 1983). Other examples of behavioral countermeasures include grooming, cleaning symbioses of marine animals, sunning of frogs and salamanders to dislodge leeches, and dust-bathing of birds (Murray, 1990).

Social Commensalism

The nests of social insects are home to many different scavengers such as mites, beetles, and millipedes. The host insects either ignore these commensals or accept them as part of

Fig. 13.2   Brood parasitism of cuckoo and cowbird. Adapted from Wilson (1975).

the colony. Some silverfish and millipedes live with army ants, participate in their forays, and even share their prey.

Among insect-eating birds, several species may flock together to better avoid predators and to forage more efficiently. Similar groups of mixed species have been reported among marine fishes, such as dolphins and whales, and African mammals. Mixed groups of animals often congregate to increase the efficiency of early warning of predators.

Social Mutualism

Mutualistic symbioses are common between ants and various insects such as aphids, scales, mealybugs, and treehoppers. The ants tend these insects like cattle, milking them for food and protecting them from predators (Davidson and McKey, 1993) (fig. 13.3).

Aphids penetrate the food-transporting system (phloem) of plants with their stylets and feed on the plant sap, which is rich in sugar and minerals. The aphids digest only a small amount of the plant sap they take in and excrete the remainder in a form called honeydew. Aphids that are tended by ants excrete the honeydew slowly so that it accumulates as drops around the anus. Mutualism between ant and aphid has evolved to such a high degree that their life cycles have become coordinated, and the aphid has lost various structures that it normally has to defend itself, such as a hard exoskeleton and jumping legs. When an ant is ready to feed, it strokes an aphid with its antennae and stimulates it to exude honeydew. Some ants carry the aphids they tend back to the nests each evening to protect them from predators and return them to the plants each morning.

To ensure a steady source of food, worker ants carry aphids and other honeydew insects to different plants if the original ones are disturbed or uprooted. Some ant species maintain aphid eggs in their nests during the winter and when the aphids hatch in the spring carry them to nearby plants. The queens of some ant species even carry scale insects during their nuptial flights. Bees and wasps also keep honeydew insects.

Some butterfly species have caterpillars that are tended by ants. The large blue butterfly *Maculinea arion* cannot complete its life cycle without help from red ants. Until the caterpillars pupate and hatch, the ants protect the caterpillars in their nests and feed on their sugary excretions. When the chrysalis opens, the butterfly leaves the ant's nest. Other symbioses between caterpillars and ants have been described (DeVries, 1992).

Some South American ant species appear to have a mutualistic relationship with each other. The ants form adjacent nests, share common trails, and collect honeydew from the same population of treehoppers. The different ants are friendly to each other and seem to thrive because of the association.

The giant cowbird (*Scaphidura oryzivora*) of Central and South America is a brood parasite that lays its eggs in the nests of oropendolas, which are members of the grackle family. Oropendolas have a natural enemy, the parasitic botfly (*Philornis*), which infests their nests and kills their young. Studies have shown that oropendola nests containing young cowbirds have fewer botflies than nests without cowbirds. Young cowbirds instinctively attack and kill maggots of the botflies and remove them not only from the nests but also from the young host birds. Benefits to

Ants tending aphids

Fig. 13.3   Ants licking honeydew from aphids. In this mutualistic symbiosis, the ants obtain food from aphids, which they protect from predators.

cowbird nestlings occur only when the host environment lacks stinging wasps and biting ants, which provide some protection against botflies (N. Smith, 1968). Cowbirds, unlike cuckoos, tolerate the eggs and young of the host birds. Thus, an association that began as one of strict parasitism has evolved into mutualism (Rothstein and Robinson, 1994).

13.3 SUMMARY

Behavior in a symbiotic association is influenced by external and internal factors. Symbiosis commonly modifies the behavior of the associating partners. Cleaning symbioses are common in tropical, marine waters, where specialized fish and shrimp clean other fish of external parasites and damaged tissues. In return for this service, the cleaners obtain food and protection from predators. Cleaners and hosts communicate with each other by means of various behavioral patterns such as posing or by means of bright colors and unusual color patterns. Some common cleaners include wrasses such as the senorita, the neon goby, remoras, Pederson shrimp, and the boxer shrimp. Experiments have shown that cleaners have an important function in tropical ecosystems by controlling the spread of parasites. Some fish have evolved color patterns similar to those of cleaner fish. This disguise protects the fish from predators and allows them to get close to prey fish. Cleaning symbioses are also present among land animals.

Clownfish live in a mutualistic association with giant anemones that grow in tropical coral reef communities. The fish acquire immunity to the anemone by means of an acclimation process during which the fish coats itself with mucus from the anemone. The mucus disguises the fish in such a way that it is not recognized as foreign by the anemone. Clownfish attract other fish to the anemone and share the prey caught by its host.

Some animal parasites such as nematodes, tapeworms, flukes, and acanthocephalan worms modify the behavior of their intermediate hosts in a manner that makes them more likely to be preyed on by the final hosts.

Social symbioses occur among insect societies and birds and include social commensalism, social mutualism, and social parasitism. In social symbioses each individual has only a limited role. Social parasitism is highly evolved among ants and may involve simple predation, slavery, or the integration of a foreign queen into an existing ant colony. According to Emery's rule, parasitic and slave-making ants evolved from closely related species that now serve as their hosts. Interestingly, a similar situation may exist among parasitic red algae, many of which occur only on other closely related red algae.

Certain beetles become accepted into ant or termite colonies because of appeasement substances produced by the beetles that appear to calm the hosts. The beetles feed on host larvae and food that is regurgitated by worker ants.

Cuckoos are highly developed social parasites that lay their eggs in the nests of other birds. When the eggs hatch, the young cuckoos kill the host's young and are raised by the host. Adaptations of the cuckoo include mimicking the eggs of the host, extendable cloacae, and eggs with thick shells that resist breaking when they are dropped into the host nest.

Silverfish and millipedes are commensals in the nest of social insects. Ants form mutualistic associations with aphids, scales, and mealybugs, milking them for the food (honeydew) they obtain from plants. Ants tend aphids like cattle, protecting them and carrying them from plant to plant. Ants have a similar relationship with the caterpillars of some butterflies.

Cowbirds and oropendolas have a mutualistic relationship in which the young cowbirds kill botfly maggots that are serious parasites of young oropendolas and thereby protect their foster nest and host.

14 SYMBIOSIS AND COEVOLUTION

Although Darwin's theory of organic evolution has provided an intellectual framework for understanding the processes of speciation and the unity of life forms, the theory has never fully satisfied evolutionary biologists. In Darwin's view, evolution is the descent, with modification, of different lineages from common ancestors. Population genetics, natural selection in gene pools, adaptation, and isolating mechanisms have traditionally been considered by many biologists as significant factors in the evolutionary process. Evolutionary ideas are now being refashioned in the language of punctuated equilibria, genetic polymorphism, molecular evolution, and sexual selection. To this list of new approaches, one can add the concepts of coevolution and symbiosis. Evolutionary ecologists are convinced that evolutionary processes cannot be understood fully until the close interactions between organisms are examined. Indeed, the term *symbiogenesis* was introduced by Mereschkovsky in 1920 to recognize a process whereby new species arise as a result of symbiotic interactions between different organisms (Margulis 1991a, 1993a, b; Khakhina, 1992).

Coevolution is a kind of evolutionary "arms race" between organisms living together in intimate associations. Adaptations that give one organism an advantage are countered by neutralizing adaptations of the other organism. For example, evolution of virulence in a pathogenic organism is often matched by a concomitant evolution of resistance of the host species. Reciprocal genetic exchange, often occurring simultaneously in interacting populations, is the basis of coevolution. Traits of one species evolve in response to those of another species, whose traits, in turn, were influenced by the first species. Evolution of the eukaryotic cell has involved a series of genetic interactions between ancient prokaryotes. Scientists have recently discovered that DNA exchange has taken place among the nuclei, mitochondria, and chloroplasts of eukaryotic cells, attesting to a complex evolutionary process. Coevolution that takes place between two different species is *pairwise coevolution.* On a larger scale, one or more species may evolve a trait in response to traits of several other interacting species, a phenomenon called *diffuse coevolution.* For example, a plant species may develop toxins or other defenses in response to herbivory, and various animal populations evolve means to overcome these defenses. The plant species then develop new defenses, and the evolutionary process is repeated. In some interacting species, a coevolutionary equilibrium appears to take place. Thompson has introduced a new "geographic mosaic theory of coevolution," which differs from approaches that focus only on local populations (Thompson, 1994).

The traditional theory of coevolution includes cospeciation as well as coadaptation among interacting organisms. Long-term, close contacts between organisms and the reciprocal influences that take place between them often result in the formation of new species. The phylogeny or evolutionary histories of organisms involved in symbioses are closely related to each other. The many relationships between insects and plants, in particular those involved in pollination, are clear examples of the adaptation and speciation that can occur in symbiotic associations. The lichen symbiosis is an example of how free-living fungi (ascomycetes) become transformed into new species during the course of symbiotic associations with algae. Parasitologists have often observed a close relationship between the phylogeny of parasites and their host species. For example, *Enterobius*, a genus of pinworms that infect primates, has a phylogeny and geographical distribution that closely parallel those of its primate hosts. Evolutionary insights have been made on the taxonomic relationships of species of bats, squirrels, and pocket gophers from the studies of their ectoparasites, such as lice and fleas.

Ecologists have applied the following rules to better understand the evolution of host parasite interactions:

- *Farenholz's rule* states that the phylogeny of many parasites mirrors that of their hosts. Because most parasites are obligate symbionts, it is assumed that parasites of present-day hosts had ancestors that were parasites of earlier hosts.
- *Szidat's rule* states that the more primitive the host, the more primitive its parasites.
- *Eichler's rule* states that because the host and parasite share a parallel evolution, parasites can be used to understand the phylogeny and origin of host species. Large host groups will have more genera of parasites than small host groups. Parasitic symbionts may evolve in two ways. Speciation may occur if some parasitic species become established in different hosts, or if the parasites invade new tissues or organs in the same host species.

Additionally, Dieter Ebert (1998) used serial passage experiments to study the evolution of parasites and their adaptation to their hosts. He found that within a host, competition among pathogens resulted in increased virulence, whereas pathogens tended to become avirulent to their former hosts.

14.1 GENETIC CONFLICTS AND SYMBIOSES

From an evolutionary perspective, symbiotic relationships are an uneasy balance between conflicting interests of host and symbiont genomes (Werren and Beukeboom, 1998). Genetic conflicts are believed to be the un-

derlying cause for major evolutionary phenomena that include (1) evolution of sex, its maintenance and its absence, (2) evolution of bacterial genomes, (3) emergence of linkage groups (chromosomes), (4) sexual selection and mate choice, (5) meiosis and crossing-over, (6) multicellularity, (7) diploidy, (8) senscence, (9) sexual and somatic incompatibility, (10) eusociality, (11) speciation and biodiversity, (12) genome size, and (13) organelle behavior (Bell, 1993; Hurst et al., 1996; Partridge and Hurst, 1998). The persistence of one-cell stages in multicellular life histories may have a role in reducing conflicts of interest among genetically different replicators within an organism (Grosberg and Strathmann, 1998). Almost 10% of present-day angiosperms show male sterility due to one or several mitochondrial genes. In *Thymus vulgaris*, all individuals contain male sterility mitochondrial genes and depend on nuclear genes to counteract their effects by restoring pollen production (Belhassen et al., 1991).

Genes, like organisms, may cooperate because they benefit from such gene-to-gene cooperation. Pathogenic virulence may be viewed as the opposite of genetic cooperation. As the transmission of pathogenic symbionts increases, they become less virulent to their hosts. In general, a pathogen that has co-transmitted with its host is selected to minimize harm to its host. Enhanced host fitness favors the symbiont's well being (Frank, 1996).

Self-promoting genetic elements, also called ultra-selfish genes, were first recognized in 1945 as parasitic chromosomes (Ostergren, 1945). Various self-promoting genetic elements are vertically transmitted, and they manipulate their host so as to increase their spread at the cost of other genes within the genome. In *Chlamydomonas reinhardtii*, mitochondria are inherited from the minus parental strain, while chloroplasts are inherited from the plus types. Likewise, in some gymnosperms, mitochondria are inherited from one parent and chloroplasts from the other. Such uniparental inheritance of mitochondria or chloroplasts may be viewed as an adaptive response to counter self-promoting cytoplasmic genes.

Mitochondria and Programmed Cell Death

Recent studies suggest that the endosymbiotic origin of mitochondria and aerobic metabolism in eukaryotes formed the basis of *apoptosis*, programmed cell death, which is common throughout metazoans (Green and Reed, 1998). Apoptosis plays a crucial role in animal development as well as in the normal functioning of the immune system. Malfunctions of apoptosis result in cancers, autoimmune diseases, and the loss of brain cells in Alzheimer's disease. In the origin of eukaryotic life, the ancestors of present-day purple bacteria that were to become mitochondria formed symbioses that benefited both the symbionts and host cell and allowed them to exploit energy resources in the emerging oxygen atmosphere. The symbiosis also carried the seeds of potential genetic conflict between the two genomes. The new mitochondrial symbionts controlled the life and death of the host cell by providing critical antioxidants. The conditions that favored early mitochondria could also have threatened the host cell's life by causing the release of mitochondria to a free-living state. The po-

tential genetic conflict and instability continued until some of the mitochondrial genes were laterally transferred to the host nucleus. In this way the symbionts became totally dependent on the host cell (Kobayashi, 1998). Mitochondrial mechanisms involved in apoptosis include (1) disruption of electron transport oxidative phosphorylation and ATP production, (2) release of proteins such as cytochrome c that trigger activation of intracellular cystein protease known as caspase, and (3) changes in cellular reduction-oxidation potential. Once cytochrome c is released, the cell becomes committed to apoptosis. A group of proteins known as Bcl-2 are channel-forming proteins that regulate mitochondrial activities (Green and Reed, 1998) (fig. 14.1).

Genetic strategies for reciprocal coadaptation have been examined for a number of host–symbiont relationships. The strategies include gene-for-gene relationships, epistatic interactions, and polymorphic variation.

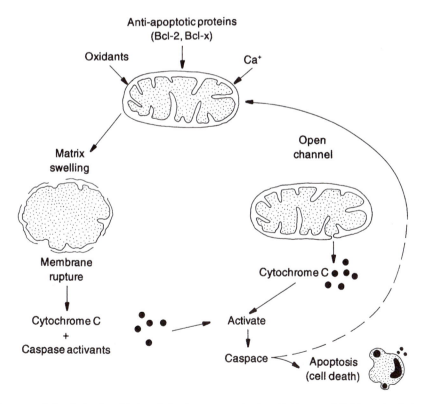

Fig. 14.1   Role of mitochondria in cell death caspase activation. Oxidants and Ca^{2+} overload trigger mitochondria to release caspase-activating proteins such as cytochrome c, which ultimately leads to apoptosis. Redrawn from Green and Reed (1998).

Gene-for-Gene Interactions

The *gene-for-gene hypothesis* was first proposed by Harold Flor and was based on his
studies with flax (*Linum usitatissimum*) and
the rust fungus *Melampsora lini*. Flor found
that two independent genes for resistance in
the host plant were matched with two genes
for virulence in the fungal pathogen (Flor,
1956). The resistance gene is inherited as a
dominant trait, whereas virulence is inherited
as a recessive trait. The host plant and its fungal pathogen have thus evolved a complementary genetic system in which each gene
determining a host response is matched by a
gene or genes for a certain behavior in the
pathogen (Innes, 1995).

Since 1942, gene-for-gene coevolution has
been documented or strongly suggested in
more than three dozen cultivated plants and
the fungi that infect them. These include
wheat–*Puccinia graminis tritici*; wheat–
Puccinia recondita; barley–*Ustilago hordei*,
barley-*Erysiphe graminis hordei*; wheat–
Erysiphe graminis tritici; apple–*Venturia in-
aequalis* (Frank, 1992). Gene-for-gene interactions have also been reported between the
snail *Biomphalaria glabrata* and the blood
fluke *Schistosoma mansoni*; resistance in the
snail results from a single, dominant gene,
which is matched with an avirulent gene of
the fluke. Resistance or susceptibility of the
individual snails is controlled by the relative
frequencies of genes. Gene-for-gene interactions are present in animal–plant symbioses
such as the cyst nematode parasitism of
potato, barley, and soybean and aphid infestation of rushberry and alfalfa. Resistance to
the Hessian fly, *Mayetiola destructor*, in
wheat plants is governed by genes that are
matched by corresponding genes for virulence in the fly (Callow et al., 1988).

Scientists suspect that there may be a clus-
tering of genes in host chromosomes that confer resistance to more than one pathogen. For
example, barley chromosome number 5 has
genes that control resistance to the powdery
mildew fungus *Erysiphe graminis hordei* and
also to the rust fungi *Puccinia hordei* and
Puccinia striiformis.

Gene-for-gene coevolution does not explain all types of host–parasite interactions
(Innes, 1995). Gurmel Sidhu (1984) has described the complex interplay of two or more
symbionts associated with a host plant. For
example, he has shown that the plant nematode *Meloidogyne incognita* alters the resistance of tomato plants to the wilt fungus
Fusarium, which alone cannot parasitize the
plant. Similar interactions between the fungal
pathogens *Fusarium* and *Verticillium* on
tomato plants have been described. Sidhu has
explained these observations in terms of the
concept of epistatic parasitism. Disease expression by the hypostatic parasite depends
on the establishment of the epistatic parasite
(Sidhu,1984).

Molecular Approaches to Plant Pathogenesis

Since 1992, disease resistance genes have
been cloned and characterized from several
plant species. Avirulence (*avr*) genes have
been isolated from several plant pathogenic
bacteria and fungi since 1984. In the presence
of an avr gene a plant pathogen is unable to
induce disease on a host plant and thus determine the host range of the pathogen. Some
avr genes, when transferred to virulent
pathogens, cause the hypersensitive response
in host plants (DeWit, 1995). For example, an
avr gene (*avrB*sT) in the bacterium *Xanthomonas compestris* causes the hypersensitive
response in peppers, and loss of the *avrB*sT
gene in the pathogen causes the disease of
bacterial leaf spot on resistant pepper varieties. *Avr* genes encode for hydrophilic proteins that lack long stretches of amino acids
known as signal sequences, which allow avr
proteins to be secreted to the outside medium.
Avr proteins are responsible for plant disease
symptoms such as angular leaf spots on cotton and citrus canker. Hypersensitivity response and pathogenicity (*Hrp*) genes that encode for the bacterial membrane pore-forming
proteins allow for the passage of the avr proteins to the outside. Resistance gene products
are able to recognize the avr proteins, and
plant defense blocks the pathogen's further
advance. Molecular biology of the host resistance gene *Hm1* was first described from corn
in 1992. Lack of *Hm1* in corn makes it susceptible to the leaf spot fungus *Cochliobolus
carbonum*. Although the fungal pathogen produces a virulence factor, the HC toxin, which
is responsible for the disease symptoms, the
host *Hm1* gene encodes for an enzyme, HC
toxin reductase, which destroys the toxin.
Since 1992, more than a dozen resistance
genes have been characterized from crop
plants. The *PTO* gene of tomato confers resistance against the bacterial speck disease
caused by *Pseudomonas syringae*, which car-

ries the avirulence *pto* gene (see fig. 5.4). The host *PTO* gene is a protein kinase, which is a signal transducer in the hypersensitive response. Other resistance genes isolated include *Cf2*, *Cf4*, and *Cf9* from tomato against the fungus *Cladosporium fulvum*, tobacco *N'* gene for resistance aginst tobacco mosaic virus; and the flax *L6* gene against the rust fungus *Melampsora lini* that carries the *avr6* gene (fig. 14.2).

Genetic Polymorphism

Genetic polymorphism is common among species that participate in symbiotic associations, particularly those involving host–parasite interactions. J.B.S. Haldane (1949) was the first to suggest that parasitic symbionts might be the reason for polymorphism in their hosts. For example, the variety of antibodies produced by a host organism has evolved in response to the many different antigens produced by trypanosome parasites. Multiple genes for resistance and virulence in host–parasite symbioses have resulted in a high degree of genetic polymorphism in the interacting species. Some scientists have suggested that parasitic associations may be the princi-

pal cause of protein polymorphism within a population. Stephen J. Gould and Richard Lewontin (1979) have argued that evolutionary changes increase polymorphism within a population. Such a view is a radical departure from Darwin's theory, which states that genetic changes of an organism are subject to selective pressures of the environment and that natural selection conserves one genetic type best suited to a particular set of environmental conditions and eliminates the others.

In humans, malaria has been responsible for the selection of three or more types of genetic variation with respect to resistance to the disease. Resistance caused by the sickle-cell hemoglobin gene (*HbS*) has been studied extensively. Red blood cells containing sickled hemoglobin collapse and assume a sickle shape under low oxygen tension. The *HbS* gene is inherited as a recessive gene, and individuals who are homozygous for it die prematurely. Individuals heterozygous for that trait survive because they contain a normal hemoglobin gene. In Africa, where malaria is endemic, heterozygous individuals have a survival advantage because *Plasmodium* is unable to develop in the sickle cells. Such an evolutionary strategy on the part of a host is

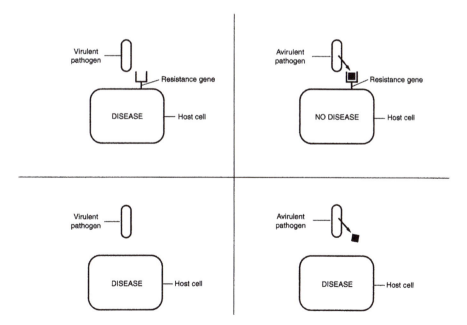

Fig. 14.2   Gene-for-gene interactions are central to plant disease resistance. Resistance is expressed only when the host plant contains a specific resistance gene product that recognizes the corresponding avirulence gene product. All other combinations fail and disease results. Adapted from Staskawicz et al. (1995).

called *heterozygote advantage* (Friedman and Trager, 1981).

Duffy blood groups, MN antigens, and glucose-6-phosphate dehydrogenase (G6PDH) deficiency are other biochemical variations in humans that have evolved in response to *Plasmodium* infection. Individuals deficient in G6PDH, whether heterozygous and homozygous, are resistant to the parasite because *Plasmodium*, an obligate intracellular symbiont of red blood cells, depends on the host to supply G6PDH. The gene for G6PDH is highly polymorphic, with more than 50 alleles coding for hundreds of variant forms of the enzyme. Individuals with low levels of this enzyme are susceptible to anemia and are also sensitive to drugs such as sulfanilamides and the antimalarial drug primaquine. The G6PDH deficiency gene occurs with high frequency among people from Africa and the Mediterranean region. The absence of Duffy antigens from African people may also be related to resistance to *Plasmodium* (Mathews and Armstrong, 1981). Duffy antigens occur on the surface of red blood cells at the sites to which *Plasmodium vivax* binds in order to invade the cell. The presence of MN antigens is also thought to be a significant example of genetic polymorphism in response to *Plasmodium* infection.

14.2 IMMUNOPARASITOLOGY: CONFLICT OR COMPROMISE?

From an evolutionary perspective, host–parasite relationships may begin in conflict but eventually they move toward compromise, and the immune system plays a central role in the complex system of checks and balances (Dawkins, 1990; Mitchell, 1991; Read, 1994). Distinctions between self and nonself exist in all animals and show phylogenetic complexity and adaptive immune responses. Mounting an immune response is metabolically expensive; a host may compromise between the available resources for its growth and development and for defense (Behnke et al., 1992). A balance between the beneficial and potential harmful effects of immune responses to infection has to be considered in terms of a series of evolutionary trade-offs. For the host it involves resistance, pathology, and loss of resources, and for the parasite it involves reproduction, immungenicity, and pathogenicity. Compromise strategies have led to stable equilibria between many parasites and their hosts (Wakelin, 1996). Genetic structure, evolution of microparasites, mechanisms of pathogenesis, and the evolution of immune response have consequences for public health, treatment, and prevention (Levin et al., 1999).

Resistance to an infection may be *nonspecific*, or *innate*, *immunity*—for example, where parasites are prevented from entering the body tissues by barriers such as the skin. Mucus secreted by epithelial cells also is a protective barrier because it prevents infectious agents from attaching to the epithelial cells. Again, secretions such as tears, saliva, nasal fluids, and milk contain antibacterial substances. *Specific acquired immunity* is another type of host resistance that involves the production of antibodies and a complex interplay of T- and B-cells. The acquired immune response results in the recovery of the host from a disease and is followed by the host acquiring a specific memory, with which it responds vigorously to an infection by the same parasite.

Immunological responses of vertebrates involve organs (thymus, spleen, and bone marrow), lymphocyte cells (T- and B-cells), and antibody molecules. T-cell receptors recognize antigens and release cytokines that coordinate the immune response. Major cytokines released by T-cells include interleukins, tumor necrosis factor, and interferons. B-cells recognize antigens and produce large amounts of immunoglobulins (Ig) that circulate throughout the body. Interactions between the host's antibody molecules and the parasite's antigens constitutes one important phase of a host's defenses.

Each Ig molecule is composed of four peptide chains: two identical heavy chains are linked by a disulfide bond to two identical light chains. There are an estimated 1 billion variations of Ig molecules in humans. Each type of Ig molecule is a product of a single clone of B-cells. Immunologists recognize five classes of Ig molecules: IgG, IgA, IgM, IgD, and IgE.

Immunoglobulin G (IgG) is the most abundant antibody in the interstitial tissues and plays a major role in destroying bacterial toxins and enhancing the phagocytosis of bacteria by lymphocytes. Because IgG can cross the placental barrier, it provides the first level of immunological defense against invading organisms during the first few weeks of an infant's life.

Immunoglobulin A (IgA) is generally found

in saliva, tears, nasal fluids, sweat, and secretions produced in the lining of the trachea, lungs, and urogenital and digestive tracts. IgA plays a vital role in defense against skin-inhabiting microorganisms and strongly inhibits the attachment of microbes to the surface of mucosal epithelial cells.

Immunoglobulin M (IgM) antibodies are large molecules that are efficient agglutinating and cytotoxic agents. These antibodies are confined mainly to the bloodstream and are an effective defense against blood parasites.

Immunoglobulin D (IgD) antibodies are unique because of their susceptibility to proteolytic breakdown and their short life span. Most IgD molecules and IgM antibodies are confined to the surface of lymphocytes. Some scientists suspect that IgD and IgM molecules interact to form receptor sites for antigens.

Immunoglobulin E (IgE) molecules remain bound mostly to the surfaces of large tissue cells (mast cells) and occasionally are found in the serum. IgE molecules protect the external mucosal surfaces of the body by causing an inflammatory reaction in response to pathogenic agents. Parasitologists have often noted high levels of IgE following infection by helminthic infections or skin response to feeding of ectoparasites. When IgE molecules beome bound to the receptors on the mast cells, biologically active mediators such as histamine and serotonin are released, which trigger hypersensitivity responses in local tissues. Increased permeability of blood vessels, changes in muscle contractions, and secretion of fluids often accompany the inflammatory hypersensitivity response. This can harm the parasites, but the host tissue may also be damaged in the process.

The adaptive immune response in a vertebrate host consists of the following stages: lymphocytes are first activated and undergo a phase of clonal proliferation, then some become effector cells, while the remainder constitute memory cells. The secondary immune response in a host originates with the activation of the memory cells. B-lymphocytes mature in the bone marrow of mammals and become the antibody-producing cells. This constitutes the humoral immunity. Lymphocytes mature under the influence of the thymus and form the basis of cellular immunity. Cellular immunity, the principal host response against intracellular symbionts, is a complicated process. In brief, cellular immunity involves the following steps. Different surface antigens of a parasite are recognized by the T- and B-cells. Macrophages bind to the antigen molecules and carry them to helper T-cells, which in turn transfer the antigens to B-lymphocytes. The activated B-cells then proliferate and mature to manufacture antibodies. Some activated T-cells kill parasites via phagocytosis from the action of destructive mediators such as amines and cytokines. The antibody–antigen combination activates complement, a complex series of proteins that operate as a cascade system and can bring about lysis on the parasite surfaces.

Interferons are antiviral molecules secreted by virus-infected host cells. Interferons attach to neighboring healthy cells and make them resistant to infection by other viruses. The phenomenon of viral interference or cross-protection has been noted in bacteria, plants, and animals. By continuously changing the structure of their surface antigens, viruses such as the influenza virus render previously manufactured antibodies ineffective. In addition to humoral antibodies, the host uses sensitized cytotoxic T cells to destroy virus-infected host cells. Examples of such responses to viral infection include mumps, measles, herpes, pox, and rabies. Children with T-cell immune deficiency are unable to recover from viral infections.

Parasitologists have often described the phenomenon of *premunition*, which is the immunity of a host to reinfection following recovery from disease. The exact mechanism of premunition is not understood. Humoral antibodies are effective against the forms of parasites that occur in the host's bloodstream (Mitchison, 1990). Some parasites, however, have evolved novel strategies to counter a host's immune system and are able to establish long-term chronic infections. A few well-known examples of such strategies are described below.

Immunity against malarial parasites is primarily a premunition, and it is effective only if a small, residual population of the parasite is present in the host individual. In endemic areas, infants are protected against malaria by maternal antibodies (Roberts and Janovy, 1996)

Trypanosomes circumvent the host's humoral antibodies by changing the structure of their surface antigens with each cycle of reproduction. When the host produces new antibodies, the parasite also changes its antigens.

Some parasitic symbionts such as *Trypanosoma cruzi*, *Leishmania*, and *Toxo-*

plasma escape the host's antibodies by taking up residence in macrophages. Adults of *Schistosoma mansoni* living in the host blood vessels disguise themselves by covering their surfaces with antigens obtained from red blood cells. This mimicry successfully masks the parasite, and the host's humoral antibodies fail to recognize it as foreign.

Sequestration is beneficial to both host and parasite. The invading organism becomes isolated and stands apart from the host. Parasites also avoid the host immune response by living in tissue or cells of immunologically isolated zones. For example, larvae of *Trichinella spiralis* reside in muscle cells, where they are protected from the host's antibodies.

14.3 ROLE OF PHYTOALEXINS AND CROSS-PROTECTION IN PLANT DEFENSES

The biochemical basis of plant resistance to fungal parasitism is not entirely understood (Kombrink and Somssich, 1995). When a fungus infects a plant, it stimulates the plant to produce *phytoalexins* that inhibit growth of the fungus (Hammerschmidt, 1999). Phytoalexins are produced in response either to complex carbohydrates that are present in the walls of the invading fungus or to compounds excreted by the pathogen. The fungal compounds are called *elicitors* because they stim-

ulate the plant cell to synthesize enzymes that bring about the production of phytoalexins. If a fungus lacks specific elicitors, then the infected plant cannot form phytoalexins and therefore has no defense against the pathogen. The relationship between phytoalexins and elicitors is extremely complex (fig. 14.3). Many different compounds and conditions stimulate phytoalexin production, and there is no clear pattern to the findings of many studies on this subject. There is also a continuing controversy over the general effectiveness of phytoalexins in determining plant resistance to pathogens. More than 300 chemicals with phytoalexinlike properties have been isolated from plants belonging to more than 30 families. Phytoalexins produced by plants in each family have similar chemical structures. For example, in most legumes the phytoalexins are isoflavonoids, whereas the members of the potato family produce terpenoids. Some well-studied phytoalexins include phaseolin in beans, pisatin in peas, glyceolin in soybeans, rishitin in potatoes, and gossypol in cotton (Kuc, 1995; Ebel, 1998).

Instead of causing severe disease, some strains of plant viruses confer immunity against more virulent strains of the same virus on a host plant. This phenomenon is called *cross-protection*. The viruses are modified strains of pathogenic forms, and their presence stimulates the host to produce un-

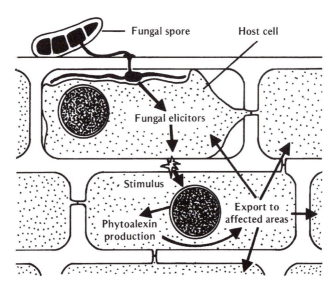

Fig. 14.3   Schematic model for phytoalexin synthesis by host cells near the infected cell. Adapted from Sequeira (1984).

known compounds that inhibit the replication of the pathogenic viruses. The resistance may spread from an infected host cell to other uninfected cells. Cross-protection has economic applications. Tomato and citrus plants can be protected from disease-causing viruses by inoculating them with less virulent strains of the same virus. For example, a common practice used to protect tomatoes against the tobacco mosaic virus (TMV) is to infect seedlings with a mutant form of TMV. The mutant virus slows the growth of the plants somewhat but does not cause the severe symptoms of the normal virus, such as leaf mosaic, growth inhibition, and fruit discoloration. The infected seedlings produce plants that are disease resistant. This method is useful for protecting plants grown in greenhouses, but scientists are hesitant to apply the technique to field crops. Under natural conditions the modified viruses might spread to different plants, causing new diseases, or they might also mutate back to the virulent strains.

14.4 PRODUCTS OF COEVOLUTIONARY SYMBIOSES

Plants are attacked by viruses, bacteria, fungi, nematodes, insects, and grazing animals, and many have developed adaptations to protect themselves from predatory and parasitic organisms. Morphological features such as spines and thorns and harmful chemical substances are examples of such adaptations (Huxley and Cutler, 1991; Bernays, 1992).

The primary chemical substances of plants are carbohydrates, lipids, and proteins. Secondary plant chemicals include phenylpropanes, acetogenins, terpenoids, steroids, and alkaloids. These secondary chemicals protect plants from being eaten by herbivores. Some insects, however, have evolved mechanisms to detoxify poisonous plant metabolites.

Passiflora adenopoda protects itself by trapping plant-eating insects by means of hooked hairs on its outer surface. Many species of *Passiflora* also have structures that resemble the eggs of *Heliconius* butterflies. These structures deter female butterflies from laying their eggs on the plant.

Ants have evolved mutualistic associations with a variety of flowering plants (Keeler, 1989). In plants such as *Acacia* the ants live in hollow areas located at the base of thorns (fig. 14.4). The ants are attracted to the host plant by secretions from glandular nectaries at the base of the leaves. The ants protect the plants from other insects and disperse the plant's seeds, and their excrement pro-

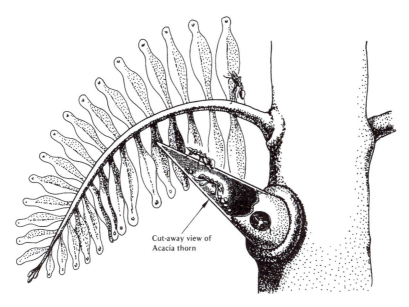

Cut-away view of Acacia thorn

Fig. 14.4   *Pseudomyrmex* ants living inside the hollow cavity of thorns of the acacia plant protect the plant from insect predators. The ants feed on glycogen-rich nodules produced by the plant. Adapted from Perry (1983).

vides nutrient to the host plant. Plants that are fed by ants include species of the genera *Hydrophton* and *Myrmecodia* in the family Rubiaceae from Southeast Asia and northeastern Australia. The plants typically produce swollen organs that the ants live in, and the mutualistic interaction is known as *myrmecotrophy* (Beattie, 1989).

Many evergreen plant species produce steroids that simulate the molting and juvenile hormones of insects. These steroids accelerate larval development and may cause premature death of an insect. Balsam fir (*Abies balsamea*) has evolved an analog of the insect juvenile hormone that arrests larval development and prevents the formation of normal adults. Analogs of insect hormones are the products of fine-tuned coevolutions between plants and insects.

Hypericin is a toxic metabolite secreted from glands of the flowering plant genus *Hypericum*. The chemical, when ingested by animals, produces skin irritation and blindness and leads to starvation; consequently, the plant is generally avoided by most grazing animals. Beetles of the genus *Chrysolina*, however, can detoxify hypericin and thus have gained access to a food supply that is generally unavailable to other animals.

All members of the mustard family, Cruciferae, produce oils that are harmful to animals and are also effective as antibiotics. Insects such as the cabbage butterfly, mustard beetle, and cabbage aphid have evolved mechanisms to detoxify the active ingredients of these oils, but in the process they have become dependent on the substances for their growth and development.

14.5 ECOLOGICAL AND EVOLUTIONARY PERSPECTIVES ON SYMBIOSIS

Ecosystems and biological communities have historical dimensions. A diverse biological community consists of many interacting organisms where the organismic characteristics reflect a unique evolutionary history. Plants, animals, fungi, protozoans, and microbial organisms have lived together in communities that reflect mutually reciprocal coevolutionary adaptations.

The Symbiotic Continuum

During the past decade, the terms symbiosis and mutualism have assumed a much broader significance. Throughout this text we have examined the mechanisms of symbiosis between interacting species. Evolutionary ecologists, however, have a different perspective of symbiosis and have emphasized the outcomes of symbiotic interactions, which they measure in terms of potential fitness. For ecologists, symbiosis includes not only intimate associations but also interspecific interactions in which the symbionts are not physically connected, for example, a diffuse mutualism such as the relationship between flowering plants and their animal pollinators.

Mortimer Starr and others have developed a classification they call the *symbiotic continuum* (Starr, 1975). This classification of symbiotic categories is based on the potential fitness, or reproductive ability, of the symbionts. Competition and mutualism are at opposite ends of the continuum, and neutralism is in the center (fig. 14.5). Competition between interacting species produces detrimental outcomes to both species, whereas mutualistic relationships increase the potential fitness of the symbionts. Thus, symbiosis means that two interacting organisms can influence each other's rates of survival and reproduction. Mutualisms are interorganismic interactions that result in improved survival or reproduction of the partners and may involve aids to dispersal, provision of nutrients, energy, shelter, or fertilization. Other continua in Starr's classification deal with the duration of the symbiosis, from transient to permanent; the physical contact between the symbionts, from incidental to close; and nutrition, from saprobic to biotrophic.

Theories of Mutualism

According to Douglas Boucher (1985), there are four types of models for ecological theo-

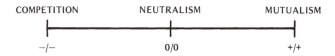

Fig. 14.5   The symbiotic continuum classification is based on the potential fitness of interacting species.

ries of mutualism: (1) individual selection model, (2) population dynamics model, (3) model of shifts of interaction, and (4) the keystone mutualist model. In the individual selection model, which is usually a fitness cost–benefit type, symbiotic associations tend to favor mutualism because the number of competitors benefited is less in a mutualistic association. As more needs of an association are met by the combined contributions of the symbionts, the pressure of competition on these symbionts from similar species will decrease.

Population dynamics models begin with the *Lotka-Volterra competition theory.* More than half a century ago, Alfred Lotka and Vito Volterra used mathematics to understand competition and predator–prey relationships (Lotka, 1932; Volterra, 1926). They devised a model that is now called a *phase-plane model* and is being used as a prototype to study mutualism. In this model, the population densities of two interacting species are plotted along vertical and horizontal axes, thus producing a plane separated by lines that form four regions. The regions show different population densities of the two species and correspond to what will happen to the populations; that is, both populations may increase or decrease, or one will increase and the other will decrease. The phase-plane model has been applied to more than 25 different case histories of mutualism with virtually identical results in the forms of the graphs. The model predicts stable equilibrium, persistence of mutualistic interaction, attainment of higher population densities in symbiosis for both species, and decrease of mutual benefit as the population grows larger.

Models of shifts of interaction are based on the symbiotic continuum model, which was described above, and takes into account fluctuations from one segment of the continuum to another, for example, from competition to mutualism. In the keystone mutualist model, removal of a particular symbiont from an association will produce significant changes in the structure of a community (J.H. Brown and Heske, 1990).

Categories of Mutualism

Boucher (1985) argues that mutualism is a major organizing principle in nature and that researchers should be able to test the following predictions.

1. Mutualists control survival and reproduction; for example, the number of seeds produced by a flower depends on the activity of its mutualist pollinators.
2. Mutualisms that do not appear to have clear values may, in fact, have definite fitness values. For example, seed dispersal by mutualists is more efficient than other means such as wind and results in a higher survival value of the seeds.
3. When mutualistic organisms are either removed from or introduced into a community, the relative abundance of other members of the community is significantly altered.
4. Interspecific competition in nature often produces indirect mutualisms, associations in which species are not in physical contact but positively affect each other's fitness or growth rates; for example, plants with small populations may mimic flowers of other plants, thereby increasing the population size recognized by pollinators.
5. Mutualistic enviroments tend to expand because mutualisms can colonize marginal habitats.
6. Antagonistic interactions stimulate mutualisms to evolve as the antagonists develop defenses against each other.
7. The biomass of natural as well as artificial communities may increase when mutualistic organisms (e.g., mycorrhizal fungi) are added to the community and decreases when they are removed.
8. Sharing of mutualists, as with pollinators, will generally produce indirect mutualisms rather than competition for mutualism. Many new examples of mutualism in nature remain to be discovered.

Four categories of mutualism in terrestrial ecosystems have been recognized: seed dispersal, pollination, resource harvest, and protection.

Seed dispersal mutualism

Seed dispersal mutualisms are diffuse and involve plants that are generally woody and broad-leaved and many different animals

(Herrera, 1995). Seed dispersal occurs in many habitats, from the Arctic to the tropics. When an animal swallows a seed and deposits it at another site, mutualism has taken place. The potential fitness of a seed is increased if it can germinate in a site that is more favorable for growth than where it was produced, that is, away from the parent population. Pine seeds are dispersed by nutcrackers and red squirrels. Nutcrackers feed on pine seeds in late summer and store many seeds in their sublingual pouches. As the birds fly to their cache sites, they bring seeds from their pouch and bury them in the soil. Nutcrackers recover these caches throughout fall, winter, and early next summer and seem to remember where each group of seeds was buried (Lanner, 1996). Plant fruits are bait, and they protect the seeds as well as provide the mutualists with their reward. Ants disperse seeds by carrying them and frequently burying them. The system provides dispersal for the seeds and food for the ants. There are also nonmutualistic means of seed dispersal such as wind, water, and explosive fruits, which forcibly disperse their seeds.

Pollination mutualism

Insects, birds, bats, some rodents, marsupials, and primates make up most of the pollinators. Many of them form diffuse types of associations, but others are obligate mutualists. The advantages for the pollinators include food, such as pollen and nectar, and for the plants the ability to produce more offspring per gram of pollen produced, more viable offspring, and minimized inbreeding (Kearns et al., 1998). For further discussion of pollination, consult chapter 12.

Harvest mutualism

In *harvest mutualisms* one symbiont gathers, or harvests, nutrients that it converts into a form the other symbiont can use. A unique feature of harvest mutualism is that one organism interacts with a geographically restricted population of the other mutualist. Examples of this type of symbiosis include the gut mutualism of herbivores, the fungus gardens of ants, mycorrhizal fungi, and lichens. Alimentary canals of all animals contain complex colonies of microorganisms. Some of these endosymbionts are mutualists or parasites; others are commensalistic. These organisms may form diffuse or obligatory mu-

tualisms. The host pays a low cost for maintenance of the symbionts and receives a high rate of return when the food it ingests is converted into a usable form. Termites and ruminants are examples of digestive mutualism. If the animal had to depend on its ability to digest, it would assimilate less material. Thus, host animals take in large amounts of potential food that cannot be directly assimilated and depend on their mutualistic microorganisms to make available molecules such as vitamins, amino acids, fatty acids, and carbohydrates.

Protection mutualism

One of the best known examples of *protection mutualism* is the association between ants and the thorns of *Acacia* plants (Janzen, 1966). In this obligate mutualism, ants live in colonies in the hollow chambers of thorns and feed on carbohydrates provided by extrafloral nectaries on the plant. The ants protect the host plant from insects and other plants such as vines (Huxley and Cutler, 1991).

Distribution of Mutualism

Although the evidence is incomplete, it appears that mutualistic associations are more common in tropical communities than elsewhere. For example, flowering plants in tropical rainforests and tropical deciduous forests are mostly animal pollinated, whereas the vast northern forests and temperate deciduous forests are mainly wind pollinated. Plant–ant relationships, including those of seed dispersal and protection, are more common in the tropics, as are mutualisms involving algae and cnidarians such as corals, cleaning symbioses, and fungus gardens of insects (Beattie, 1985; Boucher, 1985). The abundance of mutualisms in the tropics may be a reflection of the greater species diversity and productivity in this part of the world. Most tropical soils and waters are poor in nutrients, a condition that favors mutualisms between autotrophic and heterotrophic organisms.

r and K Selection: Life-History Patterns and Symbiosis

Robert H. MacArthur and Edward O. Wilson (1967) proposed a concept to elucidate the connection between the abundance of an organism and evolution. According to these scientists, density-dependent selection causes the evolution of high-*K traits*, whereas *r selection* is characteristic of an expanding population, which is density independent (fig. 14.6). In environments that are constantly changing physically, individuals with high birthrates are often favored by natural selection because their descendants can colonize empty habitats quickly. These organisms usually experience a high degree of mortality among juveniles, and the populations are generally below their saturation levels with little or no competition. The individuals are short lived and are usually small, grow fast, and mature early. Organisms such as insects and annual plants have high r values. K-selected species inhabit relatively stable environments in which population levels are consistently high. In these situations, fitness depends more on features that allow individuals to compete for resources, often at the expense of their ability to produce offspring. Populations are at the saturation level most of the time and are regulated by intense competition within and among the species. The individuals are long lived, grow slowly, and mature late. The most obvious examples of this strategy are found in tropical forests, coral reefs, and specialized environments such as the termite intestine or the rumen of herbivores (MacArthur and Wilson, 1967; Pianka, 1970).

Organisms under r selection are often described as physically controlled, and the K-selected species are biologically accommodated. The r and K selection model offers a broad framework within which to consider the forces of natural selection that regulate symbiotic organisms and their hosts. Gerald Esch and colleagues (1977) have examined the concept of r and K selection within the context of animal parasitism. They have shown that characteristics of morphology,

physiology, and life history of parasitic symbionts can be used to determine population density dynamics.

A biological community shows the range of the r and K continuum, the r end representing organisms that expend most of their energy on reproduction and little on each individual. K selection produces stabilized symbioses, with efficient use of the environment and with the organisms saturating the environment; competition among individuals is intense. Application of the r and K selection concept to the study of symbiosis is relatively new (Andrews and Harris, 1986).

Evolution in a Symbiotic Environment

The evolution of a species in the absence of interspecific interactions is largely determined by forces of natural selection. Ecologists realize, however, that a population of any given species is embedded in the matrix of a community, and interactions with other species will produce genetic changes resulting in an evolutionary feedback. Interspecific dynamics such as those involving predator–prey, host–pathogen, and competitive relationships are antagonistic interactions and therefore cause deterioration of the evolutionary environment.

Populations of mutualists will bring about an evolutionary change in other mutualists, with the following predictable results: (1) genetic change in the mutualists will be slow, (2) there will be tendency toward a lack of specificity between the interacting populations, and (3) there will be a reduction in the importance of sexual reproduction. Intracellular symbioses that involve two or more species have had a profound influence on the evolution of the symbionts as well as on the environments in which they grow. Examples of such symbioses are those of reef-building

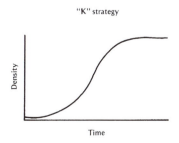
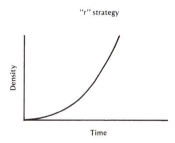

Fig. 14.6   Population growth curves for organisms that use r and k strategies. Adapted from Esch et al. (1977).

corals and foraminifera, which are described in chapter 10. In both cases, symbiosis with unicellular algae has resulted in changes in size and shape of the hosts and an increased number of host species.

One of the unresolved mysteries of evolutionary biology is the role of selection that favors the maintenance of sex in natural populations. Scientists have suggested that antagonistic forces in an environment favor the evolution of sexual reproduction. Conversely, in environments where mutualistic interactions predominate, asexual reproduction is the favored strategy because it results in continued genetic homogeneity of the mutualists, thereby ensuring continuity of the symbiosis.

Traditionally, scientists have considered the environment a product of geological and physico-chemical processes to which living organisms have to adapt or perish. This view has been challenged by James Lovelock and Lynn Margulis (1974), who have proposed the *Gaia hypothesis*. According to this hypothesis, the composition of the atmosphere, sediments, and aquatic environments is controlled by living organisms in their interactions with the environment. Present levels of atmospheric oxygen, carbon dioxide, and nitrogen and temperature ranges are the product of life mechanisms. Thus, in the broadest sense, there is a coevolution of life forms and environmental conditions, and each depends on the other for its existence and maintenance (Joseph, 1990; Levine, 1993; Myers, 1993; Lovelock, 1995; G.R. Williams, 1996; Volk, 1997).

Hypersea, like the Gaia hypothesis, requires an imaginative stretch. According to its authors, *hypersea* represents the body fluids of land eukaryotes that are involved in symbiotic associations, including mutualism and parasitism (McMenamin and McMenamin 1993, 1996). Hypersea is a new type of ocean which first became established in the Paleozoic (Zimmer, 1991).

14.6 SYMBIOSIS IN THE ORIGIN OF AGRICULTURE IN ANTS AND HUMAN SOCIETIES

David Rindos (1984) framed the origin of agriculture within the broad scope of evolutionary theory. He argues that the domestication of plants and animals is not merely an invention or a historical event but rather a mutualistic relationship of varying degrees between

different species that arise under a specific set of conditions (Rindos, 1984). Incidental domestication is the result of human dispersal and protection of wild plants and animals in the general environment. Over time this relationship will select for morphological changes in plants and animals for further domestication. Specialized domestication is mediated by the environmental impact of humans in localized areas. Agroecology is not merely a theater in which people perform their agricultural tasks, but humans are directly interacting with plants on which they feed and with which they establish a symbiotic relationship. Agricultural domestication is the climax of two processes within agroecology where humans, by means of environmental manipulation, increase the rates of evolution of domesticated plants. Weeds are opportunistic plants that parasitize the symbiosis between people and their coevolved domesticates. Most successful weeds are those that mimic cultivated plants in morphology and life cycle events (Gould, 1991).

In a recent study on the evolution of agriculture in ants, which originated some 50 million years ago, there are some striking parallels with human agriculture. Agricultural ants of the tribe Attini (Formicidae), which includes the famous leaf cutters, show extensive coevolutionary symbiosis between the "attine" ant farmers and their fungal cultivars. All agricultural ants grow their fungal crops asexually as clones. Through molecular genetic analysis it has been shown that domestication occurred on a number of occasions and that new cultivars are acquired in an ongoing evolutionary process. In some cases, a single ant species may farm a diversity of fungi, and at the same time a single cultivar is grown by ants of distinct lineages. A given ant nest may contain a single crop, but different ant nests of the same ant species alternate between as many as eight different cultivars. Horizontal transfer, like crop sharing, is practiced by ant species that borrow a crop from another species' nest. Biochemical modifications of fungal cultivars are matched with olfactory and gustatory adaptations of the ants (Mueller et al., 1998).

Although human farmers have domesticated only 14 significant species of large mammalian herbivores, they have successfully domesticated hundreds of crop plants ranging from cereals and legumes to root crops and nut trees. Sharing of domesticates such as maize and potatoes followed the

European expansion into the New World. Evidence shows that human farmers were also responsible for the "domestication" of specialized human pathogens that cause diseases such as tuberculosis, measles, and flu, which evolved from pathogens of cattle and pigs (Diamond, 1997, 1998).

14.7 SYMBIOSIS AND HUMAN AFFAIRS

According to the late Rene Dubos, the renowned microbiologist, all living things are mutually interdependent (Dubos, 1959). Earlier, the Russian philosopher Peter Kropotkin wrote a series of essays, *Mutual Aid: A Factor of Evolution* (1902), which provided an alternative explanation to the "struggle for existence" and "survival of the fittest" concepts. Kropotkin believed that cooperation and mutual help, rather than struggle, were the principal forces of evolutionary change. Cooperation resulted from an instinct of preservation which gave to species and to individuals a better chance of survival in stressful environments. Cooperation and symbiosis are an integral part of our biological heritage. Charles Darwin himself was aware of the role of interorganismic associations as he eloquently explores the "entangled bank" and the "grandeur in this view of life." In *The Origin of Species* (1859), Darwin wrote:

> It is interesting to contemplate an entangled bank, clothed with many plants of many kinds, with birds singing on the bushes, with various insects flitting about, and with worms crawling through the damp earth, and to reflect that these elaborated constructed forms, so different from each other, and dependent on each other in so complex a manner, have all been produced by laws acting around us. . . . Thus from the war of nature, from famine and death, the most exalted object which we are capable of conceiving, namely, the production of the higher animals, directly follows. There is grandeur in this view of life, with its powers, having been originally breathed into a few forms or into one; and that, whilst this planet has gone cycling on according to the fixed laws of gravity, from so simple a beginning endless forms most beautiful and most wonderful have been, and are being evolved. (pp. 489–490)

The germ theory of disease, developed during the height of the social Darwinian revolution, provided supportive evidence for the struggle for survival. With much determination, civilization declared war against microbial infections. Ecological equilibrium between host and microbes was not then a popular view among scientists. As the twenty-first century approaches, humans have become a dominant force on earth, but they still struggle with symbionts that cause disease in people, crops, and domestic animals. Human diseases such as leprosy, the black plague of the Middle Ages, typhus, malaria, and yellow fever have had consequences of historic proportions, as have plant disease epidemics such as the potato blight in Ireland (Nikiforuk, 1991).

Tremendous progress has been made in curbing catastrophic diseases. But mass-scale applications of pesticides have polluted our soils, water, and air, and at the same time the pest organisms with millions of years of evolutionary history have successfully resisted man-made challenges to their survival. We have learned several lessons in the past three decades: (1) a symbiont can rapidly develop tolerance to man-made poisons; (2) instability in the ecological equilibrium of the interacting species is an important factor in an outbreak of disease epidemics; (3) genetic uniformity in crop plants and domestic animals is an invitation for epidemics; and (4) a multifaceted approach to pest management is the best means of pest control.

With the advent of air travel, symbionts can move across geographical barriers with ease. Containing epidemics such as AIDS or Mediterranean fruit fly infestations of citrus fruit challenges the resourcefulness of the scientific community.

In agriculture, concepts of symbiosis have been applied to the biological control of insects, increasing nitrogen fixation by symbiotic cyanobacteria, and applications of mycorrhizal symbionts to forest seedlings in order to promote their growth and thus stop soil erosion. Symbiosis provides the conceptual tool for the management of an ecosystem. Its alternative has been the application of chemicals that eliminate symbioses dependent on natural biological interactions. Ever since it became known that *Rhizobium* fixes atmospheric nitrogen within legume root nodules, the application of this symbiosis has become part of modern agricultural practices. Legume crops such as alfalfa have been used as green manure, in crop rotation, and in mixed cropping. Artificial inoculation of legume seeds with strains of *Rhizobium* is a regular practice in many countries. The *Azolla–Anabaena* symbiosis is most active in flooded rice fields, where populations of the symbiotic fern can fix up to 3 kg of atmospheric nitrogen per hectare per day. In rice paddies of China and Vietnam, *Azolla* has been used for centuries as a source of nitrogen, with the result that Asian farmers can harvest up to 2 tons of rice per hectare without using chemical fertilizers. Modern rice varieties need up to 20 kg of nitrogen per hectare for each ton of rice produced. Scientists are exploring the possibility of improving the use of this symbiosis by manipulating cultural practices.

Black locust, alder, and *Casuarina* are trees with nitrogen-fixing *Frankia* symbionts. These trees have been used successfully in the reforestation of barren land.

Almost all higher plants form mycorrhizal associations. Mycorrhizas are increasingly being recognized as an important factor in the uptake of nutrients, especially phosphorus, by plants. Donald Marx has demonstrated that pine seedlings inoculated with ectomycorrhizal fungi grow faster and survive better than uninoculated seedlings (Marx, 1991). Commercial application of vesicular arbuscular mycorrhizas has yet to become establisbed. These mycorrhizas are also known to increase nitrogen fixation in nodulated legume plants.

The alimentary canals of all animals are unique ecosystems that contain large numbers of interdependent microbial species. Ruminants, with their modified stomachs, have successfully exploited the energy of cellulose and lignin by fermentation symbioses.

Social scientists have found symbiosis to be a useful explanation for a variety of phenomena. For example, Peter Corning has applied the general principles of mutualistic symbiosis to the evolution of human social dominance. According to his synergistic hypothesis, cooperation enhances survival and reproduction of humans (Corning, 1983). Similarly, psychologists are using symbiosis as a conceptual framework to study interactions between parents and their offspring.

The increasing ability of humans to manipulate natural symbiotic associations for their own advantage and create artificial symbioses through genetic engineering will have far-reaching consequences within the next few decades. Corn and soybeans are being used to manufacture human antibodies ("plantibodies") that have shown promise as drugs against human diseases such as cancer (Gibbs, 1997). Disease-resistance genes have been introduced into rice plants (Ronald, 1997). Whether all consequences will be positive is difficult to predict. Parasitic symbionts have a long history of survival and despite our best efforts may still invent ways to circumvent our attempts to eradicate them.

14.8 SUMMARY

The concepts of symbiosis and coevolution have attracted much attention from evolutionary biologists and ecologists who are not satisfied with the classical Darwinian approaches to organic evolution. Pairwise coevolution occurs between two species, whereas diffuse coevolution involves larger numbers of species. Coevolution involves reciprocal genetic changes in the interacting populations and includes coadaptation as well as cospeciation. Genetic conflicts have been responsible for a number of evolutionary phenomena, including the evolution of sex. Ultra-selfish genes are self-promoting genetic elements. Mitochondria are involved in apoptosis.

Gene-for-gene relationships between parasites and hosts have been found in both plant and animal symbioses. Plants may have clusters of genes that confer resistance to more than one pathogen. In some host–parasite interactions the expression of disease by one parasite depends on the presence of another parasite.

Parasitic symbioses are thought to be the principal causes of genetic polymorphism within a population. Resistance to pathogenic symbionts by vertebrate hosts involves immune mechanisms, cross-protection, and sequestration of the parasites. New advances in the molecular biology of immune responses are offering unique opportunities for combating infectious illnesses. Some parasites have evolved unique ways to protect themselves from the host's immune systems.

Plant responses to parasites include hypersensitivity, cross-protection, and the production of phytoalexins. Plants have evolved morphological and chemical adaptations to protect them from parasitic organisms and herbivores. Some insects have evolved mechanisms to detoxify poisonous compounds produced by plants. The immune system helps to regulate the interactions between a host and its symbiont.

By measuring the potential fitness of a species, ecologists are reexamining the role of competition and mutualism in the coevolution of interacting species. The symbiotic continuum is a classification based on the potential fitness of the partners of a symbiotic association. Competition between interacting species produces decreased fitness, whereas mutualisms result in enhanced potential fitness of both partners.

There are four main models of mutualism: individual selection model, population dynamics model, shifts of interaction model, and keystone mutualist model. Some scientists believe that mutualisms are an important organizing principle in nature and have a significant influence on the environment and community. Mutualisms result in the evolution of stable ecosystems with increased parasexual forms of reproduction. Mutualism in terrestrial ecosystems includes seed dispersal, pollination, resource harvest, and protection.

Although the phenomenon of mutualism has been recognized for more than 100 years, it has not been part of the development of modern ecology, despite its prevalence in many biological communities. Only recently have ecologists recognized that the subject of mutualism is worthy of serious study. Such interest is welcome, since ecologists, with their theoretical perspectives, bring a new dimension to a subject that has long fascinated descriptive biologists.

Symbiosis can be viewed in terms of r and K selection. Most parasites demonstrate the r-selection strategy; K-selected species are those that occur in stable or specialized environments such as those of tropical forests or the rumen of herbivores.

According to the Gaia hypothesis, the earth's environment is a product of interactions between life and nonlife molecules, and its fragile equilibrium is ultimately controlled by the dynamics of living organisms. Hypersea is a new type of ocean which consists of the body fluids of eukaryotes involved in symbiotic association. Gaia and Hypersea are ultimate manifestations of symbiotic interactions. Agricultural ants grow fungi for food. The evolution of agriculture in ants such as the leaf-cutters shows striking similarities to human agriculture.

The concept of symbiosis has been applied to human affairs in which cooperation and mutual aid have been the principal forces of evolutionary change. The importance of studying symbiosis is underscored by our increased ability to control pathogenic organisms by understanding the molecular mechanisms of biological interactions. In agriculture, our understanding of symbiotic systems has been of great value in the biological control of insect pests, the increase of nitrogen production by natural means, and the artificial inoculation of forest trees and crop plants with mycorrhizae and nitrogen-fixing symbionts.

APPENDIX 1

Heinrich Anton de Bary (1831–1888)

a phenomenon in which dissimilar organisms live together, or symbiosis

Die Erscheinung der Symbiose
(1879)

In an address to the 1878 meeting of German naturalists, Professor Heinrich Anton de Bary of the University of Strassbourg proposed the term "symbiosis" to designate phenomena such as parasitism and mutualism that were found in the plant kingdom.[1] The need for such clarification arose because earlier that year Professor Pierre-Joseph van Beneden, a renowned authority on tapeworms, had published a book titled *Commensalism and Parasites.* Professor de Bary felt that plant associations had unique properties and therefore required new definitions. He distinguished between two main categories of symbiosis: "antagonistic," which involved a struggle between the symbionts, and "mutualistic," in which the participating organisms benefited from the association. Thus began the birth of a concept which this book adopts—namely, the broad definition of the word symbiosis. During the twentieth century, symbiosis was a popular subject in biology, as well as in society at large. Curiously, for many people symbiosis became synonymous with mutualism. When he coined the term "symbiosis," de Bary was studying lichens, and his original paper on symbiosis used lichens as common examples.

The exact nature of lichens was a highly contentious issue in mid-nineteenth century German botany. Lichens were considered to be the "neglected plants" of botany until the birth of the dual hypothesis by Simon Schwendener, a Swiss botanist. Schwendener, and to some extent de Bary, provided evidence that lichens were associations of fungi and algae.[2] This view was bitterly opposed by leading botanists, including William Nylander, who regarded lichens as autonomous plants like fungi and mosses. The opponents of the dual hypothesis feared that lichens would lose their special identity once they were recognized as fungi that imprisoned algal cells. There was an excitement in the air.

In his 1878 address to the German naturalists, de Bary described lichens as follows:

> The mass of green cells which characterize Lichens have had the most chequered fate in the history of science. Until it was shown some ten years ago, lichens are not truly part of the plant having fungus-like fruiting body, but they are algae which live and grow on or in certain fungi and do not exist outside of this special association. A specific fungus and a specific alga form a specific lichen through their association and without this association, there would be no lichen. (p. 305)

Later de Bary stated that although the algal partners of lichens could live on their own when artificially isolated, the fungal partner died without its algae. He stated:

> The host depends on the tenant in order to live. The tenant is, as a result well treated; not only its growth not impeded but is favored in comparison with those grown independently. The algae remain in harmony with the fungus. The fungus

Fig. A.1   Anton de Bary.

penetrates deep into the hard rock to obtain elements necessary to form the association.

. . . many phenomena similar to parasitism, mutualism etc occur in associations of organisms of different species. Parasitism, mutualism, lichenism are each a special case of a tendency to associate for which the term symbiosis is proposed as a general term. (p. 307)

As a student, de Bary was greatly influenced by Alexander Braun, one of the most celebrated mid-nineteenth century German botanists, and studied algae, parasitic fungi, and water molds. At the age of 24, de Bary succeeded Carl Nageli as Extraordinary Professor at Freiburg and immediately began to attract research students worldwide. "The Professor" de Bary considered teaching research students to be one of his most important responsibilities, and his instructional method included teaching students to be self-reliant, to think critically, and to overcome difficulties and errors. Posing the proper question was important. Between 1855 and 1888 de Bary trained more than 100 scientists. Many of his students became distinguished scientists and used de Baryan experimental approaches to study the morphology, physiology, and life history of organisms. De Bary's most productive years (1872–1888) were at Strassbourg.

The spirit of de Baryan laboratory life was described in an essay, "Strassburg and its botanical laboratory" by William Dudley in 1888.[3] De Bary and his family lived in one wing of the ground floor of The Botanical Institute, which also housed research laboratories, instrument rooms, a library, rooms for two assistants, a lecture room, and a museum. The new campus of Kaiser Wilhelm University, built in 1871, consisted of several modern buildings around a quadrangle that provided lecture halls for philosophy, literature, and mathematics on one side and the chemical, physical, and botanical institutes and the observatory on other sides. The German model of a research university was emulated in the United States by Johns Hopkins University (1876) in Baltimore, Maryland and by Clark University (1887) in Worcester, Massachusetts.

Asa Gray, a Harvard botanist and one of Charles Darwin's American supporters, asked his student, William Gilson Farlow, to go to Germany and study the new experimental approaches to morphology and development of plants under de Bary. In a letter to Farlow in January 1873, Gray outlined what was expected of him when he returned from his studies:

We are going to have the Cryptogamic Flora of the U.S. done up, pari passu with the Phanerog. Sullivant is *engaged* to do the *Mosses*, in a volume—to which I hope to have Austin do the Hepaticae.

I want you to come home prepared to do—

1. The Algae—one vol.
2. The Fungi—
3.—since they are only *Algae & Fungi!* either dwelling together—or the lamb inside of the lion—you will have to do lichenes! Unless we can get them out of Tuckerman, in an intelligible form—which is doubtful. He is somewhere among your friends the *Germans*.[4]

Farlow, on his return from Germany, established a research laboratory at Harvard University and trained a generation of mycologists in the United States.

De Bary died on January 19, 1888 at the age of 57 but his legacy endures, and his brilliant career as an experimental botanist encompassed several firsts in the history of life sciences.[5] At the age of 22 he published his book, *Die Brandpilze*, in which he challenged the idea of spontaneous generation and described the life cycle of the rust fungus from summer to winter spores, from wheat to barberry hosts, and back to wheat again. With regard to the emerging cell theory, de Bary showed that slime molds do not have cell walls. He discovered the zoospores of the potato blight fungus, which he renamed *Phytophthora infestans*. In 1866, the first edition of his classic book *Morphologie und Physiologie der Pilze, Flechten und Myxomyceten* was published. Marshall Ward of Cambridge University, who studied in de Bary's laboratory, wrote the following about this book in 1888:

He gave definiteness to the scattered knowledge (of these organisms) and enabled the scientific world to see clearly the remarkable power of the man. His un-flinching honesty and rigorous self-criticism and modesty had already attracted the attention of all who came into contact with him or his work. Now, however, was seen the marvelous grasp of details, the power of logical generalization which he possessed, and thenceforward the name of de Bary was associated with the leadership of the modern school of biologists, he was himself creating.[6]

De Bary is regarded as a founder of modern mycology and plant pathology. De Bary, along with Wilhem Hofmister, Hugo von Mohl, Nageli, Schwendener, and Julius Sachs were part of a mid-nineteenth century German scientific movement that brought about a paradigm shift to botany as an inductive science.[7] What started out as a way to describe "the lichen situation" in 1878 has come to define our age where life is a complex web of interdependent interactions. Symbioses range widely from those formed by viruses, as evolutionary brewmeisters, to the Gaian view of the unity of life and global interdependence. There is a conceptual congruence of symbiosis, life, science, and religion that de Bary could not have anticipated.

NOTES

1. De Bary's address to the German naturalists was published in 1879 (A. de Bary, "Die Erscheinung der Symbiose." Vortag, Gehalten and versammlung Deutscher Naturforscher and Aerzte zu Cassel. Strassburg: Verlag von Karl J. Thubner). The French translation was published in 1879 in *Revue Internationale des Sciences*, edited by J.L. DeLanessan and Tome Troisieme (Paris: Octave Dion). Professor David Blitz, Department of Philosophy, Central Connecticut State University, in New Britain, kindly provided the English translation.

2. De Bary. (1879). "De la symbiose." *Revue Internationale des Sciences* 3: 301–309.

3. William R. Dudley, "Strassburg and its botanical laboratory." *Botanical Gazette*, 1888, 13(12): 305–311. J.T. Rothrock discussed the laboratory environment under Professor De Bary's guidance in *Botanical Gazette* (1881, 1888).

4. Hilda F. Harris, "The Correspondence of William G. Farlow during his student days at Strasbourg." *Farlowia*, 1945, (2) 1: 9–27.

5. Some significant essays on De Bary include Frederick K. Sparrow. (1978) "Professor Anton De Bary," *Mycologia*, 1978, 70: 222–252; G. Murry, "Heinrich Antone De Bary," *Journal of Botany*, 1988, 4: 65–67; James G. Horsfall and Stephen Wilhelm, "Heinrich Anton De Bary: Nach Einhundertfunfzig Jahren," *Annual Review of Phytopathology*, 1982, 20: 27–32.

6. Marshall Ward, "Anton De Bary," *Nature*, 1888, 37: 297–299.

7. Eugene Cittadino, *Nature as the laboratory. Darwinian Plant Ecology in the German Empire, 1880–1900* (Cambridge: Cambridge University Press, 1990).

APPENDIX 2
HISTORICAL LANDMARKS OF SYMBIOSIS

1500 BC   Papyrus Ebers describes the human intestinal parasite *Ascaris lumbricoides* and the tissue parasite *Dracunculus medinensis*.

AD 1200   Albertus Magnus describes a parasitic symbiont of green plants—a mistletoe.

1674–81   Antonie van Leeuwenhoek describes microscopic protozoans such as *Emeria* and *Giardia*.

1684   Francesco Redi, father of parasitology, was the first scientist to search for parasitic symbionts of fish, birds, mammals, and humans and report the presence of these organisms in intestines, air sacs, and swim bladders. Redi refutes the idea of the spontaneous generation of life in 1668 by demonstrating that maggots develop from the eggs of flies.

1743   Turbevill Needham finds the nematode *Anguina tritici* in cockled wheat seeds.

1755   Matlieu Tillet, a pioneer experimentalist, proves that bunt disease of cereals is contagious.

1796   Edward Jenner develops the first method of inoculation against an infectious disease (vaccination against smallpox).

1807   Isaac-Benedict Prevost, a founding father of plant pathology, experimentally proves that a plant disease (bunt disease of wheat) is caused by a fungus.

1835   Agostino Bassi demonstrates that a fungus, *Beauveria bassiana*, causes muscardine disease of silkworm larvae.

This was the first experimental proof that microbes cause animal disease.

1836–44   J. Schmidberger and T. Hartig coin the term "ambrosia" and recognize its fungal nature.

1840–44   David Gruby, founder of medical mycology, experimentally shows that fungi cause diseases in human and animals.

1842   Johann J. Steenstrup demonstrates the penetration of a snail's body by cercariae and elucidates the life cycle of a liver fluke. His idea of alternation of generations greatly accelerated research on the life histories of flukes, tapeworms, and nematodes.

1853   Heinrich Anton de Bary unravels the complex life cycle of wheat stem rust and lays the foundation for the experimental study of plant diseases. In 1879, de Bary coins the term "symbiosis."

1857–60   Louis Pasteur discovers fermentation as a microbial process and produces artificial immunity to anthrax, chicken pox, cholera, and rabies.

1858   J.G. Kuhn writes one of the early textbooks on the theme that pathogens cause disease (the germ theory of disease).

1858   Rudolf Virchow writes a classic book, *Cellular Pathology,* in which disease processes are viewed as abnormalities at the cellular level.

1860   Claude Bernard formulates the concept of homeostasis. In later decades the internal environment of the host is recognized as a specialized habitat for symbionts.

1867   Simon Schwendener proposes the dual hypothesis to explain the nature of lichens and states that lichens are fungi that live parasitically on algal hosts.

1875   William G. Farlow, after completing studies under de Bary, begins the study of fungi in the United States.

1876   P. van Beneden coins the terms "commensalism" and "mutualism."

1876–84   Robert Koch proves that in anthrax disease a specific symbiont produces a specific disease and develops criteria for establishing microbes as disease agents (Koch's postulates).

1878   Thomas J. Burrill was the first scientist to show that bacteria cause plant disease by demonstrating the cause of fire blight of pears.

1878   Patrick Manson observes the development of *Wuchereria bancrofti* in the body of female mosquitoes and demonstrates the importance of the intermediate host in the life history of a pathogenic symbiont.

1878   M. Woronin describes the nature of legume nodules.

1880   Alphonse Laveran describes the malarial organism *Plasmodium malariae* in the red blood cells of humans; he wins a Nobel Prize in 1907.

1881–82   Rudolph Leuckart and Algernon Thomas describe the life cycle of a digenetic trematode *Fasciola hepatica*.

1883   A.E.W. Schimper suggests that chloroplasts may have evolved from blue-green bacteria.

1884   Elie Mechnikov discovers the phenomenon of phagocytosis and pioneers studies on cellular immunity; he wins a Nobel Prize in 1908.

1885   A.B. Frank coins the term "mycorrhiza" to describe the fungus growth around the roots of woody plants.

1888   Hermann Hellriegel, H. Wilfarth, and Martinus W. Beijerinck discover the role of root nodule bacteria in the legume symbiosis and their significance in the nitrogen cycle.

1888   Roland Thaxter begins his lifelong studies on the Laboulbeniales.

1890   S. Kitasato and Emil von Behring discover antibodies for diphtheria and tetanus antitoxins.

1891   Theobald Smith successfully transmits the cattle fever organism by using ticks as vectors.

1891–1903   Dimitri Iwanovski describes the first plant virus disease, tobacco mosaic virus, and demonstrates its filterability. He made microscopic observations and described crystal line inclusions of the virus in diseased tobacco leaves.

1894   Richard Altmann reports on the discovery of mitochondria.

1895   David Bruce discovers the role of the tsetse fly in the transmission of African sleeping sickness.

1897   Friedrich Loeffler and Paul Frosch discover the first animal virus, the causal agent of foot-and-mouth disease.

1897   Ronald Ross discovers the sexual stages of the malarial organism in mosquitoes and was awarded a Nobel Prize in 1902.

1898   Martinus W. Beijerinck experimentally isolates the tobacco mosaic virus and describes it as *contagium vivum fluidum* because of its exclusively intracellular mode of reproduction.

1905   K.C. Mereschovsky proposes the bacterial origin of mitochondria.

1915   Frederick W. Twort discovers bacteriophages.

1921   Paul Buchner, a pioneer of research on mutualism, writes a classic textbook, *Endosymbiosis of Animals with Plant Microorganisms*.

1921   Elias Melin synthesizes a mycorrhizal association and describes its physiological relationships.

1923–25   Lemuel R. Cleveland explores the relationship of symbiotic intestinal flagellates with their termite and cockroach hosts.

1924   W. Goetsch reports on the mutualistic association between hydra and unicellular algae.

1927   Ivan E. Wallin presents evidence for the bacterial nature of mitochondria and the role such bacteria may have played in the origin of species.

1928   Frederick Griffith discovers the transformation of pneumococci from avirulent to virulent strains following exposure to extracts of killed virulent bacteria. The nature of the "transformation factor" became a subject of intense research activity.

1929   Alexander Fleming reports on the discovery of the antibiotic effect of penicillin.

1933   Charles Drechsler, a pioneer in the study of nematode-trapping fungi in the United States, begins a long and productive career.

1935–37   Wendell M. Stanley isolates and crystallizes the tobacco mosaic virus, which earned him a Nobel Prize in 1946.

1938   T.M. Sonneborn reports on the "killer factor" in *Paramecium*.

1944   Oswald T. Avery, Colin M. MacLeod, and Maclyn McCarty identify Griffith's transforming factor as being nucleic acid with little or no protein. Nucleic acid is later recognized as the hereditary material that makes up genes.

1949   J.B.S. Haldane suggests that pathogens and disease can be strong selective agents maintaining genetic diversity within host populations.

1950   Andrew Lwoff and A. Gutman demonstrate that a lysogenic strain of *Bacillus megatherium* carries a noninfectious form of virus (prophage) and that ultraviolet light induces the prophage to produce active viral particles.

1952   Joshua Lederberg coins the term "plasmid" to describe extranuclear genetic structures that can reproduce independently.

1952   Norton D. Zinder and Joshua Lederberg discover transduction, the transfer of genes by viruses.

1954   A.K. Sarkisov publishes the first major textbook on mycotoxicoses and ergotism.

1956   Harold Flor proposes the gene-for-gene hypothesis to explain the nature of the host–parasite relationship of the flax rust *Melampsora lini*.

1955–56   Heinz Fraenkel-Conrat successfully reconstitutes an active tobacco mosaic virus from protein and nucleic acid components derived from separate sources and demonstrates that RNA is the infective component of the virus.

1957   A. Isaacs and D. Lindemann discover interferon.

1962   Heinz Stolp discovers *Bdellovibrio*, a parasitic symbiont of bacteria.

1962   T.O. Diener discovers the viroid, a new type of infectious agent that consists only of RNA.

1965   Paul R. Ehrlich and Peter H. Raven present the concept of coevolution in an article titled "Butterflies and plants: A study of coevolution."

1964   William Hamilton explores the evolutionary connection between sex, genetic polymorphism, and symbionts and develops computer models such as the "prisoner's dilemma" and a theory of mutualism and cooperation.

1966   Daniel Jensen describes the coevolution between ants and acacias and successfully brings the coevolutionary process and symbiosis into the mainstream of biological thought.

1970   Lynn Margulis uses information from cellular and molecular biology to promote the serial endosymbiotic theory for the origin of the eukaryotic cell.

1960–70   Life-history studies by R.R. Askew and the work of P.W. Price draw attention to the study of parasitoid symbioses.

1970–present   Stanley Falkow and his co-workers bring about a merger of molecular biology and microbial pathogenesis to provide new evolutionary insights into the host–pathogen symbiosis.

1972   Andrew Wyllie and his colleagues open a new discipline by discussing the role of apoptosis (programmed cell death) with wide-ranging implications.

1973   Van Valen's "Red Queen hypothesis" as an influential evolutionary theory opens a new perspective on the escalating arms race between parasitic and pathogenic symbionts and their hosts.

1973   Herbert Boyer, Robert Helling, Annie Chang, and Stanley Cohen insert DNA into a plasmid and propagate it in the host *Escherichia coli* and recombinant DNA technology is born.

1974   R.D. Alexander describes the origin and evolution of cooperation as a cost–benefit dynamics between the biological interactions.

1975   Lynn Margulis and James Lovelock propose the Gaia hypothesis, according to which the planet Earth is a complex self-regulatory, resilient living system.

1977   A.M. Maxam and W. Gilbert discover a new method for sequencing DNA.

1981   Vernon Ahmadjian and his co-workers report separation and artificial synthesis of lichens and introduce the concept of "controlled parasitism" to describe the nature of lichen symbioses.

1982   W.D. Hamilton and M. Zuk suggest that defense against parasites is involved in the evolution of bird coloration. Accordingly, the birds should choose mates with bright ornamental plumage and complex songs because these characteristics signal resistance to parasites.

1982   Human insulin is the first human gene product manufactured using recombinant DNA and is licensed for therapeutic use.

1985   First genetically modified crop plants such as corn, cotton, and tobacco are produced by using Ti plasmid of *Agrobacterium tumefaciens* for glyphosate-resistance to a herbicide.

1985−95   Birth of the International Symbiosis Society, journals

(*Endocytobiosis and Cell Research*, *Symbiosis*) and international conferences to promote research on and teaching of symbiosis.

1992   H. Mason and C. Arntzen produce hepatitis B vaccine in tobacco as a host plant by using *Agrobacterium*-based vectors. Similarly, alfalfa plants are used as hosts for the vaccine against cholera toxin, opening a new age of molecular symbiosis that spans across all kingdoms of life.

1992   Scott O'Neill and his colleagues open a new field of reproductive symbionts of arthropods such as *Wolbachia* that distort the sex ratio of the host by feminization or male killing.

1995   R.D. Fleischmann and his colleagues sequence the first complete genome sequence of *Haemophilus influenzae*. Since then, complete sequences of several other prokaryotes and the yeast *Saccharomyces cerevisiae*, the first eukaryotic organism, have become known.

1997   Stanley Prusiner is awarded a Nobel Prize for his research on the nature of prions as infectious proteins.

REFERENCES

Abelson, P.H. (1985) "Plant-fungus symbiosis." *Science* 229: 617. Editorial.

Abrahamson, W.G., ed. (1989) *Plant-Animal Interactions.* New York: McGraw-Hill.

Ackerman, H.W. and M.S. DuBow. (1987) *Viruses of Prokaryotes.* Boca Raton, FL: CRC Press.

Ackerman, J. (1997) "Parasites: Looking for a free lunch." *National Geographic* 192: 74–91.

Ackerveken, G.V. and U. Bonas. (1997) "Bacterial avirulence proteins as triggers of plant disease resistance." *Trends in Microbiology* 5: 394–398.

Adam, R.D. (1991) "The Biology of *Giardia* spp." *Microbiological Reviews* 55: 706–732.

Adamec, L. (1997) "Mineral nutrition of carnivorous plants: a review." *Botanical Review* 63: 273–299.

Addicott, J.E., J.L. Bronstein, and F. Kjellberg. (1990) "Evolution of mutualistic life cycles: yucca moths and fig wasps." In *Insect Life Cycles,* ed. F. Gilbert, pp.143–161. New York: Springer-Verlag.

Adey, W.H. (1998) "Coral reefs: algal structured and mediated ecosystems in shallow turbulent, alkaline waters." *Journal of Phycology* 34: 393–406.

Agrios, G.N. (1997) *Plant Pathology,* 4th ed. New York: Academic Press.

Aguilar, F.S., P. Hussain, and P. Cerutti. (1993) "Aflatoxin B1 induces the transversion of G-T in codon 249 of the p53 tumor suppressor gene in human hepatocytes." *Proceedings National Academy of Sciences, USA* 90: 8586–8590.

Ahmadjian, V. (1993) *The Lichen Symbiosis.* New York: John Wiley.

Ahmadjian, V. (1995a) "Lichens are more important than you think." *BioScience* 45: 124.

Ahmadjian, V. (1995b) "Lichens—specialized groups of parasitic fungi." In *Pathogenesis and Host Specificity in Plant Diseases,* eds. K. Kohmoto, U.S. Singh, and R.P. Singh, pp. 277–288. New York: Elsevier.

Aikawa, M. (1988) "Fine structure of malaria parasites in the various stages of development." In *Malaria,* eds. W.H. Wernsdorfer and I. McGregor, pp. 97–129. Edinburgh: Churchill Livingstone.

Alderman, D.J. and J.L. Polglase. (1988) "Pathogens, parasites and commensals." In *Freshwater Crayfish: Biology, Management and Exploitation,* eds. D.M. Holdich and R.S. Lowery. London: Croon Helm. pp. 167–212.

Alexopoulos, C.J., C.W. Mims, and M. Blackwell. (1996) *Introductory Mycology,* 4th ed. New York: John Wiley.

Allee, W.C. (1931) *Animal Aggregation.* Chicago: University of Chicago Press.

Alle, W.C. (1938) *The Social Life of Animals.* New York: Henry Schuman.

Allee, W.C. (1951) *Cooperation Among Animals.* New York: Henry Schuman.

Allen, M.F. (1991) *The Ecology of Mycorrhizae.* New York: Cambridge University Press.

Allen, M.F., ed. (1992) *Mycorrhizal Functioning: An Integrative Plant-Fungal Process.* London: Chapman and Hall.

Amabile-Cuevas, C. and M. Chicurel. (1993) "Horizontal gene transfer." *American Scientist* 81: 332–341.

Amman, R.I., W. Ludwig, and K-H. Schleifer. (1995) "Phylogenetic Identification and *in situ* detection of individual microbial cells without cultivation." *Microbiological Reviews* 59: 143–149.

Anagnostakis, S.L. (1995) "The pathogens and pests of chestnuts." *Advances in Botanical Research* 21: 125–145.

Andersen, A. (1991) "Seed harvesting by ants in Australia." In *Ant-Plant Interactions,* eds. C. Huxley and D. Cutler, pp. 493–503. Oxford: Oxford University Press.

Anderson, D.M. (1994) "Red tides." *Scientific American* 271: 62–68.

Anderson, O.R. (1992) "Radiolarian algal symbioses." In *Algae and Symbioses,* ed. W. Reisser, pp. 93–109. Bristol: Biopress Ltd.

Anderson, R. and R. May. (1978) "Regulation and stability of host parasite population interactions: I. Regulatory processes." *Journal of Animal Ecology* 47: 219–247.

Andersson, G.E.S. and C.G. Kurkland. (1999) "Reductive evolution of resident genomes." *Trends in Microbiology* 6: 263–268.

Andrews, J.H. and R.F. Harris. (1986) "r- and K- selection and microbial ecology." *Advances in Microbial Ecology* 9: 99–147.

Anstett, M.C., M. Hossaert-McKey, and F. Kjellberg. (1997) "Figs and fig pollinators: evolutionary conflicts in a coevolved mutualism." *Trends in Ecology and Evolution* 12: 94–99.

Appleton, J.A. and D.D. McGregor. (1984) "Rapid expulsion of *Trichinella spiralis* in suckling rats." *Science* 226: 70–72.

Applemelk, B.J., R.N. Negrini, A.P. Moran, and E.J. Kuipers. (1997) "Molecular mimicry between *Helicobacter pylori* and the host." *Trends in Microbiology* 5: 70–71.

Arai, H., ed. (1980) *Biology of the Tapeworm Hymenolepis diminuta.* London: Academic Press.

Aravalli, R.N., Q. She, and R.A. Garrett. (1998) "Archaea and the new age of microorganisms." *Trends in Ecology and Evolution* 13: 190–194.

Armaleo, D. (1995) "Factors affecting depside and depsidone biosynthesis in a cultured lichen fungus." *Cryptogamic Botany* 5: 14–21.

Atkinson, H.J. (1995) "Plant-nematode interactions: molecular and genetic basis." In *Pathogenesis and Host Specificity in Plant Diseases,* eds. K. Kohmoto, U.S. Singh, and R.P. Singh, pp. 355–369. New York: Elsevier.

Atlas, R.M. (1997) *Principles of Microbiology,* 2nd ed. Dubuque, IA: Wm. C. Brown.

Atsatt, P. (1973) "Parasitic flowering plants: How did they evolve?" *American Naturalist* 107: 579–586.

Atsatt, P.R. (1988) "Are vascular plants "inside-out lichens?" *Ecology* 69: 17–23.

Atsatt, P. (1991) "Fungi and the origin of land plants." In *Symbiosis as a Source of Evolutionary Innovation: Speciation and Morphogenesis,* eds. L. Margulis and R. Fester, pp. 301–315. Cambridge, MA: MIT Press.

Attenborough, D. (1995) "Living together." In *The Private Life of Plants,* pp. 199–243. Princeton, NJ: Princeton University Press.

Axelrod, R. (1984) *The Evolution of Cooperation.* New York: Basic Books.

Axelrod, R. and W.D. Hamilton. (1981) "The evolution of cooperation." *Science* 211: 1390–1396.

Baertschi, S.W., K.D. Raney, T.S. Himada, T.H. Harris, and F.P. Guengerich. (1989) "Comparison of rates of enzymatic oxidation of aflatoxin B1, aflatoxin G1, and sterigmatocystin and activities of the epoxides in forming guanyl-N7 adducts and inducing different genetic responses." *Chemical Research Toxicology* 2: 114–120.

Bailie, B.K., V. Monje, V. Silvestre, M. Sison, and C. Belda-Baillie. (1998) "Allozyme electrophoresis as a tool for distinguishing different zooxanthellae symbiotic with giant clams." *Proceedings of the Royal Society London B* 265: 1949–1956.

Baker, H.F. and R.M. Ridley, eds. (1996) *Prion Diseases. Methods in Molecular Medicine.* Totowa, NJ: Humana Press.

Baldauf, S.L. and J.D. Palmer. (1993) "Animals and fungi are each other's closest relations: congruent evolution from multiple proteins." *Proceedings National Academy of Sciences. USA* 90: 11558–11562.

Ball, D.M., J.F. Pedersen, and G.D. Lacefield. (1993) "The tall-fescue endophyte." *American Scientist* 81: 370–379.

Balows, A., H.G. Truper, M. Dworkin, Wm. Harder, and K.H. Schleifer, eds. (1992) *The Prokaryotes.* 2nd edition. New York: Springer-Verlag.

Baltimore, D. (1985) "Retroviruses and retrotransposons." *Cell* 40: 481–482.

Banuett, F. (1992) "*Ustilago maydis,* the delightful blight." *Trends in Genetics* 8: 174–179.

Banuett, F. (1995) "Genetics of *Ustilago maydis,* a fungal pathogen that induces tumors in maize." *Annual Review of Genetics* 29: 179–208.

Barbosa, P., V.A. Krischik, and C.G. Jones, eds. *Microbial Mediation of Plant-Herbivore Interactions.* New York: Wiley and Sons.

Barbosa, P. and D.K. Letourneau. (1988) *Novel Aspects of Insect-Plant Interactions.* New York: John Wiley and Sons.

Barker, J.M., R.W. Brown, P.J. Collier, I. Farrell, and P. Gilbert. (1992) "Relationship between *Legionella pneumophila* and *Acanthamoeba polyphaga.*" *Applied Environmental Microbiology* 58: 2420–2425.

Barnard, C. and J. Behnke, eds. (1990) *Parasitism and Host Behaviour.* London: Taylor & Francis.

Baron, C. and P. Zambryski. (1995a) "The plant response in pathogenesis, symbiosis, and wounding: variations on a common theme?" *Annual Review of Genetics* 29: 107–129.

Baron, C. and P.C. Zambryski. (1995b) "Notes from the underground: highlights from plant-microbe interactions." *Trends in Biotechnology* 13: 356–362.

Barron, G.L. (1977) *The Nematode-destroying Fungi.* Guelph, Ontario: Canadian Publications.

Barron, G. (1992) "Jekyll-hyde mushrooms." *Natural History* 46–53.

Barron, G. (1981) "Predators and parasites of microscopic animals." In *Biology of Conidial Fungi,* eds. G.L. Cole and B. Kendrick, pp. 176–200. New York: Academic Press.

Barth, F.G. (1991) *Insects and Flowers: The Biology of a Partnership,* trans. M.A. Biederman-Thorson. Princeton, NJ: Princeton University Press.

Batra, L.R. (1979) *Insect-Fungus Symbiosis: Nutrition, Mutualism, and Commensalism.* Allenheld. Montclair, NJ.

Batra, L.R., and Batra, S.W.T. (1985) "Floral mimicry induced by mummy-berry fungus exploits host's pollinators as vectors." *Science* 228: 1011–1013.

Batra, L.R. (1985) "Ambrosia beetles and their associated fungi: research and techniques." *Proceedings Indian Academy of Sciences* 94: 137–148.

Bauchop, T. (1989) "Biology of gut anaerobic fungi." *BioSystems* 23: 53–64.

Baumann, P., L. Baumann, C. Lai, D. Rouhbakhsh, N.A. Moran, and M.A. Clark. (1995) "Genetics, physiology, and evolutionary relationships of the genus *Buchnera*: intracellular symbionts of aphids." *Annual Review of Microbiology* 49: 55–94.

Baumann, P., N.A. Moran, and L. Baumann. (1997) "The evolution and genetics of aphid endosymbionts." *BioScience* 47: 12–20.

Baumler, A.J. (1997) "The record of horizontal gene transfer in *Salmonella*." *Trends in Microbiology* 5: 318–322.

Beard, C.B., S.L. O'Neill, R.B. Tesh, F.F. Richards, and S. Aksoy. (1993) "Modification of arthropod vector competence via symbiotic bacteria." *Parasitology Today* 9: 179–183.

Beattie, A. (1985) *The Evolutionary Ecology of Ant-Plant Mutualisms.* Cambridge: Cambridge University Press.

Beattie, A. (1989) "Myrmecotrophy: plants fed by ants." *Trends in Ecology and Evolution* 4: 172–176.

Beaver, R.A. (1989) "Insect-fungus relationships in the bark and ambrosia beetles." In *Insect-Fungus Interactions,* eds. N. Wilding, N.M. Collins, P.M. Hammond, and J.F. Webber, pp. 121–143. London: Academic Press.

Beckage, N.E., ed. (1997a) *Parasites and Pathogens: Effects on Host Hormones and Behavior.* New York: Chapman and Hall.

Beckage, N.E. (1997b) "The parasitic wasp's secret weapon." *Scientific American* 272: 82–87.

Becker, Y. (1996a) " A short introduction to the origin and molecular evolution of viruses." *Virus Genes* 11: 73–77.

Becker, Y. (1996b) "Molecular evolution of viruses: an interim summary." *Virus Genes* 11: 299–302.

Behnke, J.M., C.J. Barnard, and D. Wakelin. (1992) "Understanding chronic nematode infections: evolutionary considerations, current hypotheses and the way forward." *International Journal of Parasitology* 22: 861–907.

Beier, J.C. (1998) "Malaria parasite development in mosquitoes." *Annual Review of Entomology* 43: 519–542.

Belfort, M. (1989) "Bacteriophage introns: parasites within parasites?" *Trends in Genetics* 5: 209–216.

Belhassen, E., B. Dommee, A. Altan, P. Gouyon, D. Pomente, M. Assouad, and D. Couvet. (1991) "Complex determination of male sterility in *Thymus vulgaris* L., genetic and molecular analysis." *Theoretical and Applied Genetics* 82: 1237–1243.

Bell, G. (1993) "Pathogen evolution within host individuals as a primary cause of senescence." *Genetica* 91: 21–34.

Bengtson, S., ed. (1994) *Early Life on Earth.* New York: Columbia University Press.

Berg, D.A., P.S. Hoffman, B.J. Appelmelk, and J.G. Kusters. (1997) "The *Helicobacter pylori* genome sequence: genetic factors for long life in the gastric mucosa." *Trends in Microbiology* 4: 468–474.

Bergman, B., C. Johansson, and E. Soderback. (1992a) "*Nostoc-Gunnera* symbiosis." *New Phytologist* 122: 379–400.

Bergman, B., A. Matveyev, and U. Rasmussen. (1996) "Chemical signalling in cyanobacterial-plant symbioses." *Trends in Plant Science* 1: 191–197.

Bergman, B., A.N. Rai, C. Johansson, and E. Soderback. (1992b) "Cyanobacterial-plant symbiosis." *Symbiosis* 14: 61–81.

Bergquist, N.R., and D.G. Colley. (1998) "Schistosomiasis vaccines: research to development." *Parasitology Today* 14:99–104.

Bermudes, D., G. Hinkle, and L. Margulis. (1994) "Do prokaryotes contain microtubules?" *Microbiological Reviews* 58: 387–400.

Bernard, H.U. (1994) "Coevolution of papillomaviruses with human population." *Trends in Microbiology* 2: 140–142.

Bernays, E.A., ed. (1992) *Insect-Plant Interactions.* Boca Raton, FL: CRC Press.

Berry, R.J., T.J. Crawford, and G.M. Hewitt, eds. (1992) *Genes in Ecology.* Oxford: Blackwell.

Bigger, C. (1988) "Historecognition and immunocompetence in selected marine invertebrates." In *Invertebrate Historecognition,* eds. R. Grosberg, D. Hedgecock, and K. Nelson. New York: Plenum Press, pp. 55–65.

Bilbrami, K.S., and K.K. Sinha. (1992) "Aflatoxins: their biological effects and ecological significance." In *Handbook of Applied Mycology,* vol. 5, *Mycotoxins in Ecological Systems,* eds. D. Bhatnagar, E.B. Lillehoj, and D.K. Arora, pp. 59–86. New York: Marcel Dekker.

Blackstone, N.W. (1995) "A unit-of-evolution perspective on the endosymbiont theory of the origin of the mitochondrion." *Evolution* 49: 785–796.

Blackwell, M. (1994) "Minute mycological mysteries: the influence of arthropods on the lives of fungi." *Mycologia* 86: 1–17.

Blackwell, M. and D. Malloch. (1991) "Life history and arthropod dispersal of a coprophilous *Stylopage.*" *Mycologia* 83: 360–366.

Blank, R. and Trench, R. (1985) "Speciation and symbiotic dinoflagellates." *Science* 229: 656–659.

Boller, T. and F. Meins, eds. (1992) *Genes Involved in Plant Defense.* Vienna: Springer-Verlag.

Bonner, J.T. (1996) *Sixty Years of Biology. Essays on Evolution and Development.* Princeton, NJ: Princeton University Press.

Bonnett, I. and T. Silvester. (1981) "Specificity in *Gunnera-Nostoc* endosymbiosis." *New Phytologist* 89: 121–128.

Bonning, B.C. and B.D. Hammock. (1996) "Development of recombinant baculoviruses for insect control." *Annual Review of Entomology* 41: 191–210.

Boothroyd, J.C. (1990) "Molecular biology of trypanosomes." In *Modern Parasite Biology,* ed. D.J. Wyler, pp. 333–347. New York: W.H. Freeman.

Borst, A. (1991) "Molecular genetics of antigenic variation." *Immunoparasitology Today,* 7: A29–A33.

Bottone, E.J., R.M. Madayag, and M.N. Qureshi. (1992) "*Acanthamoeba keratitis*: synergy between amebic and bacterial co-contaminants in contact lens care system as a prelude to infection." *Journal of Clinical Microbiology* 30: 2447–2450.

Boucher, D.H., ed. (1985) *The Biology of Mutualism.* New York: Oxford University Press.

Bourke, A.F.G. and N.R. Franks. (1995) *Social Evolution in Ants.* Princeton, NJ: Princeton University Press.

Bourtzis, K. and S. O'Neill. (1998) "*Wolbachia* infections and arthropod reproduction." *BioScience* 48: 287–293.

Bowen, D., T.A. Rocheleau, M. Blackburn, O. Andreev, E. Golubeva, R. Bhartia, and R.H. ffrench-Constant. (1998) "Insecticidal toxins from the bacterium *Photorhabdus luminescens.*" *Science* 280: 2129–2132.

Bracho, A.M., D. Martinez-Torres, A. Moya, and A. Latorre. (1995) "Discovery and molecular characterization of a plasmid localized in *Buchnera* sp. bacterial symbiont of the aphid *Rhopalosiphum padi.*" *Journal of Molecular Evolution* 41: 67–73.

Bradley, D.W. (1993) "Introduction: the diversity of human hepatitis viruses." *Seminars in Virology* 4: 269–271.

Brandt, P. (1991) *Evolution der Eukaryotischen Zelle.* Stuttgart: G. Thieme Verlag.

Braun-Howland, E.B. and Nierzwicki-Bauer, S.A. (1990) "*Azolla-Anabaena* symbiosis: Biochemistry, physiology, ultrastructure, and molecular biology." In *CRC Handbook of Symbiotic Cyanobacteria,* ed. A.N. Rai. Boca Raton, FL: CRC Press, pp. 65–117.

Breitwisch, R. (1992) "Tickling for ticks." *Natural History* (March): 56–61.

Brennicke, A., M. Klein, S. Binder, V. Knoop, L. Grohmann, O. Malek, J. Marchfelder, J. Marienfeld, and M. Unseld. (1996) "Molecular biology of plant mitochondria." *Naturwissenschaften* 83: 339–346.

Brewin, N.J. (1993) "The *Rhizobium*-legume symbiosis: plant morphogenesis in a nodule." *Cell Biology* 4: 149–156.

Breznak, J. (2000) Termite gut microbiology. *Annual Review of Microbiology.* In press.

Broda, P.M.A., S.G. Oliver, and P.F.G. Sims, eds. (1993) *The Eukaryotic Genome-Organization and Regulation,* vol. 50. Cambridge: Cambridge University Press.

Bronstein, J.L. (1994a) "Our current understanding of mutualism." *The Quarterly Review of Biology* 69: 31–51.

Bronstein, J.L. (1994b) "Conditional outcomes in mutualistic interactions." *Trends in Ecology and Evolution* 9: 214–217.

Brosius, J. and H. Tiedge. (1996) "Reverse transcriptase." *Virus Genes* 11: 163–179.

Brown, B.E. and J.C. Ogden. (1993) "Coral bleaching." *Scientific American* 268: (January): 64–70.

Brown, J. (1983) "Cooperation—a biologist's dilemma." In *Advances in The Study of Behaviour,* ed. J. Rosenblatt, pp. 1–37. New York: Academic Press.

Brown, J.H. and E.J. Heske. (1990) "Control of a desert-grassland transition by a keystone rodent guild." *Science* 250: 1705–1707.

Brown, J.R. and W.F. Doolittle. (1995) "Root of the universal tree of life based on ancient aminoacyl-tRNA synthetase gene duplications." *Proceedings National Academy of Sciences, USA* 92: 2441–2445.

Brown, J.R. and W.F. Doolittle. (1997) "Archaea and the prokaryote-to-eukaryote transition." *Microbiology and Molecular Biology Reviews* 61: 456–502.

Brown, V.K. and J.H. Lawton. (1991) "Herbivory and the evolution of leaf size and shape." *Philosophical Transactions of Royal Society, London B* 333: 265–272.

Brul, S. and C.K. Stumm. (1994) "Symbionts and organelles in anaerobic protozoa and fungi." *Trends in Ecology and Evolution* 9: 319–324.

Brundrett, M. (1991) "Mycorrhizas in natural ecosystems." *Advances in Ecological Research* 21: 171–313.

Brune, A. (1998) "Termite guts: the world's smallest bioreactors." *Trends in Biotechnology* 16: 16–21.

Bryant, D.A. (1992) "Puzzles of chloroplast ancestry." *Current Biology* 2: 240–242.

Bryant, D.A., ed. (1995) *The Molecular Biology of Cyanobacteria.* Dordrecht: Kluwer.

Buchmann, S.L. and G.P. Nabhan. (1996) *The Forgotten Pollinators.* Washington, DC: Island Press.

Buchner, P. (1965) *Endosymbiosis of Animals with Plant Microorganisms.* New York: Interscience.

Buddemeier, R.W. and D.G. Fautin. (1993) "Coral bleaching as an adaptive mechanism." *BioScience* 43: 320–326.

Bui, E.T.N., P.J. Bradley, and P.J. Johnson. (1996) "A common evolutionary origin for mitochondria and hydrogenosomes." *Proceedings National Academy of Sciences, USA* 93: 9651–9656.

Bull, J., I. Molineux, and W.R. Rice. (1991) "Selection of benevolence in a host-parasite system." *Evolution* 45: 875–882.

Buller, R.M. and G. Palumbo. (1991) "Poxvirus pathogenesis." *Microbiological Reviews* 55: 80–122.

Burkholder, J.M. (1999) "The lurking perils of *Pfiesteria.*" *Scientific American* 281: 42–49.

Burleigh, B.A. and N.W. Andrews. (1995) "The mechanism of *Trypanosoma cruzi* invasion of mammalian cells." *Annual Review of Microbiology* 49: 175–200.

Caetano-Anolles, G., P.A. Goshi, and P.M. Gresshoff. (1993) "Nodule morphogenesis in the absence of *Rhizobium.*" In *New Horizons in Nitrogen Fixation,* eds. R. Palacio, J. Mora, and W.E. Newton, pp. 297–302. Dordrecht: Kluwer Academic Publishers.

Caiola, M.G. (1992) "Cyanobacteria in symbioses with bryophytes and tracheophytes." In *Algae and Symbioses,* ed. W. Reisser, pp. 231–254. Bristol: BioPress Ltd.

Callow, J.A., T. Ray, T.M. Estrada-Garcia, and J.R. Green. (1988) "Molecular signals in plant cell recognition." In *Cell to Cell Signals,* eds. S. Scannerini, D. Smith, P. Bonfante-Fasola, and V. Gianinazzi-Pearson, pp. 167–182. Berlin: Springer-Verlag.

Cann, A.J. (1997) *Principles of Molecular Virology,* 2nd ed. London: Academic Press.

Cano, R.J. and M.K. Borucki. (1995) "Revival and identification of bacterial spores in 25- to 40-million year old Dominican amber." *Science* 268: 1060–1064.

Capron, M. and A. Capron. (1994) "Immunoglobulin E and effector cells in schistosomiasis." *Science* 264: 1876–1877.

Carrington, M., B. Allsopp, H. Baylis, N. Malu, Y. Shochat, and S. Sohal. (1995) "Lymphoproliferation caused by *Theileria parava* and *Theileria anulata.*" In *Molecular Approaches to Parasitology,* eds. J.C. Boothroyd and R. Komuniecki, pp. 43–56. New York: Wiley-Liss.

Carroll, G. (1988) "Fungal endophytes in stems and leaves from latent pathogen to mutualistic symbiont." *Ecology* 69: 2–9.

Cavalier-Smith, T. (1987a) "Eukaryotes with no mitochondria." *Nature* 326: 332–333.

Cavalier-Smith, T. (1987b) "The simultaneous symbiotic origin of mitochondria, chloroplasts and microbodies." In *Endocytobiology III,* eds. J.L. Lee and J.F. Frederick, pp. 55–71.

Cavalier-Smith, T. (1991) "The evolution of cells." In *Evolution of Life: Fossils, Molecules, and Culture,* eds. S. Osawa and T. Honjo, pp. 271–304. Tokyo: Springer-Verlag.

Cavalier-Smith, T. (1992a) "Origin of the cytoskeleton." In *The Origin and Evolution of the Cell,* eds. H. Hartman and K. Matsuno, pp. 79–83. Singapore: World Scientific.

Cavalier-Smith, T. (1992b) "The number of symbiotic origins of organelles." *BioSystems* 28: 91–106.

Cavalier-Smith, T. (1993) "Kingdom Protozoa and its 18 phyla." *Microbiological Reviews* 57: 953–994.

Cavalier-Smith, T. (1997) "Cell and genome coevolution: facultative anaerobiosis, glycosomes and kinetoplastan RNA editing." *Trends in Genetics* 13: 6–9.

Cavanaugh, C.M. (1994) "Microbial symbiosis: patterns of diversity in the marine environment." *American Zoologist* 34: 79–89.

Cavedon, K. and E. Canale-Parola. (1992) "Physiological interactions between a mesophilic cellulolytic *Clostridium* and a non-cellulolytic bacterium." *FEMS Microbial Ecology* 86: 237–245.

Chang, K.-P. (1995) "Cell biology of *Leishmania.*" In *Modern Parasite Biology,* ed. D.J. Wyler, pp. 79–90. New York: W.H. Freeman.

Chanway, C.P., R. Turkington, and F.B. Holl. (1991) "Ecological implications of specificity between plants and rhizosphere microorganisms." *Advances in Ecological Research* 21: 121–169.

Chao, L. (1992) "Evolution of sex in RNA viruses." *Trends in Ecology and Evolution* 7: 147–151.

Chapela, I.H., S.A. Rehner, T.R. Schultz, and U.G. Mueller. (1994) "Evolutionary history of the symbiosis between fungus-growing ants and their fungi." *Science* 266: 1691–1694.

Chapman, R.L. and D.A. Waters. (1992) "Epi- and endobiotic chloroplasts." In *Algae and Symbioses,* ed. W. Reisser, pp. 619–639. Bristol: Biopress Ltd.

Cherfas, J. (1991) "New hope for vaccine against schistosomiasis." *Science* 251: 630–631.

Cherrett, J.M., R.J. Powell, and D.J. Stradling. (1989) "The mutualism between leaf-cutting ants and their fungus." In *Insect-Fungus Interactions,* eds. N. Wilding, N.M. Collins, P.M. Hammond, and J.F. Webber, pp. 93–120. London: Academic Press.

Childress, J.J. (1995) "Life in sulfidic environments: historical perspective and current research trends." *American Zoologist* 35: 83–90.

Childress, J.J., H. Felbeck, and G.N. Somero. (1987) "Symbiosis in the deep sea." *Scientific American* 255: 114–120.

Childress, J.J. and C.R. Fisher. (1992) "The biology of hydrothermal vent animals: physiology, biochemistry and autotrophic symbiosis." *Oceanographic Marine Biological Annual Review* 30: 337–441.

Chisholm, S.W., R.J. Olsen, E.R. Zettler, R. Goericke, J.B. Waterberg, and N.A. Welschmeyer. (1988) "A novel free living prochlorophyte abundant in the oceanic euphotic zone." *Nature* 334: 340–343.

Choe, J.C. and B.J. Crespi, eds. (1997) *The Evolution of Social Behaviour in Insects and Arachnids.* Cambridge: Cambridge University Press.

Choi, E.Y., G.S. Ahn, and K.W. Jeon. (1991) "Elevated levels of stress proteins associated with bacterial symbiosis in *Amoeba proteus* and soybean root nodule cells." *BioSystems* 25: 205–212.

Chusman, J.H. and J.F. Addicott. (1991) "Conditional interactions in ant-plant-herbivore mutualism." In *Ant-Plant Interactions,* eds. C.R. Huxley and D.F. Cutler, pp. 92–103. Oxford: Oxford University Press.

Cirillo, J.D. (1999) "Exploring a novel perspective on pathogenic relationships." *Trends in Microbiology* 6: 96–98.

Clark, C.G. and A.J. Roger. (1995) "Direct evidence for secondary loss of mitochondria in *Entamoeba histolytica.*" *Proceedings National Academy of Sciences, USA* 92: 6518–6521.

Clarke, B.C. (1979) "The evolution of genetic diversity." *Proceedings of the Royal Society London B* 205: 453–474.

Clarkson, J.M. and A.K. Charnley. (1996) "New insights into the mechanisms of fungal pathogenesis in insects." *Trends in Microbiology* 4: 197–203.

Clay, K. (1988) "Fungal endophytes of grasses: a defensive mutualism between plants and fungi." *Ecology* 69: 10–16.

Clay, K. (1990) "Fungal endophytes of plants." *Annual Review of Ecology and Systematics.* 21: 275–297.

Clay, K. (1991) "Parasitic castration of plants." *Trends in Ecology and Evolution* 6: 162–166.

Clay, K. and P. Kover. (1996) "Evolution and stasis in plant-pathogen associations." *Ecology* 77: 997–1003.

Cleveland, L.R. and A.V. Grimstone. (1964) The fine structure of flagellate *Mixotricha paradoxa* and its associated microorganisms. *Proceedings of the Royal Society of London B* 159: 668–686.

Coles, G. (1984) "Recent advances in schistosome biochemistry." *Parasitology* 89: 603–637.

Coley, P.D. and J.A. Barone. (1996) "Herbivory

and plant defenses in tropical forests." *Annual Review of Ecology and Systematics* 27: 305–335.

Collins, F.M. (1993) "Tuberculosis: The return of an old enemy." *Critical Reviews in Microbiology* 19: 1–16.

Collins, F.H. and S.M. Paskewitz. (1995) "Malaria: current and future prospects for control." *Annual Review of Entomology* 40: 195–219.

Colwell, R.R. (1997) "Global climate and infectious disease: the cholera paradigm." *Science* 274: 2025–2031.

Connor, R. (1986) "Peudoreciprocity: investing in mutualism." *Animal Behaviour* 34: 1652–1654.

Connor, R. (1995) "The benefits of mutualism: a conceptual framework." *Biological Reviews* 70: 427–457.

Cooke, R. (1977) *The Biology of Symbiotic Fungi.* London: John Wiley.

Corliss, J.O. (1981) "What are the taxonomic and evolutionary relationships of the protozoa to other Protista?" *BioSystems* 14: 445–459.

Corliss, J.O. (1994) "An interim utilitarian (user-friendly) hierarchial classification and characterization of the protists." *Acta Protozoologica* 33: 1–51.

Corning, Peter. (1983) *The synergism hypothesis: a theory of progressive evolution.* New York: McGraw-Hill.

Couch, J.N. (1938) *The Genus Septobasidium.* Chapel Hill, NC: University of North Carolina Press.

Covacci, A., S. Falkow, D.E. Berg, and R. Rappuoli. (1997) "Did the inheritance of a pathogenicity island modify the virulence of *Helicobacter pylori*?" *Trends in Microbiology* 5: 205–208.

Crampton, J.M. (1994) "Molecular studies of insect vectors of malaria." *Advances in Parasitology* 34: 1–31.

Crofton, H. (1966) *Nematodes.* London: Hutchinson.

Crompton, D. (1988) "The prevalence of ascariasis." *Parasitology Today* 4: 162–169.

Crozier, R.H. and P. Pamilo. (1996) *Evolution of Social Insect Colonies. Sex Allocation and Kin Selection.* New York: Oxford University Press.

Cundell, D.A. and E. Tomanen. (1995) "Bacterial adherence, colonization, and invasion of mucosal surfaces." In *Virulence Mechanisms of Bacterial Pathogens,* eds. Roth, J.A., C.A. Bolin, K.A. Brogden, F.C. Minion, and M.J. Wannemuehler, pp. 3–20. St. Paul, MN: American Society for Microbiology Press.

Curio, E. (1988) "Behavior and parasitism." In *Parasitology in Focus. Facts and Trends,* ed. H. Mehlhorn, pp. 149–158. Berlin: Springer-Verlag.

Cushman, J.H. and J. Addicott. (1991a) "Conditional interactions in ant-plant-herbivore mutualism." In *Ant-Plant Interactions,* eds. C. Huxley and D. Cutler, pp. 92–103. Oxford: Oxford University Press.

Cushman, J.H. and A.J. Beattie. (1991b) "Mutualisms: assessing the benefits to hosts and visitors." *Trends in Ecology and Evolution* 6: 193–195.

Damian, R.T. (1964) "Molecular mimicry: antigen sharing by parasite and host and its consequences." *American Naturalist* 98: 129–149.

Damian, R.T. (1987) "Presidential address— The exploitation of host immune responses by parasites." *Journal of Parasitology* 73: 1–13.

Damian, R. (1989) "Molecular mimicry: parasite evasion and host defense." *Current Topics in Microbiology and Immunology* 145: 101–115.

Dangl, J.L., R.A. Dietrich, and M.H. Richberg. (1996) "Death don't have no mercy: cell death programs in plant-microbe interactions." *The Plant Cell* 8: 1793–1807.

Darwin, C. (1859) *On the Origin of Species by Means of Natural Selection, or the Preservation of Favoured Races in the Struggle for Life.* London: John Murray.

Darwin, C. (1862) *The Various Contrivances by which Orchids are Fertilized by Insects.* London: John Murray.

Darwin, C. (1871) *The Descent of Man and Selection in Relation to Sex.* London: John Murray.

Darwin, C. (1872) *The Expression of the Emotion in Man and Animals.* London: John Murray.

Darwin, C. (1875) *Insectivorous Plants.* London: John Murray.

Darwin, C. (1876) *The Effects of Cross and Self-Fertilization in the Vegetable Kingdom.* London: John Murray.

Darwin, C. (1877) *The Different Forms of Flowers on Plants of the Same Species.* London: John Murray.

Darwin, C. (1881) *The Formation of Vegetable Mould through the action of Worms, with observations on their habits.* London: Murray.

Davidson, D.W. and D. McKey. (1993) "Ant-plant symbioses: stalking the Chuyachaqui." *Trends in Ecology and Evolution* 8: 326–332.

Davies, N.B. and M. Brooke. (1991) "Coevolution of the cuckoo and its hosts." *Scientific American* 264: 92–98.

Dawkins, R. (1990) "Parasites, desiderata lists and the paradox of the organism." *Parasitology* 100: S63–S73.

Dawkins, R. and J.R. Krebs. (1979) "Arms races within and between species." *Proceedings Royal Society London B* 205: 489–511.

Deacon, J.W. (1980) *Introduction to Modern Mycology.* New York: Halsted Press.

De Bary, A. (1879) *Die Erscheinung der Symbiose.* Strassburg: Verlag Von Karl J. Trubner.

DeDuve, C. (1995) *Vital Dust. Life as a Cosmic Imperative.* New York: Basic Books.

Deitsch, K.W., E.R. Moxon, and T.E. Wellems. (1997) "Shared themes of antigenic variation and virulence in bacterial, protozoal, and fungal infections." *Microbiology and Molecular Biology Reviews* 61: 281–293.

Denarie, J. and J. Cullimore. (1993) "Lipo-oligosaccharide nodulation factors: a minireview of a new class of signaling molecules mediating recognition and morphogenesis." *Cell* 74: 951–954.

DePriest, P.T. (1993) "Molecular innovations in lichen systematics: the use of ribosomal and intron nucleotide sequences in the *Cladonia chlorophaea* complex." *Bryologist* 96: 314–325.

DePriest, P.T. (1995) "Phylogenetic analyses of the variable ribosomal DNA of the *Cladonia chlorophaea* complex." *Cryptogamic Botany* 5: 60–70.

Derbyshire, V., M.M. Parker, and M. Belfort. (1995) "Homing sweet homing: mobile introns in bacterial viruses." *Seminars in Virology* 6: 65–73.

Deretic, V.M., M.J. Schurr, J.C. Boucher, and D.W. Martin. (1994) "Conversion of *Pseudomonas aeruginosa* to mucoidy in cystic fibrosis: environmental stress and regulation of bacterial virulence by alternative sigma factors." *Journal of Bacteriology* 176: 2773–2780.

De Souza, W. (1984) "Cell biology of *Trypanosoma cruzi.*" *International Review of Cytology* 86: 197–283.

Despommier, D. (1990) "*Trichinella spiralis*: the worm that would be a virus." *Parasitology Today* 6: 193–196.

Despommier, D. (1993) "*Trichinella spiralis* and the concept of niche." *Journal of Parasitology* 79: 472–482.

DeVay, J.E., C.F. Gonzalez, and R.J. Wakeman. (1978) "Comparison of the biocidal activities of syringomycin and syringotoxin and the characterization of isolates of *Pseudomonas syringae* from citrus hosts." *Proceedings of Fourth International Conference on Plant Pathological Bacteriology* 4: 643.

DeVries, P.J. (1992) "Singing caterpillars, ants and symbiosis." *Scientific American* 267: 76–82.

DeWit, P.J.G.M. (1995) "Fungal avirulence genes and plant resistance genes: unraveling the molecular basis of gene-for-gene interactions." *Advances in Botanical Research* 21: 147–185.

Diamond, J. (1997) *Guns, Germs, and Steel: The Fates of Human Societies.* New York: Norton.

Diamond, J. (1998) "Ants, crops, and history." *Science* 281: 1974–1975.

Diaz, C.L., L.S. Melchers, P.J.J. Hooykaas, B.J.J. Lugtenberg, and J.W. Kijne. (1989) "Root lectin as a determinant of host plant specificity in the *Rhizobium*-legume symbiosis." *Nature* 338: 579–581.

Dickman, C.R. (1992) "Commensal and mutualistic interactions among terrestrial vertebrates." *Trends in Ecology and Evolution* 7: 194–197.

Diener, T.O. (1996) "Origin and evolution of viroids and viroid-like satellite RNAs." *Virus Genes* 11: 119–131.

Dijksterhuis, J., M. Veenhuis, Wm. Harder, and B. Nordbring-Hertz. (1994) "Nematophagus fungi: physiological aspects and structure-function relationships." *Advances in Microbial Physiology* 36: 111–143.

Dimmock, N.J. and S.B. Primrose. (1994) *Introduction to Modern Virology,* 4th ed. Oxford: Blackwell.

Distel, D.L. (1998) "Evolution of chemoautotrophic endosymbioses in bivalves." *BioScience* 48: 277–286.

Distel, D.L. and C.M. Cavanaugh. (1994) "Independent phylogenetic origins of methanotrophic and chemoautotrophic bacterial endosymbiosis in marine bivalves." *Journal of Bacteriology* 176: 1932–1938.

Dobson, A. and P. Hudson. (1986) "Parasites, disease, and the structure of ecological communities," *Trends in Ecology and Evolution* 1: 11–14.

Docamp, R. and S.N. Moreno. (1996) "The role of Ca^{2+} in the process of cell invasion by intracellular parasites." *Parasitology Today* 12: 61–65.

Doeller, J.E. (1995) "Cellular energetics of animals from high sulfide environments." *American Zoologist* 35: 154–165.

Dollet, M. (1984) "Plant diseases caused by flagellated protozoa (*Phytomonas*)." *Annual Review of Phytopathology* 22: 115–132.

Domnas, A.J., J.P. Srebro, and B.F. Hicks.

(1977) "Sterol requirement for zoospore formation in the mosquito-parasitizing fungus *Lagenidium*." *Mycologia* 69: 875–886.

Donnenberg, M.S., J.B. Kaper, and B.B. Finlay. (1997) "Interactions between enteropathogenic *Escherichia coli* and host epithelial cells." *Trends in Microbiology* 5: 109–114.

Doolittle, W.F. (1995) "Of archae and eo: what's in a name?" *Proceedings National Academy of Sciences, USA* 92: 2421–2423.

Doolittle, W.F. (1996) "Some aspects of the biology of cells and their possible evolutionary significance." In *54th Symposium of the Society for General Microbiology*, eds. D. Mcl. Roberts, P. Sharp, G. Alderson, and M.A. Collins, pp. 1–21. Cambridge: Cambridge University Press.

Doolittle, W.F. (1998a) "A paradigm gets shifty." *Nature* 392: 15–16.

Doolittle, W.F. (1998b) "You are what you eat: a gene transfer ratchet could account for bacterial genes in eukaryotic nuclear genomes." *Trends in Genetics* 14: 307–311.

Doolittle, W.F. (1999) "Phylogenetic classification and the universal tree." *Science* 284: 2124–2128.

Doolittle, W.F. and J.R. Brown. (1994) "Tempo, mode, the progenote and the universal root." *Proceedings of the National Academy of Sciences, USA* 91: 6721–6728.

Douglas, A.E. (1988) "Specificity in the *Convoluta roscoffensis/Tetraselmis* symbiosis." In *Cell to Cell Signals*, eds. S. Scannerini, D. Smith, P. Bonfante-Fasola, and V. Gianinazzi-Pearson, pp. 131–142. Berlin: Springer-Verlag.

Douglas, A.E. (1989) "Mycetocyte symbiosis in insects." *Biological Reviews* 64: 409–434.

Douglas, A.E. (1992) "Symbiosis in evolution." In *Oxford Surveys in Environmental Biology*, eds. D. Futuyma and J. Antonovics, pp. 347–382. Oxford: Oxford University Press.

Douglas, A.E. (1994a) *Symbiotic Interactions*. New York: Oxford University Press.

Douglas, A.E. (1994b) "Chloroplast origins and evolution." In *Molecular Biology of Cyanobacteria*, ed. D.A. Bryant, pp. 91–118. Dordrecht: Kluwer.

Douglas, A.E. (1995) "The ecology of symbiotic microorganisms." *Advances in Ecological Research* 26: 69–103.

Douglas, A.E. (1996) "Microorganisms in symbiosis: adaptation and specialization." In *Evolution of Microbial Life*, eds. D. McL. Roberts, P. Sharp, G. Alderson, and M.A.

Collins, pp. 225–241. Cambridge: Cambridge University Press.

Douglas, A.E. (1998) "Nutritional interactions in insect-microbial symbioses: aphids and their symbiotic bacteria *Buchnera*." *Annual Review of Entomology* 43: 17–37.

Douglas, A.E. and W.A. Prosser. (1992) "Synthesis of the essential amino acid tryptophan in the pea aphid (*Acyrthosiphon pisum*) symbiosis." *Journal of Insect Physiology* 38: 566–568.

Douglas, A.E. and D.C. Smith. (1984) "The green hydra symbiosis. VIII. Mechanisms in symbiont regulation." *Proceedings Royal Society London B* 221: 291–319.

Douglas, A.E. and S. Turner. (1991) "Molecular evidence for the origin of plastids from a cyanobacterium-like ancestor." *Journal of Molecular Evolution* 33: 267–273.

Downie, J.A. (1994) "Signaling strategies for nodulation of legumes by rhizobia." *Trends in Microbiology* 2: 318–324.

Doyle, J. J. (1998) " Phylogenetic perspectives on nodulation: evolving views of plants and symbiotic bacteria. " *Trends in Plant Science 3*: 473–478.

Dramsi, S. and P. Cossart. (1998) "Intercellular pathogens and the actin cytoskeleton." *Annual Review of Cell and Developmental Biology* 14: 137–166.

Dreschler, C. (1934) "Organs of capture in some fungi preying on nematodes." *Mycologia* 26: 135–144.

Dubnau, D. (1999) "DNA uptake in bacteria." *Annual Review of Microbiology* 53: 217–244.

Dubos, Rene. (1959) *Mirage of Health. Utopias, Progress, and Biological Change.* New York: Harper and Brothers. Reissued by Rutgers University Press, 1987.

Dubos, Rene. (1976) "Symbiosis between the earth and humankind." *Science* 193: 459–462.

Dubremetz, J.F. (1998) "Host cell invasion by *Toxoplasma gondii*." *Trends in Microbiology* 6: 27–30.

Ducklow, H. and R. Mitchell. (1979) "Bacterial populations and adaptations in the mucus layers of living corals." *Limnology and Oceanography* 24: 715–725.

Duddington, C.L. (1957) *The Friendly Fungi*. London: Faber and Faber.

Dugatkin, L. (1997) *Cooperation among Animals*. New York: Oxford University Press.

Duncan, H.E. and S.C. Edberg. (1995) "Host-microbe interaction in the gastrointestinal tract." *Critical Reviews in Microbiology* 21: 85–100.

Dybvig, K. and L.L. Voelker. (1996) "Molecular biology of mycoplasmas." *Annual Review of Microbiology* 50: 25–57.

Dyer, B.D. (1989) "Symbiosis and organismal boundaries." *American Zoologist* 29: 1085–1093.

Dyer, B.D. and R.A. Obar. (1994) *Tracing the History of Eukaryotic Cells. The enigmatic smile.* New York: Columbia University Press.

Dyer, B.D. and R. Obar, eds. (1985) *The Origin of Eukaryotic Cells.* New York: Van Nostrand Reinhold.

Dyer, B.D. and R.A. Obar. (1994) *Tracing the History of Eukaryotic Cells. The Enigmatic Smile.* New York: Columbia University Press.

Dyer, M. and A. Tait. (1987) "Control of lymphoproliferation by *Theileria annulata.*" *Parasitology Today* 3: 3309–3311.

Ebbole, D.J. (1997) "Hydrophobins and fungal infections of plants and animals." *Trends in Microbiology* 5: 405–408.

Ebel, J. (1998) "Oligoglucoside elicitor-mediated activation of plant defense." *BioEssays* 20: 569–576.

Ebert, D. (1998) "Experimental evolution of parasites." *Science* 282: 1432–1435.

Ebert, D. and E.A. Herre. (1996) "The evolution of parasitic diseases." *Parasitology Today* 12: 96–100.

Ebringer, L. and J. Krajcovic. (1994) *Cell Origin and Evolution.* Bratislava: Veda Publishing House of the Slovak Academy of Sciences.

Edward, R. (1996) "Lessons from cooperative bacteria-animal association: the *Vibrio fischeri-Euprymna scolopes* light organ symbiosis." *Annual Review of Microbiology* 50: 591–624.

Edwards, C. (1990) *Microbiology of Extreme Environments.* New York: McGraw Hill.

Ehrlich, P.R. and P. Raven. (1965) "Butterflies and plants: a study in coevolution." *Evolution* 18: 596–604.

Eichinger, D. (1997) "Encystation of *Entamoeba* parasites." *BioEssays* 19: 633–639.

Elad, Y. (1995) "Mycoparasitism." In *Pathogenesis and Host Specificity in Plant Diseases,* vol. 2, eds. K. Kohmoto, U.S. Singh, and R.P. Singh, pp. 289–307. New York: Elsevier.

Eldredge, N. (1997) *Fossils. The Evolution and Extinction of Species.* Princeton, NJ: Princeton University Press.

Embley, T.M., D.A. Horner, and R.P. Hirt. (1997) "Anaerobic eukaryote evolution: hydrogenosomes as biochemically modified mitochondria?" *Trends in Ecology and Evolution* 12: 437–441.

Emerson, A.E. (1939) "Social coordination and the superorganism." *American Midland Naturalist* 21: 182–209.

Emerson, A.E. (1946) "The biological basis of social cooperation." *Illinois Academy of Sciences Transactions* 39: 9–18.

Emery, Carlo. (1909) "Uber den ursprung der-dulotischen, parasitischen und myrome-kophilen Amenisen." *Biologisches Centralblatt* 29: 352–362.

Erickson, H.P. (1997) "FTsZ, a tubulin homologue in prokaryote cell division." *Trends in Cell Biology* 7: 362–367.

Esch, G.W., T.C. Hazen, and J.M. Aho. (1977) "Parasitism and r- and k-selection." In *Regulation of Parasite Populations,* ed. G.W. Esch, pp. 9–62. New York: Academic Press.

Espinas, A.V. (1878) *Des Societes Animales.* Paris: Bailliere.

Evans, D. and D. Jamison. (1994) "Economics and the argument for parasitic disease control." *Science* 264: 1866–1867.

Evered, D. and S. Clark, eds. (1987) *Filariasis.* Chichester, UK: John Wiley & Sons.

Ewald, P.W. (1983) "Host-parasite relations, vectors, and the evolution of disease severity." *Annual Review of Ecology and Systematics* 14: 465–485.

Ewald, P.W. (1994) *Evolution of Infectious Disease.* New York: Oxford University Press.

Ewald, P.W. (1993) "The evolution of virulence." *Scientific American* 268: 86–93.

Ewald, P.W. (1995) "The evolution of virulence: a unifying link between parasitology and ecology." *Journal of Parasitology* 81: 659–669.

Facer, C.A. and M. Tanner. (1997) "Clinical trials of malarial vaccine: progress and prospects." *Advances in Parasitology* 39: 1–68.

Falkow, S. (1988) "Molecular Koch's postulates applied to microbial pathogenicity." *Review of Infectious Diseases* 10: S274–S276.

Falkow, S. (1990) "The Zen of bacterial pathogenicity." In *Molecular Basis of Bacterial Pathogenesis,* eds. B.H. Iglewski and V.L. Clark, pp. 3–9. San Diego, CA: Academic Press.

Farber, J. and P. Peterkin. (1991) "*Listeria monocytogenes,* a food-borne pathogen." *Microbiological Reviews* 55: 476–511.

Fautin, D.G. (1991) "The anemone fish symbiosis: what is known and what is not." *Symbiosis* 10: 23–46.

Fautin, D.G. and G.R. Allen. (1994) *Anemone Fishes and their Host Sea Anemones.* Melle, Germany: Tetra Werke.

Feagin, J.E. (1994) "The extrachromosomal DNAs of apicomplexican parasites." *Annual Review of Microbiology* 48: 81–104.

Federici, B.A. (1981) "Mosquito control by the fungi *Culicinomyces, Lagenidium* and *Coelomomyces.*" In *Microbial Control of Pests and Plant Diseases 1970–1980*, ed. H.D. Burges, pp. 555–572. New York: Academic Press.

Federici, B.A. and J.V. Maddox. (1996) "Host specificity in microbe-insect interactions." *BioScience* 46: 410–421.

Feener Jr., D.H. and B.V. Brown. (1997) "Diptera as parasitoids." *Annual Review of Entomology* 42: 73–97.

Felbeck, H. and D.L. Distel. (1992) "Prokaryotic symbionts of marine invertebrates." In *The Prokaryotes*, 2nd ed., vol. 4, eds. A. Balows, H.G. Truper, M. Dworkin, W. Harder, and K.H. Schleifer, pp. 3891–3906. New York: Springer-Verlag.

Fenchel, T. and B.J. Finlay. (1994) "The evolution of life without oxygen." *American Scientist* 82: 22–29.

Fenchel, T. and B.J. Finlay. (1995) *Ecology and Evolution in Anoxic Worlds*. New York: Oxford University Press.

Fenner, F. (1983) "Biological control, as exemplified by smallpox eradication and myxomatosis." *Proceedings of the Royal Society of London B* 218: 259–285.

Fernholm, B., K. Bremer, and H. Jornvall, eds. (1989) *The Hierarchy of Life*. New York: Elsevier.

Ferrante, A. (1986) "Discovery and control of primary amoebic meningoencephalitis (PAM)." *Parasitology Today* 2: S10.

Ferry, J.G. (1993) *Methanogenesis: Ecology, Physiology, Biochemistry and Genetics*. New York: Chapman and Hall.

Fiala, B., U. Maschwitz, and T. Pong. (1991) "The association between Macaranga trees and ants in South-east Asia." In *Ant-Plant Interactions*, eds. C. Huxley and D. Cutler, pp. 263–270. Oxford: Oxford University Press, 1991.

Fields, B., D.M. Knipe, and P.M. Howley. (1996) *Fundamental Virology*, 3rd ed. New York: Lippincott-Raven.

Finlay, B.B. and P. Cossart. (1997) "Exploitation of mammalian host cell functions by bacterial pathogens." *Science* 276: 718–725.

Finlay, B. and S. Falkow. (1989) "Common themes in microbial pathogenesis." *Microbiological Reviews* 53: 210–230.

Finlay, B. and S. Falkow. (1997) "Common themes in microbial pathogenicity revisited." *Microbiology and Molecular Biological Reviews* 61: 136–169.

Finlay, B.J. and T. Fenchel. (1993) "Methanogens and other bacteria as symbionts of free-living anaerobic ciliates." *Symbiosis* 14: 375–390.

Fisher, A. (1989) "The wheels within wheels in the superkingdom Eucaryotae." *Mosaic* 20: 2–13.

Fisher, C.R. (1990) "Chemoautotrophic and methanotrophic symbioses in marine invertebrates." *Critical Review Aquatic Sciences* 2: 399–436.

Fisher, R. (1930) *The Genetic Theory of Natural Selection*. Oxford: Clarendon Press.

Fisher, R.F. and S.R. Long. (1992) "*Rhizobium*-plant signal exchange." *Nature* 357: 655–659.

Fitter, A.H. (1992) "Costs and benefits of mycorrhizae: implications for functioning under natural conditions." *Experientia* 47: 350–354.

Fitter, A.H. and D.P. Stribley, eds. (1996) *Plant-Microbe Symbiosis: Molecular Approaches*. New York: Cambridge University Press, 1996.

Flemming, J.G.W. (1992) "Polydnaviruses: mutualists and pathogens." *Annual Review of Entomology* 37: 401–426.

Flint, H.J. (1997) "The rumen microbial ecosystem-some recent developments." *Trends in Microbiology* 5: 483–488.

Flor, H.H. (1956) "The complementary genetic systems in flax and flax rust." *Advances in Genetics* 8: 29–54.

Forst, S., B. Dowds, N. Boemare, and E. Stackebrandt. (1997) "*Xenorhabdus* and *Photorhabdus* spp.: bugs that kill bugs." *Annual Review of Microbiology* 51: 47–72.

Forst, S. and K. Nealson. (1996) "Molecular biology of the symbiotic-pathogenic bacteria *Xenorhabdus* spp. and *Photorhabdus* spp." *Microbiological Reviews*. 60: 21–43.

Forterre, P. (1997) "Archaea: what can we learn from their sequences?" *Current Opinion in Genetics and Development* 7: 764–770.

Fraenkel-Conrat, H. and P.C. Kimball (1982) *Virology*. Englewood Cliffs, NJ: Prentice-Hall.

Fraenkel-Conrat, H. and R.C. Williams. (1955) "Reconstitution of active tobacco mosaic virus from its inactive protein and nucleic acid components." *Proceedings of the National Academy of Sciences USA* 41: 690–698.

Franches, C., L. Laplaze, E. Duhoux, and D. Bogusz. (1998) "Actinorhizal symbioses: Recent advances in plant molecular and genetic transformation studies." *Critical Reviews in Plant Sciences* 17: 1–28.

Frank, S.A. (1992) "Models of plant-pathogen coevolution." *Trends in Genetics* 8: 213–219.

Frank, S.A. (1996) "Models of parasitic virulence." *The Quarterly Review of Biology* 71: 37–78.

Franssen, H.J., I. Vijn, W.C. Yang, and T. Bisseling. (1992) "Developmental aspects of the *Rhizobium*-legume symbiosis." *Plant Molecular Biology* 19: 89–107.

Franz Lang, B., G. Burger, and M.W. Gray. (1999) "Evolution of mitochondrial genomes." *Annual Review of Genetics* 33. In press.

Friedl, T. and C. Rokitta. (1997) "Species relationships in the lichen alga *Trebouxia* (Chlorophyta, Trebouxiophyceae): molecular phylogenetic analyses of nuclear-encoded large subunit rRNA gene sequences." *Symbiosis* 23: 125–148.

Friedman, M.J. and W. Trager. (1981) "The biochemistry of resistance to malaria." *Scientific American* 244: 154–164.

Frisch, K. von. (1950) *Bees, their vision, chemical senses, and language.* Ithaca, NY: Cornell University Press.

Furness, D.N. and R.D. Butler. (1983) "The cytology of sheep rumen ciliates. I. Ultrastructure of *Epidinium caudatum*." *Journal of Protozoology* 30: 676–687.

Futuyma, D.J. (1998) *Evolutionary Biology,* 3rd ed. Sunderland, MA: Sinauer Associates.

Futuyma, D.J. and M. Slatkin, eds. (1983) *Coevolution.* Sunderland, MA: Sinauer Associates.

Fyson, A. and J.I. Sprent. (1980) "A light and scanning microscope study of stem nodules in *Vicia faba* L." *Journal of Experimental Botany* 31: 1101–1106.

Gadagkar, R. (1997) *Survival Strategies: Cooperation and Conflict in Animal Societies.* Cambridge, MA: Harvard University Press.

Galan, J.E. and J.B. Bliska. (1996) "Cross-talk between bacterial pathogens and their host cells." *Annual Review of Cell and Developmental Biology* 12: 221–255.

Galun, M., ed. (1988) *Handbook of Lichenology.* Boca Raton, FL: CRC Press.

Gandon, S. (1998) "Local adaptation and host-parasite interactions." *Trends in Ecology and Evolution* 13: 214–216.

Garbary, D.J. and F.J. London. (1995) "The *Ascophyllum, Polysiphonia, Mycosphaerella* symbiosis. Fungal infection protects *A. nodosum* from desiccation." *Botanica Marina* 38: 529–533.

Gargas, A., P.T. DePriest, M. Grube, and A. Tehler. (1995) "Multiple origins of lichen symbioses in fungi suggested by SSU rDNA phylogeny." *Science* 268 (1995): 1492–1495.

Garnett, G.P. and E.C. Holms. (1996) "The ecology of emergent infectious disease." *BioScience* 46: 127–135.

Gaume, L., D. McKey, and S. Terrin. (1998) "Ant-plant-homopteran mutualism: how the third partner affects the interaction between a plant-specialist ant and its myrmecophyte host." *Proceedings of the Royal Society London B* 265: 569–575.

Ge, Zhangming, and D.E. Taylor. (1999) Contributions of genome sequencing to understand the biology of *Helicobacter pylori. Annual Review of Microbiology* 53: 353–387.

Gehrig, H.H., A. Schussler, and M. Kluge. (1996) "*Geosiphon pyriforme,* a fungus forming endocytobiosis with *Nostoc* (cyanobacteria), is an ancestral member of the Glomales-evidence by SSU rRNA analysis." *Journal of Molecular Evolution* 43: 71–81.

Gelvin, S.B. (2000) "*Agrobacterium* and plant genes involved in T-DNA transfer and integration." *Annual Review of Plant Physiology and Plant Molecular Biology* (in press).

Germot, A., H. Philippe, and H. LeGuyader. (1996) "Presence of a mitochondrial-type 70-KDNA heat shock protein in *Trichomonas vaginalis* suggests a very early mitochondrial endosymbiosis in eukaryotes." *Proceedings National Academy of Sciences, USA* 93: 14614–14617.

Ghabrial, S.A. (1994) "New developments in fungal virology." *Advances in Virus Research* 43: 303–388.

Gianinazzi, S. and H. Schuepp, eds. (1994) *Impact of Arbuscular Mycorrhizas on Sustainable Agriculture and Natural Ecosystems.* Basel: Birkhuser Verlag.

Gibbs, A.J., C.H. Calisher, and F. Garcia-Arenal, eds. (1995) *Molecular Basis of Virus Evolution.* New York: Cambridge University Press.

Gibbs, S.P. (1992) "The evolution of algal chloroplasts." In *Origins of Plastids,* ed. R.A. Lewin, pp. 107–121. New York: Chapman Hall.

Gibbs, W.W. (1997) "Plantibodies." *Scientific American* 278: 44.

Gibson, D. and G. Mani. (1984) "An experimental investigation of the effects of selective predation by birds and parasitoid attack on the butterfly *Danaus chrysippus*." *Proceedings Royal Society London B* 221: 31–51.

Gibson, E.K., D.S. McKay, K. Thomas-Keptra, and C.S. Romanek. (1997) "The case for relic life on Mars." *Scientific American* 276 (1997): 58–65.

Gilbert, F., ed. (1990) *Insect Life Cycles. Genetics, Evolution and Co-ordination.* London: Springer-Verlag.

Gilchrist, D.G. (1998) "Programmed cell death in plant disease: the purpose and promise of cellular suicide." *Annual Review of Phytopathology* 36: 393–414.

Giles, N. (1983) "Behavioural effects of the parasite *Schistocephalus solidus* (Cestoda) on an intermediate host, the three-spined stickleback, *Gasterosteus aculeatus*." *Animal Behavior* 31: 1192–1194.

Gilles, H. and P. Ball, eds. (1991) *Hookworm Infections.* Amsterdam Elsevier.

Gillham, N.W. (1994) *Organelle Genes and Genomes.* New York: Oxford University Press.

Gillin, F.D., D.S. Reiner, and J.M. McCaffery. (1996) "Cell biology of the primitive eukaryote, *Giardia lamblia*." *Annual Review of Microbiology* 50: 679–705.

Gil-Turnes, M.S., M.F. Hay, and W. Fenical. (1989) "Symbiotic marine bacteria chemically defend crustacean embryos from a pathogenic fungus." *Science* 246: 116–118.

Giovannoni, S.J., S. Turner, G.J. Olsen, S. Barns, D.J. Lane, and N.R. Pace. (1988) "Evolutionary relationships among cyanobacteria and green chloroplasts." *Journal of Bacteriology* 170: 3584–3592.

Gitler, C. and D. Mirelman. (1986) "Factors contributing to the pathogenic behavior of *Entamoeba histolytica*." *Annual Review of Microbiology* 40: 237–262.

Glynn, P.W. (1993) "Coral reef bleaching: ecological perspectives." *Coral Reefs* 12: 1–7.

Godfrey, H.C.J. (1994) *Parasitoids: Behavioral and Evolutionary Ecology.* Princeton, NJ: Princeton University Press.

Goff, L. (1983) *Algal Symbiosis.* Cambridge: Cambridge University Press.

Goff, L.J., J. Ashen, and D. Moon. (1997) "The evolution of parasites from their hosts: a case study in the parasitic red algae." *Evolution* 51: 1068–1078.

Goff, L.J., D.A. Moon, P. Nyvall, B. Stache, K. Mangin, and G. Zuccarello. (1996) "The evolution of parasitism in the red algae: molecular comparisons of adelphoparasites and their hosts." *Journal of Phycology* 32: 297–312.

Golding, G.B., and R.S. Gupta. (1995) "Protein-based phylogenies support a chimeric origin for the eukaryotic genome." *Molecular Biological Evolution* 12: 1–6.

Goldsmith, D. (1997) *The Hunt for Life on Mars.* New York: Dutton Publishers.

Good, M.F., D.C. Kaslow, and L.H. Miller. (1998) "Pathways and strategies for developing a malaria blood-stage vaccine." *Annual Review of Immunology* 16: 57–87.

Goodenough, U.W. (1991) "Deception by pathogens." *American Scientist* 79: 344–355.

Gordon, M.S. and E.C. Olson. (1995) *Invasions of the Land. The Transitions of Organisms from Aquatic to Terrestrial Life.* New York: Columbia University Press.

Gottfert, M. (1993) "Regulation and function of rhizobial nodulation genes." *FEMS Microbiological Reviews* 104: 39–64.

Gould, F. (1991) "The evolutionary potential of crop pests." *American Scientist* 79: 496–507.

Gould, S.J. and R.C. Lewontin. (1979) "The spandrels of San Marco and the Panglossian paradigm. A critique of the adaptationist programme." *Proceedings Royal Society London B* 205: 581–598.

Graf, J. and E. Ruby. (1998) "Host-derived amino acids support the proliferation of symbiotic bacteria." *Proceedings National Academy of Sciences, USA* 95: 1818–1822.

Grajal, A. and S.D. Strahl. (1991) "A bird with the guts to eat leaves: bizarre in appearance and diet, the hoatzin may well be the world's oldest birds." *Natural History* August (8): 48–54.

Gray, K.M. (1997) "Intercellular communication and group behavior in bacteria." *Trends in Microbiology* 5: 184–188.

Gray, M.W. (1989) "The evolutionary origins of organelles." *Trends in Genetics* 5: 294–299.

Gray, M.W. (1991) "Origin and evolution of plastid genomes and genes." In *The Molecular Biology of Plastids,* eds. L. Bogorad and I.K. Vasil, pp. 303–330. San Diego, CA: Academic Press.

Gray, M.W. (1992) "The endosymbiont hypothesis revisited." *International Review of Cytology* 141: 233–357.

Gray, M.W. (1993) "Origin and evolution of organelle genomes." *Current Opinion in Genetics and Development* 3: 884–890.

Gray, M.W. (1995) "Mitochondrial evolution." In *The Molecular Biology of Plant Mitochondria,* eds. C.S. Levings and I.K. Vasil, pp. 635–659. Dordrecht: Kluwer.

Gray, M.W. and D.F. Spencer. (1996) "Organellar evolution." In *Evolution of Microbial Life,* 54th Symposium of the Society for General Microbiology, eds. D. McL. Roberts, P. Sharp, G. Alderson, and M.A. Collins, pp. 109–126. Cambridge: Cambridge University Press.

Gray, M.W., G. Burger, and B. Franz Lang. (1999) "Mitochondrial evolution." *Science* 283: 1476–1481.

Gray, N.F. (1987) "Nematophagous fungi with particular reference to their ecology." *Biological Reviews* 62: 245–304.

Green, D. and J. Reed. (1998) "Mitochondria and apoptosis." *Science* 281: 1309–1312.

Green, T.G.A., B. Budel, U. Heber, A. Meyer, H. Zellner, and O.L. Lange. (1993) "Differences in photosynthetic performance between cyanobacterial and green algal components of lichen photosymbiodemes measured in the field." *New Phytologist* 125: 723–731.

Greenberg, J.T. (1997) "Programmed cell death in plant-pathogen interactions." *Annual Review of Plant Physiology and Plant Molecular Biology* 48: 525–545.

Gresshoff, P.M., ed. (1990) *Molecular Biology of Symbiotic Nitrogen Fixation.* Boca Raton, FL: CRC Press.

Griffin, D.E. (1994) "Introduction: cytokines in viral infections." *Seminars in Virology* 5: 403–404.

Groisman, E.A. and H. Ochman. (1997) "How *Salmonella* became a pathogen." *Trends in Microbiology* 5: 343–349.

Grosberg, R.K. and E.R. Strathmann. (1998) "One cell, two cell, red cell, blue cell: the persistence of a unicellular stage in multicellular life histories." *Trends in Ecology and Evolution* 13: 112–116.

Guerrero, R. (1991) "Predation as prerequisite to organelle origin: *Daptobacter* as example." In *Symbiosis as a Source of Evolutionary Innovation,* eds. L. Margulis and R. Fester, pp. 106–117. Cambridge: MIT Press.

Guillen, N. (1996) "Role of signaling and cytoskeletal rearrangements in the pathogenesis of *Entamoeba histolytica*." *Trends in Microbiology* 5: 191–196.

Guo, P. (1994) "Introduction: principles, perspectives and potential applications in virus assembly." *Seminars in Virology* 5: 1–3.

Gupta, R.S. (1995a) "Phylogenetic analysis of the 90KD heat shock family of protein sequences and an examination of the relationship among animals, plants and fungi species." *Molecular Biological Evolution* 12: 1063–1073.

Gupta, R.S. (1995b) "Evolution of the chaperonin families (Hsp60, Hsp10 and Tcp-1) of proteins and the origin of eukaryotic cells." *Molecular Microbiology* 15: 1–11.

Gupta, R.S., K. Aitken, M. Falah, and B. Singh. (1994) "Cloning of *Giardia lamblia* heat shock protein HSP70 homologs: implications regarding origin of eukaryotic cells and of endoplasmic reticulum." *Proceedings of the National Academy of Sciences, USA* 91: 2895–2899.

Gupta, R.S. and G.B. Golding. (1996) "The origin of the eukaryotic cell." *Trends in Biochemical Sciences* 21: 166–171.

Gutnick, D.L. (1992) "Prokaryotic symbionts of the aphid." In *The Prokaryotes,* 2nd ed., vol. 4, eds. A. Balows, H.G. Truper, M. Dworkin, W. Harder, and K.H. Schleifer, pp. 3907–3913. New York: Springer-Verlag.

Hagan, P. and H. Wilkins. (1993) "Concomitant immunity in schistosomiasis." *Parasitology Today* 9: 3–6.

Hajek, A.E. and R.J. St. Leger. (1994) "Interactions between fungal pathogens and insect hosts." *Annual Review of Entomology* 39: 293–322.

Haldane, J.B.S. (1949) "Disease and evolution." *La Scientifica Ricerca Scienti* (supplement) 19: 68–76.

Hall, B.F. and K.A. Joiner. (1991) "Strategies of obligate intracellular parasites for evading host defenses." *Parasitology Today* 7: A22–A27.

Hamada, N., H. Miyagawa, H. Miyawaki, and M. Inoue. (1996) "Lichen substances in mycobionts of crustose lichens cultured on media with extra sucrose." *Bryologist* 99: 71–74.

Hamilton, W.D. (1963) "The evolution of altruistic behavior." *American Naturalist* 97: 354–356.

Hamilton, W.D. (1982) "Pathogens as causes of genetic diversity in their host populations." In *Population Biology of Infectious Diseases,* eds. Anderson, R. and R. May, pp. 269–296. New York: Springer-Verlag.

Hamilton, W.D. and M. Zuk. (1982) "Heritable true fitness and bright birds: a role for parasites?" *Science* 218: 384–387.

Hammerschmidt, R. (1999) "Phytoalexins: what we have learned after 60 years?" *Annual Review of Phytopathology* 37: 285–306.

Handel, S.N. and A.J. Beatie. (1990) "Seed dispersal by ants." *Scientific American* 263: 76–83.

Hanley, K., J. Biardi, C. Greene, T. Markowitz, C. O'Connell, and J. Hornberger. (1996) "The behavioral ecology of host-parasite interactions: an interdisciplinary challenge." *Trends in Ecology and Evolution* 12: 371–373.

Hargreaves, J.A., A.M. Bailey, and J.P.R. Keon. (1995) "Determinants of parasitism in smut fungi." In *Pathogenesis and Host Specificity in Plant Diseases,* Vol 2, eds. K. Kohmoto, U.S. Singh, and R.P. Singh, pp. 189–201. New York: Elsevier.

Harley, J.L. and S.E. Smith. (1983) *Mycor-rhizal Symbiosis.* New York: Academic Press.

Harper, G.H. (1985) "Teaching Symbiosis." *Journal of Biological Education (UK)* 19: 219–223.

Harrison, M.J. (1998) "Development of the arbuscular mycorrhizal symbiosis." *Current Opinion in Plant Biology* 1: 360–365.

Harrison, M.J. (1999) " Molecular and cellular aspects of the arbuscular mycorrhizal symbiosis." *Annual Review of Plant Physiology and Plant Molecular Biology* 50: 361–389.

Hartman, H. (1992) "The eukaryotic cell: evolution and development." In *The Origin and Evolution of the Cell,* eds. H. Hartman and K. Matsuno, pp. 3–11. Singapore: World Scientific Publishing.

Hartman, H. and K. Matsuno, eds. (1992) *The Origin and Evolution of the Cell.* Singapore: World Scientific Publishing.

Harvell, C.D. (1990) "The evolution of inducible defence." *Parasitology* 100: S53–S61.

Hashimoto, T., L.B. Sanchez, T. Shirakura, M. Muller, and M. Hasegawa. (1998) "Secondary absence of mitochondria in *Giardia lamblia* and *Trichomonas vaginalis* revealed by valyl-t RNA synthetase phylogeny." *Proceedings of the National Academy of Sciences USA* 95: 6860–6865.

Hausen, H.Z. (1991) "Viruses in human cancers." *Science* 254: 1167–1172.

Hausfater, G. and D. Watson. (1976) "Social and reproductive correlates of parasite ova emissions by baboons." *Nature* 262: 688–689.

Hawksworth, D.L. (1988) "The variety of fungal-algal symbioses, their evolutionary significance, and the nature of lichens." *Botanical Journal of the Linnean Society* 96: 3–20.

Hawksworth, D.L. and A.Y. Rossman. (1997) "Where are all the undescribed fungi?" *Phytopathology* 87: 888–891.

Hawksworth, F.G. and D. Wiens. (1972) *Biology and Classification of Dwarf Mistletoes* (Arceuthobium). Agriculture Handbook no. 401. Washington, DC: U.S. Forest Service.

Haygood, M.G. (1993) "Light organ symbioses in fishes." *Critical Reviews in Microbiology* 19: 191–216.

Heckmann, K. and H.D. Gortz. (1992) "Prokaryotic symbionts of ciliates." In *The Prokaryotes,* 2nd ed., vol. 4, eds. A. Balows, H.G. Truper, M. Sworkin, W. Harder, and K.H. Schleifer, pp. 3865–3890. New York: Springer-Verlag.

Hedge, R.S., J.A. Mastrianni, M.R. Scott, K.A. DeFea, P. Tremblay, M. Torchia, S.J. DeArmond, S.B. Prusiner, and V.R. Lingappa. (1998) "A transmembrane form of the prion protein in neurodegenerative disease." *Science* 279: 827–834.

Heinemann, J.A. (1991) "Genetics of gene transfer between species." *Trends in Genetics* 7: 181–185.

Henig, R.M. (1993) *A Dancing Matrix—Voyages along the Viral Frontier.* New York: Alfred Knopf.

Hennecke, H. and D.P.S. Verma, eds. (1991) *Advances in Molecular Genetics of Plant-Microbe Interactions.* Dordrecht: Kluwer.

Henze, K., A. Badr, M. Wettern, R. Cerff, and W. Martin. (1995) "A nuclear gene of eubacterial origin in *Euglena gracilis* reflects cryptic endosymbioses during protist evolution." *Proceedings of the National Academy of Sciences, USA* 92: 9122–9126.

Herms, D. and W. Mattson. (1992) "The dilemma of plants: to grow or defend." *The Quarterly Review of Biology* 67: 283–335.

Herre, E.A. (1993) "Population structure and the evolution of virulence in nematode parasites of fig wasps." *Science* 259: 1442–1445.

Herre, E.A., N. Knowlton, U.G. Mueller, and S.A. Rehner. (1999) "The evolution of mutualisms." *Trends in Ecology and Evolution* 13: 49–53.

Herrera, C.M. (1995) "Plant-vertebrate seed dispersal systems in the Mediterranean; ecological, evolutionary, and historical determinants." *Annual Review of Ecology and Systematics* 26: 705–727.

Heslinga, G.A. (1986) "Biology and culture of the giant clam." In *Clam Mariculture in North America,* eds. J. Manzi and M. Castagna, pp. 299–322. Amsterdam: Elsevier.

Heslinga, G.A. and W.K. Fitt. (1987) "The domestication of reef-dwelling clams." *BioScience* 37: 332–339.

Heslinga, G.A., and T.C. Watson. (1985) "Recent advances in giant clam mariculture." *Proceedings of the Fifth International Coral Reef Congress. Papeete, Tahiti.* pp. 531–537.

Hessler, R.R. and V.A. Kaharl. (1995) "The deep-sea hydrothermal vent community: an overview." In *Seafloor Hydrothermal Systems: Physical, Chemical, Biological and Geological Interactions,* eds. S.E. Humphries, R.A. Zierenberg, L.S. Mullineau, and R.E. Thomson, Geophysical Monograph 91. Washington, DC: American Geophysical Union.

Hill, D.J. (1994) "The nature of the symbiotic relationship in lichens." *Endeavour* 18: 96–103.

Hinkle, G. (1991) "Status of the theory of the symbiotic origin of undulopodia (cilia)." In *Symbiosis as a Source of Evolutionary Innovation: Speciation and Morphogenesis,* eds. L. Margulis and R. Fester. 135–142. Cambridge: M.I.T.

Hinkle, G., J.K. Wetterer, T.R. Schultz, and M.L. Sogin. (1994) "Phylogeny of the attine ant fungi based on analysis of small subunit ribosomal RNA gene sequences." *Science* 266: 1695–1697.

Hirsch, A.M. (1992) "Developmental biology of legume nodulation." *New Phytologist* 122: 211–237.

Hirsch, A.M. and T.A. LaRue. (1997) "Is the legume nodule a modified root or stem or an organ *sui generis?*" *Critical Reviews in Plant Sciences* 16: 361–392.

Hively, W. (1997) "Looking for life in all the wrong places." *Discover Magazine* 18: 76–85.

Hobson, P.N. and C.S. Stewart, eds. (1997) *The Rumen Microbial Ecosystem,* 2nd ed., London: Blackie.

Hoek, C. van den, D.G. Mann, and H.M. Jahns. (1995) *Algae: An Introduction to Phycology.* Cambridge: Cambridge University Press.

Holldobler, B. and E.O. Wilson. (1990) *The Ants.* Cambridge: Harvard University Press.

Hofmann, D.K. and U. Brand. (1987) "Induction of metamorphis in the symbiotic *Cassiopea andromeda*: role of marine bacteria and of biochemicals." *Symbiosis* 4: 99–116.

Holland, C., D. Crompton, S. Assolu, W. Crichton, S. Torimiro, and D. Walters. (1992) "A possible genetic factor influencing protection from infection with *Ascaris lumbricoides* in Nigerian children." *Journal of Parasitology* 78: 915–916.

Holmes, J. (1983) "Evolutionary relationships between parasitic helminths and their hosts." In *Coevolution,* eds. F. Futuyma and M. Slatkin, pp. 161–185. Sunderland, MA: Sinauer Associates.

Holmes, J. and W. Bethel. (1972) "Modification of intermediate host behaviour by parasites." In *Behavioural Aspects of Parasitic Transmission,* eds. E. Canning and C. Wright, pp. 123–149. London: Academic Press.

Honegger, R. (1991) "Functional aspects of the lichen symbiosis." *Annual Review of Plant Physiology and Plant Molecular Biology* 42: 553–578.

Honegger, R. (1992) "Lichens: mycobiont-photobiont relationships." In *Algae and Symbioses,* ed. W. Reisser, pp. 255–275. Bristol, UK: Biopress Ltd.

Honegger, P., M. Peter, and S. Scherrer. (1996) "Drought-induced structural alterations at the mycobiont-photobiont interface in a range of foliose macrolichens." *Protoplasma* 190: 212–232.

Honigberg, B.M., ed. (1989) *Trichomonads Parasitic in Humans.* New York: Springer-Verlag.

Hooykaas, P.J.J. and R.A. Schilperoort. (1992) "*Agrobacterium* and plant genetic engineering." *Plant Molecular Biology* 19: 15–38.

Hopkins, D. (1990) "Dracunculiasis." In *Tropical and Geographical Medicine,* eds. K. Warren and A. Mahmoud, pp. 439–442. New York: McGraw-Hill.

Hopwood, D.A. and K.E. Chater, eds. (1989) *Genetics of Bacterial Diversity.* London: Academic Press.

Horikoshi, K., and W.D. Grant, eds. (1998) *Extremophiles: Microbial Life in Extreme Environments.* New York: Wiley-Liss.

Horn, B.W. and R.W. Lichtwardt. (1981) "Studies of nutritional relationships of larval *Aedes aegypti* (Diptera: Culicidae) with *Smittium culisetae.*" *Mycologia* 73: 724–740.

Horner, D.S., R.P. Hirt, S. Kilvington, D. Lloyd, and T.M. Embley. (1996) "Molecular data suggest an early acquisition of the mitochondrion endosymbiont." *Proceedings of the Royal Society of London B* 263: 1053–1059.

Howard, R.J. and B. Valent. (1996) "Breaking and entering: host penetration by the fungal rice blast pathogen *Magnaporthe grisea.*" *Annual Review of Microbiology* 50: 491–512.

Howe, C.J., T.J. Beanland, A.W.D. Larkhum, and P.J. Lockhart. (1992) "Plastid origins." *Trends in Ecology and Evolution* 7: 378–383.

Hudson, P. and J. Greenman. (1998) "Competition mediated by parasites: biological and theoretical progress." *Trends in Ecology and Evolution* 13: 387–390.

Hugouvieux-Cotte-Pattat, N., G. Condemine, W. Nasser, and S. Reverchon. (1996) "Regulation of pectinolysis in *Erwinia chrysanthemi.*" *Annual Review of Microbiology.* 50: 213–257.

Hull, R. and S.N. Covey. (1996) "Retroelements: propagation and adaptation." *Virus Genes* 11: 105–118.

Hungate, R.E. (1950) "The anaerobic, mesophilic, cellulolytic bacteria." *Bacteriological Reviews* 14: 1–49.

Hungate, R.E. (1966) *The Rumen and its Microbes.* New York: Academic Press.

Hurdler, G.W. (1998) *Magical mushrooms, mischievous molds.* Princeton, NJ: Princeton University Press.

Hurst, G.D.D., L.D. Hurst, and M.E.N. Majerus. (1993) "Altering sex ratios: the games microbes play." *BioEssays* 15: 695–697.

Hurst, L., A. Atlan, and B. Bengtsson. (1996) "Genetic conflicts." *The Quarterly Review of Biology* 71: 317–364.

Huss-Danell, K. (1997) "Actinorhizal symbioses and their N_2 fixation." *New Phytologist* 136: 375–405.

Hutson, V. and R. Law. (1993) "Four steps to two sexes." *Proceedings of the Royal Society of London B* 253: 43–51.

Huxley, C.R. and D.F. Cutler, eds. (1991) *Ant-Plant Interactions.* Oxford: Oxford University Press.

Iglewski, B.H. and V.L. Clark, eds. (1990) *Molecular Basis of Bacterial Pathogenesis.* San Diego, CA: Academic Press.

Innes, R.W. (1995) "Plant-parasite interactions: has the gene for gene model become outdated?" *Trends in Microbiology* 3: 483–485.

Ireton, K. and P. Cossart. (1997) "Host-pathogen interactions during entry and actin-based movement of *Listeria monocytogenes.*" *Annual Review of Genetics* 31: 113–138.

Isaac, S. (1991) *Fungal-Plant Interactions.* London: Chapman and Hall.

Isberg, R. (1991) "Discrimination between intracellular uptake and surface adhesion of bacterial pathogens." *Science* 252: 934–938.

Jacobson, R.L. and R.J. Doyle. (1996) "Lectin-parasite interactions." *Parasitology Today* 12: 55–60.

Janerette, C.A. (1991) "An introduction to mycorrhizae." *American Biological Teacher* 53: 13–19.

Janzen, D.H. (1966) "Coevolution of mutualism between ants and acacias in Central America." *Evolution* 20: 249–275.

Janzen, D.H. (1967) "Interaction of the bull's-horn acacia (*Acacia cornigera* L.) with an ant inhabitant (*Pseudomyrmex ferruginea* F. Smith) in eastern Mexico." *Kansas University Science Bulletin* 47: 315–558.

Janzen, D.H. (1985) "The natural history of mutualisms." In *The Biology of Mutualism,* ed. D.H. Boucher, pp. 40–99. New York: Oxford University Press.

Jarrell, K.F., D.P. Bayley, J.D. Correia, and N.A. Thomas. (1998) "Recent excitement about archaea." *BioScience* 49: 530–541.

Jarroll, E.L., P. Manning, A. Berrada, D. Hare, and D.G. Lindmark. (1989) "Biochemistry and metabolism of *Giardia.*" *Journal of Protozoology* 36: 190–192.

Jeffries, P. (1997) "Mycoparasitism." In *The Mycota: A Comprehensive Treatise on Fungi as Experimental System for Basic and Applied Research. Environmental and Microbial Relationships,* vol. 4, eds. D.T. Wicklow, K. Esser, P.A. Lemke, and B.E. Soderstrom, pp. 149–164. Berlin: Springer-Verlag.

Jeffries, P. and T.W.K. Young (1994) *Interfungal Parasitic Relationships.* Wallingford, Oxon, UK: CAB International.

Jensen, T.E. (1994) "Alternative pathway (cyanobacteria to eukaryota)." In *Evolutionary Pathways and Enigmatic Algae: Cyanidium caldarium (Rhodophyta) and Related Cells,* ed. J. Seckbach, pp. 53–66. Dordrecht: Kluwer.

Jenzen, D. (1988) "Ecological characterization of a Costa Rican dry forest caterpillar fauna." *Biotropica* 20: 120–135.

Jeon, K. (1991) "*Amoeba* and X-bacteria: symbiont acquisition and possible species change." In *Symbiosis as a Source of Evolutionary Innovation: Speciation and Morphogenesis,* eds. L. Margulis and R. Fester, pp. 118–131. Cambridge, MA: MIT Press.

Jeon, K.W. (1992) "Prokaryotic symbionts of amoebae and flagellates." In *The Prokaryotes. A Handbook on the Biology of Bacteria: Ecophysiology, Isolation, Identification, Applications,* 2nd ed., vol. 4, eds. A. Balows, H.G. Truper, M. Dworkin, Wm. Harder, and K.H. Schleifer, pp. 3855–3864. New York: Springer-Verlag.

Jeon, K.W. (1995) "The large, free living amoebae: cells for biological studies." *Journal Eukaryotic Microbiology* 42: 1–7.

Jepson, M.A. and M. Ann Clark. (1998) "Studying M cells and their role in infection." *Trends in Microbiology* 6: 359–365.

Jindal, S. and M. Malkovsky. (1994) "Stress responses to viral infection." *Trends in Microbiology* 2: 89–90.

Joel, D.M. (1988) "Mimicry and mutualism in carnivorous pitcher plants (Sarraceniaceae, Nepenthaceae, Cephalotaceae, Bromeliaceae). "*Biological Journal of the Linnean Society* 35:185–197.

John, D.M., S.J. Hawkins, and J.H. Price, eds. (1992) *Plant-Animal Interactions in the Marine Benthos.* Oxford: Clarendon Press.

John, D.T., T.B. Cole, and R.A. Bruner. (1985) "Amebastomes of *Naegleria fowleri.*" *Journal of Protozoology* 32: 12–19.

Johnsgard, P. (1997) *The Avian Brood*

Parasites: Deception at the Nest. New York: Oxford University Press.

Johnson, H.M., J.K. Russell, and C.H. Pontzer. (1992) "Superantigens in human disease." *Scientific American* 266: 20–26.

Jolivet, P. (1996) *Ants and Plants: An Example of Coevolution.* Leiden: Backhuys.

Jones, C.G. and R.D. Firn. (1991) "On the evolution of plant secondary chemical diversity." *Philosophical Transactions of the Royal Society London B* 333: 273–280.

Joseph, L.E. (1990) *Gaia: The Growth of an Idea.* New York: St. Martin.

Juniper, B.E., R.J. Robins, and D.M. Joel (1989) *The Carnivorous Plants.* London: Academic Press.

Kabnick, K.S. and D.A. Peattie. (1990) "*In situ* analyses reveal that two nuclei of *Giardia lamblia* are equivalent." *Journal Cell Science* 95: 353–360.

Kabnick, K.S. and D.A. Peattie. (1991) "*Giardia: a missing link between prokaryotes and eukaryotes.*" *American Scientist* 79: 34–43.

Kado, C.I. (1991) "Molecular mechanisms of crown gall tumorigenesis." *Critical Reviews in Plant Sciences* 10: 1–32.

Kaharl, V.A. (1990) *Water Baby: The Story of ALVIN.* New York: Oxford University Press.

Kaiser, D. (1993) "Rolland Thaxter's legacy and the origins of multicellular development." *Genetics* 135: 249–254.

Kandler, O. and W. Zillig, eds. (1985) *Archaebacteria' 85. Proceedings of the EMBO Workshop on Molecular Genetics of Archaebacteria and the International Workshop on Biology and Biochemistry of Archaebacteria.* Munich: Gustav Fischer Verlag.

Kannenberg, E.L. and N.J. Brewin. (1994) "Host-plant invasion by *Rhizobium*: the role of cell-surface components." *Trends in Microbiology* 2: 277–283.

Kasamatsu, H. and A. Nakanishi. (1998) "How do animal DNA viruses get to nucleus?" *Annual Review of Microbiology* 52: 627–686.

Kasper, L.H. and Buzoni-Gatel. (1998) "Some opportunistic parasitic infections in AIDS: candidiasis, pneumocystis, cryptosporidiosis, toxoplasmosis." *Parasitology Today* 14: 150–156.

Kassai, T. (1982) *Handbook of Nippostrongylus brasiliensis (Nematoda).* Slough: Commonwealth Agricultural Bureaux.

Kates, M., D.J. Kushner, and A.T. Matheson, eds. (1993) *The Biochemistry of Archaea (Archaebacteria).* Amsterdam: Elsevier.

Katz, L.A. (1998) "Changing perspectives on the origin of eukaryotes." *Trends in Ecology and Evolution* 13: 493–497.

Katz, M., D.D. Despommier, and R.W. Gwadz. (1982) *Parasitic Diseases.* New York: Springer-Verlag.

Kaya, H.K. and R. Gaugler. (1993) "Entomopathogenic nematodes." *Annual Review of Entomology* 38: 181–206.

Kazura, J., T. Nutman, and B. Greene. (1993) "Filariasis." In *Immunology and Molecular Biology of Parasitic Infections,* ed. K. Warren, pp. 473–495. Oxford: Blackwell.

Kearns, C.A. and D.W. Inouye. (1997) "Pollinators, flowering plants and conservation biology." *BioScience* 47: 297–307.

Kearns, C.A., D.W. Inouye, and N.M. Waser. (1998) "Endangered mutualisms: The conservation of plant-pollinator interactions." *Annual Review of Ecology and Systematics* 29: 83–112.

Keeble, K. (1910) *Plant-Animals. A Study in Symbiosis.* Cambridge: Cambridge University Press.

Keeler, K.H. (1985) "Cost:benefit models of mutualism." In *The Biology of Mutualism, Ecology and Evolution,* ed. D.H. Boucher, pp. 100–127. New York: Oxford University Press.

Keeler, K.H. (1989) "Ant-plant interactions." In *Plant-Animal Interactions,* ed. W.G. Abrahamson, pp. 207–242. New York: McGraw Hill.

Keeling, P.J. (1998) "A Kingdom's progress: Archezoa and the origin of eukaryotes." *BioEssays* 20: 87–95.

Keeling, P.J. and W.F. Doolittle. (1995) "Archaea: Narrowing the gap between prokaryotes and eukaryotes." *Proceedings of the National Academy of Sciences, USA* 92: 5761–5764.

Kelly, C.K. (1992) "Resource choice in *Cuscuta europaea.*" *Proceedings of the National Academy of Sciences USA* 89: 12194–12197.

Kelly, J.M. (1997) "Genetic transformation of parasitic protozoa." *Advances in Parasitology* 39: 227–270.

Kerr, R.A. (1997) "Life goes to extremes in the deep earth and elsewhere?" *Science* 276: 703–704.

Khakhina, L.N. (1992) "Evolutionary significance of symbiosis: development of the symbiogenesis concept." *Symbiosis* 14: 217–228.

Khakhina, L.N. and M. McMenamin, eds. (1992) *Concepts of Symbiogenesis: a Historical and Critical Study of the Research of Russian Botanists,* trans. S. Merkel and R. Coalson. New Haven, CT: Yale University Press.

Kimura, J. and T. Nakano. (1990) "Reconstitution of a *Blasia-Nostoc* symbiotic association under axenic conditions." *Nova Hedwigia* 50: 191–200.

Kluger, M. (1979) *Fever: Its Biology, Evolution and Function.* Princeton, NJ: Princeton University Press.

Knoop, V. and A. Brennicke. (1994) "Promiscuous mitochondrial group II intron sequences in plant nuclear genomes." *Journal of Molecular Evolution* 39: 144–150.

Knop, D. (1996) *Giant Clams. A Comprehensive Guide to the Identification and Care of Tridacnid Clams.* Ettlingen, Germany: Dahne Verlag.

Knowlton, D.M. (1998) "The evolution of interspecific mutualisms." *Proceedings of the National Academy of Sciences USA* 95: 8676–8680.

Kobayashi, I. (1998) "Selfishness and death: rasion d'etre of restriction, recombination and mitochondria." *Trends in Genetics* 14: 368–374.

Koch, A.L. (1995) "The origin of intracellular and intercellular pathogens." *The Quarterly Review of Biology* 70: 423–437.

Kohler, S., C.F. Delwiche, P.W. Denny, L.G. Tilney, P. Webster, R.J.M. Wilson, J.D. Palmer, and D.S. Roos. (1997) "A plastid of probable green algal origin in Apicomplexan parasites." *Science* 275: 1485–1489.

Kohlmeyer, J. and E. Kohlmeyer. (1979) "Submarine lichens and lichenlike associations." In *Marine Mycology: The Higher Fungi,* pp. 70–78. New York: Academic Press.

Kombrink, E. and I.E. Somssich. (1995) "Defense responses of plants to pathogens." *Advances in Botanical Research* 21: 1–34.

Komiya, Y. (1966) "*Clonorchis* and clonorichiasis." *Advances in Parasitology* 4: 53–106.

Kowallik, K.V. (1993) "From endosymbionts to chloroplasts: evidence for a single prokaryotic/eukaryotic endocytobiosis." *Endocytobiosis and Cell Research* 10: 137–149.

Krause, D.C. (1998) "*Mycoplasma pneumoniae* cytoadherence: organization and assembly of the attachment organelle." *Trends in Microbiology* 6: 15–18.

Krebs, J.R. and N.B. Davies, eds. (1991) *Behavioural Ecology: An Evolutionary Approach.* Oxford: Blackwell.

Kropotkin, P. (1902) *Mutual Aid: A Factor of Evolution.* London: William Heinemann.

Krumholz, L.R., C.W. Forsberg, and D.M. Veira. (1983) "Association of methanogenic bacteria with rumen protozoa." *Canadian Journal of Microbiology* 29: 676–680.

Kuc, J. (1995) "Phytoalexins, stress metabolism, and disease resistance." *Annual Review of Phytopathology* 33: 275–297.

Kuijt, J. (1969) *The Biology of Parasitic Flowering Plants.* Berkeley: University of California Press.

Kung, S. and C.J. Arntzen, eds. (1989) *Plant Biotechnology.* Boston: Butterworth Publishers.

Kunoh, H. (1995) "Host-parasite specificity in powdery mildews." In *Pathogenesis and Host Specificity in Plant Diseases,* eds. K. Kohmoto, U.S. Singh, and R.P. Singh, pp. 239–250. New York: Elsevier.

Kunze, R., H. Saedler, and W.E. Lonnig. (1997) "Plant transposable elements." *Advances in Botanical Research* 27: 331–470.

Kuti, J.O., B.B. Jarvis, N. Mokhti-Rejal, and G.A. Bean. (1990) "Allelochemical regulation of reproduction and seed germination of two Brazilian *Baccharis* species by phytotoxic trichothecenes." *Journal Chemical Ecology* 16: 3441–3453.

Labrador, M. and V.G. Corces. (1997) "Transposable elements-host interactions: regulation of insertion and excision." *Annual Review of Genetics* 31: 381–404.

Ladle, R., R. Johnstone, and O. Judson. (1993) "Coevolution dynamics of sex in a metapopulation: escaping the Red Queen." *Proceedings of the Royal Society of London B* 253: 155–160.

Lake, J.A. (1991) "Tracing origins with molecular sequences: metazoan and eukaryotic beginnings." *Trends in Biochemical Sciences* 16: 46–50.

Lake, J.A. and M.C. Rivera. (1994) "Was the nucleus the first endosymbiont?" *Proceedings of the National Academy of Sciences, USA* 91: 2880–2881.

Lake, J.A. and M.C. Rivera. (1996) "The prokaryotic ancestry of eukaryotes." In *Evolution of Microbial Life,* eds. D. Roberts, P. Sharp, G. Alderson, and M.A. Collins, pp. 87–108. Cambridge: Cambridge University Press.

Lamb, C. (1996) "A ligand-receptor mechanism in plant pathogen recognition." *Science* 274: 2038–2039.

Lancaster, J.R. (1992) "Nitric oxide in cells." *American Scientist* 80: 248–259.

Landman, O.E. (1993) "Inheritance of acquired characteristics revisited." *BioScience* 43: 696–705.

Lanner, R.M. (1996) *Made for Each Other: A Symbiosis of Birds and Pines.* New York: Oxford University Press.

LaRue, T. and N.F. Weeden. (1994) "The symbiosis genes of the host." In *Proceedings First European Nitrogen Fixation Conference,* eds. G.B. Kiss and G. Endre, pp. 147–151. Szeged: Officina Press.

Latch, G.C.M. (1995) "Endophytic fungi of grasses." In *Pathogenesis and Host Specificity in Plant Diseases,* eds. K. Kohmoto, U.S. Singh, and R.P. Singh, pp. 265–276. New York: Pergamon/Elsevier.

Laval-Peuto, M. (1992) "Plastidic protozoa." In *Algae and Symbioses,* ed. W. Reisser, pp. 471–499. Bristol, UK: Biopress Ltd.

Lavine, M.D. and N.E. Beckage. (1995) "Polydnaviruses: potent mediators of host insect immune dysfunction." *Parasitology Today* 11: 368–378.

Law, R. (1989) "New phenotypes from symbiosis." *Trends in Ecology and Evolution* 4: 334–335.

Law, R. (1991) "The symbiotic phenotype: origins and evolution." In *Symbiosis as a Source of Evolutionary Innovation,* eds. L. Margulis and R. Fester, pp. 57–71. Cambridge, MA: The MIT Press.

Lederberg, J. (1952) "Cell Genetics and Hereditary Symbiosis." *Physiological Reviews* 32: 403–430.

Lederberg, J. (1993) "Viruses and humankind: Intracellular symbiosis and evolutionary competition." In *Emerging Viruses,* ed. S.S. Morse, pp. 3–9. New York: Oxford University Press.

Lee, C. (1997) "Type III secretion systems: machines to deliver bacterial proteins into eukaryotic cells?" *Trends in Microbiology* 5: 149–156.

Lee, J.J., A.T. Soldo, W. Reisser, M.J. Lee, K.W. Jeon, and H.D. Gortz. (1985) "The extent of algal and bacterial endosymbioses in protozoa." *Journal of Protozoology* 32: 391–403.

Lee, J.J. (1995) "Living sands." *BioScience* 45: 252–261.

Leib-Mosch, C. and W. Seifarth. (1996) "Evolution and biological significance of human retroelements." *Virus Genes* 11: 133–145.

Lemke, P.A., ed. (1979) *Viruses and Plasmids in Fungi.* New York: Marcel Dekker.

Lerdau, M., M. Litvak, and R. Monson. (1994) "Plant chemical defense: monoterpenes and the growth-differentiation balance hypothesis." *Trends in Ecology and Evolution* 9: 58–61.

Leschine, S. (1995) "Cellulase degradation in anaerobic environments." *Annual Review of Microbiology* 49: 399–426.

Letourneau, D. (1991) "Parasitism of ant-plant mutualisms and the novel case of *Piper.*" In *Ant–Plant Interactions,* eds. C. Huxley and D. Cutler, pp. 390–396. Oxford: Oxford University Press.

Levin, B.R., M. Lipsitch, and S. Bonhoeffer. (1999) "Population biology, evolution, and infectious disease: convergence and synthesis." *Science* 283: 806–809.

Levin, B.R. and J.J. Bull. (1994) "Short-sighted evolution and the virulence of pathogenic microorganisms." *Trends in Microbiology* 2: 76–81.

Levin, M.A., R.J. Seidler, and M. Rogul, eds. (1992) *Microbial Ecology: Principles, Methods, and Applications.* New York: McGraw Hill.

Levine, L. (1993) "Gaia; goddess and idea." *BioSystems* 31: 85–92.

Levine, N.D., J.O. Corliss, F.E.G. Cox, G. Deroux, J. Grain, B.M. Honigberg, G.F. Leedale, A.R. Loeblich, J. Lom, D. Lynn, E.G. Merinfeldt, C.F. Page, G. Poljanski, V. Sprague, J. Vavra, and F.G. Wallace. (1980) "A newly revised classification of the Protozoa." *Journal of Protozoology* 27: 37–57.

Levings, C.S. and I.K. Vasil, eds. (1995) *The Molecular Biology of Plant Mitochondria.* Dordrecht: Kluwer.

Levinton, J.S. (1995) *Marine Biology: Function, Biodiversity, Ecology.* New York: Oxford University Press.

Levy, J.A. (1993) "Pathogenesis of human immunodeficiency virus infection." *Microbiological Reviews* 57: 183–289.

Lewin, R.A,. ed. (1993) *Origins of Plastids: Symbiogenesis, Prochlorophytes, and the Origin of Chloroplasts.* New York: Chapman and Hall.

Lewin, R.A. (1995) "Symbiotic algae: definitions, quantification and evolution." *Symbiosis* 19: 31–37.

Lewin, R.A. and L. Cheng, eds. (1989) *Prochloron: A Microbial Enigma.* New York: Chapman and Hall.

Li, W.H. and D. Graur. (1991) *Fundamentals of Molecular Evolution.* Sunderland, MA: Sinauer Associates.

Lichtwardt, R.W. (1986) *The Trichomycetes.* New York: Springer-Verlag.

Limbaugh, C. (1961) "Cleaning symbiosis." *Scientific American* 205: 42–49.

Linblad, P. and B. Bergman. (1990) "The cycad-cyanobacterial symbiosis." In *Handbook of Symbiotic Cyanobacteria,* ed. A.N. Rai, pp. 137–159. Boca Raton, FL: CRC Press.

Lingelbach, K. and K.A. Joiner. (1998) "The parasitophorous vacuole membrane surrounding *Plasmodium* and *Toxoplasma*: an unusual compartment in infected cells." *Journal of Cell Science* 111: 1467–1475.

Lipps, J.H. and P.W. Signor, eds. (1992) *Origin and Early Evolution of the Metazoa.* New York: Plenum Press.

Lipsitch, M., S. Siller, and M.A. Nowak. (1996) "The evolution of virulence in pathogens with vertical and horizontal transmission." *Evolution* 50: 1729–1741.

Lively, C.M. (1996) "Host-parasite coevolution and sex." *BioScience* 46: 107–114.

Lloyd, F.E. (1942) *The Carnivorous Plants.* New York: Ronald Press.

Lloyd, S. and E. Soulsby. (1987) "Immunobiology of gastro-intestinal nematodes of ruminants." In *Immune Responses in Parasitic Infections: Immunology, Immunopathology, and Immunoprophylaxis,* ed. E. Soulsby, pp. 1–41. Boca Raton, FL: CRC Press.

Lockhart, P.J., T.J. Beanland, C.J. Howe, and A.W.D. Larkhum. (1992a) "Sequence of *Prochloron didemni* atpBE and the influence of chloroplast origins." *Proceedings of the National Academy of Sciences, USA* 89: 2742–2746.

Lockhart, P.J., D. Penny, M.D. Hendy, C.J. Howe, T.J. Beanland, and A.W.D. Larkum. (1992b) "Controversy on chloroplast origins." *Federation European Biological Society Letter* 301: 127–131.

Long, S.R. (1989) "*Rhizobium*-legume nodulation: life together in the underground." *The Cell* 56: 203–214.

Long, S.R. (1996) "*Rhizobium* symbiosis: nod factors in perspective." *The Plant Cell* 8: 1885–1898.

Long, S.R. and B.J. Staskawicz. (1993) "Prokaryotic plant parasites." *The Cell* 73: 921–935.

Lopez, S. (1998) "Acquired resistance affects male sexual display and female choice in guppies." *Proceedings of the Royal Society of London B* 265: 717–723.

Lopez-Garcia, P. and D. Moreira. (1999) "Metabolic symbiosis at the origin of eukaryotes." *Trends in Biochemical Sciences* 24: 88–93.

Losey, G.S. (1987) "Cleaning symbiosis." *Symbiosis* 4: 229–258.

Lotem, A. and S.I. Rothstein. (1995) "Cuckoo-host coevolution: from snapshots of an arms race to the documentation of microevolution." *Trends in Ecology and Evolution* 10: 436–437.

Lotka, A.J. (1932) "The growth of mixed populations: two species competing for a common food supply." *Journal of Washington Academy of Science* 22: 461–469.

Lovelock, J.E. (1995) *The Ages of Gaia: A Biography of Our Living Earth.* New York: W.W. Norton.

Lovelock, J.E. and L. Margulis. (1974) "Atmospheric homeostasis by and for the biosphere: the Gaia hypothesis." *Tellus* 26: 2–10.

Lucas, J. (1994) "The biology, exploitation and mariculture of giant clams (Tridacnidae)." *Review of Fisheries Science* 2: 181–223.

Lugtenberg, B., ed. (1986) *Recognition in Microbe-Plant Symbiotic and Pathogenic Interactions.* Berlin: Springer-Verlag.

Lugtenberg, B.J., ed. (1989) *Signal Molecules in Plant and Plant-Microbe Interactions.* Berlin: Springer-Verlag.

Luria, S.E., J.E. Darell, D. Baltimore, and A. Campbell. (1978) *General Virology,* 3rd ed. New York: John Wiley.

Lutzoni, F. and M. Pagel. (1997) "Accelerated evolution as a consequence of transitions to mutualism." *Proceedings of the National Academy of Sciences, USA* 94: 11422–11427.

Lynch, J.M., ed. (1990) *The Rhizosphere.* New York: John Wiley.

Lynch, M. (1996) "Mutation accumulation in transfer RNAs: molecular evidence for Muller's ratchet in mitochondrial genomes." *Evolution* 13: 209–220.

Lyon, B.E. (1993) "Conspecific brood parasitism as a flexible female reproductive tactic in American coots." *Animal Behaviour* 46: 911–928.

MacArthur, R. and E. Wilson. (1967) *The Theory of Island Biogeography.* Princeton, NJ: Princeton University Press.

MacDonald, J.R. (1990) "Macroevolution and retroviral elements." *BioScience* 40: 183–191.

MacDonough, K.A., Y. Kress, and B. Bloom. (1993) "Pathogenesis of tuberculosis: interaction of *Mycobacterium tuberculosis* with macrophages." *Infection and Immunology* 61: 2763–2773.

MacFadden, B. (1997) "Origin and evolution of the grazing guild in New World terrestrial mammals." *Trends in Ecology and Evolution* 12: 182–187.

Madigan, M.T. and B.L. Marrs. (1997) "Extremophiles." *Scientific American* 272: 82–87.

Madigan, M.T., J.M. Martinko, and J. Parker. (1997) *Brock Biology of Microorganisms,* 8th ed. Upper Saddle River, NJ: Prentice Hall.

Maggenti, A. (1981) *General Nematology.* New York: Springer-Verlag.

Maier, R.J. and E.W. Triplett. (1996) "Toward more productive, efficient and competitive nitrogen-fixing symbiotic bacteria." *Critical Reviews in Plant Sciences* 15: 191–234.

Maizels, R. and R. Lawrence. (1991) "Immunological tolerance: the key feature in human filariasis." *Parasitology Today* 7: 271–276.

Mansfield, J.M. (1995) "Immunology of african trypanosomiasis." In *Modern Parasite Biology.* ed. D.J. Wyler, pp. 222–246. New York: W.H. Freeman.

Marasas, W.F.O. and P.E. Nelson. (1987) *Mycotoxicology.* University Park, PA: The Pennsylvania State University Press.

Marciano-Cabral, F. (1988) "Biology of *Naeglaria* spp." *Microbiological Reviews* 52: 114–133.

Margulis, L. (1991a) "Symbiogenesis and symbionticism." In *Symbiosis as a Source of Evolutionary Innovation,* eds. L. Margulis and R. Fester, pp. 2–14. Cambridge, MA: MIT Press.

Margulis, L. (1991b) "Symbiosis in evolution: origins of cell motility." In *Evolution of Life, Fossils, Molecules and Culture,* ed. T. Honjo, pp. 305–324. Tokyo: Springer-Verlag.

Margulis, L. (1992a) "Biodiversity: molecular biological domains, symbiosis and kingdom origins." *BioSystems* 27: 39–51.

Margulis, L. (1992b) "Protoctists and polyphyly: comment on The number of symbiotic . . . by T. Cavalier Smith." *BioSystems* 28: 107–108.

Margulis, L. (1993a) *Symbiosis in Cell Evolution,* 2nd ed. New York: W.H. Freeman.

Margulis, L. (1993b) "Origins of species: acquired genomes and individuality." *BioSystems* 31: 121–125.

Margulis, L. (1996) "Archael-eubacterial mergers in the origin of eukarya: phylogenetic classification of life." *Proceedings of the National Academy of Sciences, USA* 93: 1071–1076.

Margulis, L. and M.J. Chapman. (1998) "Endosymbiosis: cyclical and permanent in evolution." *Trends in Microbiology* 6: 342–345.

Margulis, L., D. Chase, and R. Guerrero. (1986) "Microbial communities." *BioScience* 36: 160–170.

Margulis, L., J.O. Corliss, M. Melkonian, and D.J. Chapman, eds. (1990) *Handbook of Protoctista: The Structure, Cultivation, Habitats and Life Histories of the Eukaryotic Microorganisms and Their Descendants Exclusive of Animals, Plants and Fungi.* Boston: Jones and Bartlett.

Margulis, L. and R. Fester, eds. (1991) *Symbiosis as a Source of Evolutionary Innovation, Speciation and Morphogenesis.* Cambridge, MA: MIT Press.

Margulis, L. and R. Guerrero. (1991) "Kingdoms in turmoil." *New Scientist* 46–50.

Margulis, L. and M.A.S. McMenamin. (1990) "Marriage of convenience." *Annals of the New York Academy of Sciences.* 30: 31–37.

Margulis, L. and D. Sagan. (1986a) *Origins of Sex.* New Haven, CT: Yale University Press.

Margulis, L. and D. Sagan. (1986b) "Strange fruit on the tree of life." *Annals of the New York Academy of Sciences* 26: 38–45.

Margulis, L. and D. Sagan. (1995) *What is Life?* New York: Simon Schuster.

Margulis, L. and D. Sagan. (1997a) *Microcosmos. Four Billion Years of Evolution from Our Microbial Ancestors.* Berkeley: University of California Press.

Margulis, L. and D. Sagan. (1997b) *Slanted Truths: Essays on Gaia, Symbiosis and Evolution.* New York: Springer-Verlag.

Margulis, L. and K.V. Schwartz. (1997) *Five Kingdoms: An Illustrated Guide to the Phyla of Life on Earth.* 3rd ed. San Francisco: W.H. Freeman.

Mariscal, R.N. (1971) "Experimental studies on the protection of anemone fishes from sea anemones." In *Aspects of the Biology of Symbiosis,* ed. T.C. Cheng, pp. 283–315. Baltimore: University Park Press.

Marsh, M. and H.T. McMahon. (1999) "The structural era of endocytosis." *Science* 285: 215–220.

Marshall, A. (1996) "Calcification in hermatypic and ahermatypic corals." *Science* 271: 1788–1792.

Martin, M.M. (1991) "The evolution of cellulose digestion in insects." *Philosophical Transactions of the Royal Society B* 333: 281–288.

Martin, M.M. (1992) "The evolution of insect-fungus associations: from contact to stable symbiosis." *American Zoologist* 32: 593–605.

Martin, W.F. (1996) "Is something wrong with the tree of life?" *BioEssays* 18: 523–527.

Martin, W. (1999) "A briefly argued case that mitochondria and plastids are descendants from endosymbionts, but that the nuclear compartment is not." *Proceedings of the Royal Society: B. Biological sciences* 266: 1387–1395.

Martin, W. and M. Muller. (1998) "The hydrogen hypothesis for the first eukaryote." *Nature* 392: 37–41.

Martin, W., B. Stoebe, V. Goremykin, S. Hansmann, M. Hasegawa, and K. Kowallik.

(1998) "Gene transfer to the nucleus and the evolution of chloroplasts." *Nature* 393: 162–165.

Marx, D.H. (1991) "Practical significance of ectomycorrhizae in forest establishment." *Proceedings of Marcus Walenberg Foundation Symposium* 7: 54–90.

Masucci, M.G. and I. Ernberg. (1994) "Epstein-Barr virus: adaptation to life within the immune system." *Trends in Microbiology* 2: 125–130.

Mathews, H.M. and J.C. Armstrong. (1981) "Duffly blood types and vivax malaria in Ethiopia." *American Journal of Tropical Medicine and Hygiene* 30: 299–303.

Mathieu, L.G. and S. Sonea. (1995) "A powerful bacterial world." *Endeavour* 19: 112–117.

Matossian, M.K. (1989) *Poisons of the Past: Molds, Epidemics, and History.* New Haven, CT: Yale University Press.

Matthews, R.E.F. (1991a) "Classification and nomenclature of viruses." *Intervirology* 17: 1–19.

Matthews, R.E.F. (1991b) *Plant Virology,* 3rd ed. New York: Academic Press.

Matthijs, H.C.P., G.W.M. Van der Staay, and L.R. Mur (1994) "Prochlorophytes: the "other" cyanobacteria?" In *The Molecular Biology of Cyanobacteria,* ed. D.A. Bryant, pp. 49–64. Amsterdam: Kluwer.

Mattox, K.R., K.D. Stewart, and G.L. Floyd. (1972) "Probable virus infections in four genera of green algae." *Canadian Journal of Microbiology* 18: 1620–1621.

Mauel, J. (1996) "Intracellular survival of protozoan parasites with special reference to *Leishmania* spp. *Toxoplasma gondii* and *Trypanosoma cruzi.*" *Advances in Parasitology* 38: 1–51.

May, R.M. and R.M. Anderson. (1983) "Epidemiology and genetics in the coevolution of parasites and hosts." *Proceedings of the Royal Society of London B* 219: 291–313.

Maynard-Smith, J. (1989) "Generating novelty by symbiosis." *Nature* 341: 284–285.

Maynard-Smith, J. (1991) "A darwinian view of symbiosis." In *Symbiosis as a Source of Evolutionary Innovation. Speciation and Morphogenesis,* eds. L. Margulis and R. Fester, pp. 26–39. Cambridge, MA: MIT Press.

Maynard-Smith, J. and E. Szathmary. (1995) *The Major Transitions in Evolution.* New York: W.H. Freeman.

Mayr, E. (1990) "A natural system of organisms." *Nature* 348: 491.

McAuley, P.J. (1985) "The cell cycle of symbiotic *Chlorella.* I. The relationship between host feeding and algal cell growth and division." *Journal of Cell Science* 77: 225–239.

McAuley, P.J., M. Dorling, and H. Hodge. (1996) "Effect of maltose release on uptake and assimilation of ammonium by symbiotic *Chlorella* (Chlorophyta)." *Journal of Phycology* 32: 839–846.

McBurney, M.I., P.J. Van Soest, and J.L. Jeraci. (1987) "Colon carcinogenesis: the microbial feast or famine mechanism." *Nutrition and Cancer* 10: 23–28.

McCutchan, T.F. (1995) "Molecular biology of *Plasmodium.*" In *Modern Parasite Biology,* ed. D.J. Wyler, pp. 317–332. New York: W.H. Freeman.

McDermott, J. (1989) "A biologist whose heresy redraws Earth's tree of life." *Smithsonian* (August): 72–81.

McFadden, G.I. (1993) "Second-hand chloroplasts: evolution of cryptomonad algae." *Advances in Botanical Research* 19: 189–230.

McFadden, G.I. and P. Gilson. (1995) "Something borrowed, something green: lateral transfer of chloroplasts by secondary endosymbiosis." *Trends in Ecology and Evolution* 10: 12–17.

McFadden, G.I., P.R. Gilson, S.E. Douglas, T. Cavalier-Smith, C.J.B. Hofmann, and U.G. Maier. (1997) "Bonsai genomics: sequencing the smallest eukaryotic genomes." *Trends in Genetics* 13: 46–49.

McFadden, G.I. and R.F. Waller. (1997) "Plastids in parasites of humans." *BioEssays* 19: 1033–1040.

McFall-Ngai, M.J. (1999) "Consequences of evolving with bacterial symbionts: lessons from the squid-*Vibrio* associations." *Annual Review of Ecology and Systematics* 30: 235–256.

McFall-Ngai, M.J. and E.G. Ruby. (1998) "Sepiolids and Vibrios: when first they meet." *BioScience* 48: 257–265.

McLaren, D. (1989) "Will the real target of immunity to schistosomiasis please stand up." *Parasitology Today* 5: 279–282.

McLaren, D. and D. Hockley. (1976) "*Schistosoma mansoni*: the occurrence of microvilli on the surface of the tegument during transformation from cercaria to schistosomulum." *Parasitology* 73: 169–187.

McLennan, D. and D. Brooks. (1991) "Parasites and sexual selection: a macroevolutionary perspective." *The Quarterly Review of Biology* 66: 255–286.

McMenamin, M.A.S. and D.L.S. McMenamin. (1993) "Hypersea and the land ecosystem." *BioSystems* 31: 145–153.

McMenamin, M.A.S. and D.L.S. McMenamin.

(1996) *Hypersea. Life on Land.* New York: Columbia University Press.

Meeks, J.C. (1990) "Cyanobacterial-bryophyte associations." In *Handbook of Symbiotic Cyanobacteria,* ed. A.N. Rai, pp. 43–63. Boca Raton, FL: CRC Press.

Meeks, J.C. (1998) "Symbiosis between nitrogen-fixing cyanobacteria and plants." *BioScience* 48: 266–276.

Meeuse, B. and S. Morris. (1984) *The Sex Life of Flowers.* New York: Facts on File.

Mehlhorn, H., ed. (1988) *Parasitology in Focus, Facts and Trends.* Berlin: Springer-Verlag.

Mehlhorn, H. and E. Schein. (1984) "The piroplasms: life cycle and sexual stages." *Advances in Parasitology* 23: 37–103.

Meier, R. and W. Wiessner. (1989) "Infection of algae-free *Paramecium bursaria* with symbiotic *Chlorella* sp. isolated from green paramecia. II. A timed study." *Journal of Cell Science* 93: 571–579.

Meighen, E.A. (1991) "Molecular biology of bacterial bioluminescence." *Microbiological Reviews* 55: 123–143.

Meighen, E.A. and P.V. Dunlap. (1993) "Physiological, biochemical and genetic control of bacterial bioluminescence." *Advances in Microbial Physiology* 34: 1–67.

Melin, E. (1962) "Physiological aspects of mycorrhizae of forest trees." In *Tree Growth,* ed. T.T. Kozlowski, pp. 247–263. New York: Ronald Press.

Milinski, M. (1990) "Parasites and host decision-making." In *Parasitism and Host Behaviour,* eds. C. Bernard and J. Behnke, pp. 95–116. London: Taylor and Frances.

Miller, J.A. (1986) "Clinical opportunities for plant and soil fungi." *BioScience* 36: 656–658.

Miller, I.M. (1990) "Bacterial leaf nodule symbiosis." *Advances in Botanical Research* 17: 163–234.

Miller, L.H., M.F. Good, and G. Milon. (1994) "Malaria pathogenesis." *Science* 264: 1878–1883.

Miller, V.A., J.B. Kaper, D.A. Portnoy, and R.R. Isberg, eds. (1994) *Molecular Genetics of Bacterial Pathogenesis.* St. Paul, MN: American Society for Microbiology Press.

Minchella, D., B. Leathers, K. Brown, and J. McNair. (1985) "Host and parasite counteradaptation: an example from freshwater snail." *American Naturalist* 118: 843–854.

Mirelman, D. (1987) "Ameba-bacterium relationships in amoebiasis." *Microbiological Reviews* 51: 272–284.

Mitchell, G.F. (1991) "Co-evolution of parasites and adaptive immune responses." *Parasitology Today* 7: A2–A5.

Mitchison, N.A. (1990) "The evolution of acquired immunity to parasites." *Parasitology* 100: S27–S34.

Miyakawa, M. and T.D. Luckey, eds. (1968) *Advances in Germ Free Research and Gnotobiology.* Boca Raton, FL: CRC Press.

Moat, A.G. and J.W. Foster. (1995) *Microbial Physiology,* 3rd ed. New York: Wiley-Liss.

Moberg, C.L. and Z.A. Cohen. (1991) "Rene Jules Dubos." *Scientific American* 263: 66–74.

Moe, R. (1997) "*Verrucaria tavaresiae* sp. nov., a marine lichen with a brown algal photobiont." *Bulletin of the California Lichen Society* 4: 7–11.

Mohan, S., C. Dow, and J.A. Cole, eds. (1992) *Prokaryotic Structure and Function: A New Perspective,* vol. 47. Society of General Microbiology. Cambridge: Cambridge University Press.

Molina, M.C. and C. Vicente. (1995) "Correlationships between enzymatic activity of lectins, putrescine content and chloroplast damage in *Xanthoria parietina* phycobionts." *Cell Adhesion and Communication* 3: 1–12.

Moller, A.P., P. Christe, and E. Lux. (1999) "Parasitism, host immune function, and sexual selection." *The Quarterly Review of Biology* 74: 3–20.

Monastersky, R. (1998) "The rise of life on earth." *National Geographic* 193: 54–81.

Montagu, A. (1952) *Darwin: Competition and Cooperation.* New York: Shuman.

Montfort, I. and R. Perez-Tamayo. (1994) "Is phagocytosis related to virulence in *Entamoeba histolytica* Schaudinn, 1903?" *Parasitology Today* 10: 271–273.

Montgomery, M.K. and M.J. McFall-Ngai. (1995) "The inductive role of bacterial symbionts in the morphogenesis of a squid light organ." *American Zoologist* 35: 372–380.

Mooi, F.R. and E.M. Bik. (1997) "The evolution of epidemic *Vibrio cholerae* strains." *Trends in Microbiology* 5: 161–165.

Moore, J. (1984a) "Altered behavioral responses in intermediate hosts- an acanthocephalan parasitic strategy." *American Naturalist* 123: 572–577.

Moore, J. (1984b) "Parasites that change the behavior of their host." *Scientific American* 250: 108–115.

Moore, J. (1995) "The behavior of parasitized animals." *BioScience* 45: 89–96.

Moore, J. and N.J. Gotelli. (1996) "Evolutionary patterns of altered behavior and susceptibility in parasitized hosts." *Evolution* 50: 807–819.

Moore-Landecker, E. (1996) *Fundamentals*

of the Fungi. Upper Saddle River, NJ: Prentice Hall.

Mor, T.S., M.A. Gomez-Lim, and K.E. Palmer. (1998) "Perspective: edible vaccines—a concept coming of age." *Trends in Microbiology* 6: 449–453.

Moran, N.A. and A. Telang. (1998) "Bacteriocyte-associated symbionts of insects." *BioScience* 48: 295–304.

Morden, C.W., C.F. Delwiche, M. Kuhsel, and J.D. Palmer. (1992) "Gene phylogenies and the endosymbiotic origin of plastids." *BioSystems* 28: 75–90.

Moreira, D. and P. Lopez-Gracia. (1998) "Symbiosis between methanogenic Archaea and proteobacteria as the origin of eukaryotes. The syntrophic hypothesis." *Journal of Molecular Evolution* 47: 517–530.

Morell, V. (1997) "Microbiology's scarred reactionary." *Science* 276: 699–702.

Morin, J. (1999) "Shedding light: distribution, diversity and functions of bioluminescence." *Annual Review of Ecology and Systematics.* In press.

Morowitz, H.J. (1992) "*Beginnings of Cellular Life: Metabolism Recapitulates Biogenesis.*" New Haven, CT: Yale University Press.

Morse, S.S., ed. (1993) *Emerging Viruses.* New York: Oxford University Press.

Morse, S.S., ed. (1994) *The Evolutionary Biology of Viruses.* New York: Raven Press.

Moss, S.T. (1979) "Commensalism of the Trichomycetes." In *Insect-Fungus Symbiosis: Nutrition, Mutualism, and Commensalism,* ed. L.R. Batra, pp. 175–227. Totowa, NJ: Allanheld, Osmun.

Mott, K.A. and J.Y. Takemoto. (1989) "Syringomycin, a bacterial phytotoxin, closes stomata." *Plant Physiology* 90: 1435–1439.

Moulder, J.W. (1985) "Comparative biology of intracellular parasites." *Microbiological Reviews* 49: 298–337.

Mueller, J. (1974) "The biology of *Spirometra.*" *Journal of Parasitology* 60: 3–14.

Mueller, U.G., S.A. Rehner, and T.R. Shultz. (1998) "The evolution of agriculture in ants." *Science* 281: 2034–2038.

Mujer, C.V., D.L. Andrews, J.R. Manhart, S.K. Pierce, and M.E. Rumpho. (1996) "Chloroplast genes are expressed during intracellular symbiotic association of *Vaucheria litorea* plastids with the sea slug *Elysia chlorotica.*" *Proceedings of the National Academy of Sciences, USA* 93: 12333–12338.

Muller, M. (1993) "The hydrogenosome." *Journal of General of Microbiology* 139: 2879–2889.

Mullineaux, C.W. (1999) "The plankton and the planet." *Science* 283: 801–802.

Murphy, F.A., C.M. Fauquet, D.H. Bishop, S.A. Ghabrial, A.W. Jarvis, G.P. Martelli, and M.A. Summers, eds. (1995) *Virus Taxonomy: Classification and Nomenclature of Viruses: Sixth Report of the International Committee on Taxonomy of Viruses.* Berlin: Springer-Verlag.

Murphy, F.A. and N. Nathanson. (1994) "The emergence of new virus diseases: an overview." *Seminars in Virology* 5: 87–102.

Murray, M.D. (1990) "Influence of host behaviour on some ectoparasites of birds and mammals." In *Parasitism and Host Behaviour,* eds. C. Barnard and J. Behnke, pp. 290–315. London: Taylor & Francis.

Muscatine, L. (1990) "The role of symbiotic algae in carbon and energy flux in reef corals." In *Coral Reefs,* ed. Z. Dubinsky, pp. 75–87. Amsterdam: Elsevier.

Muscatine, L. and P.L. McNeil. (1989) "Endosymbiosis in *Hydra* and the evolution of internal defense systems." *American Zoologist* 29: 371–386.

Musselman, L.J. and M.C. Press. (1995) "Introduction to parasitic plants." In *Parasitic Plants,* eds. M.C. Press and J.D. Graves, pp. 1–13. London: Chapman and Hall.

Myers, N., ed. (1993) "Gaia: An Atlas of Planet Management. New York: Anchor Books.

Nadler, S. (1987) "Biochemical and immunological systematics of some ascaridoid nematodes: genetic divergence between congeners." *Journal of Parasitology* 73: 811–816.

Nap, J.P. and T. Bisseling. (1990) "Developmental biology of a plant-pathogen symbiosis: the legume root nodule." *Science* 250: 948–954.

Nardon, P. (1988) "Cell to cell interactions in insect endocytobiosis." In *Cell to Cell Signals in Plant, Animal, and Microbial Symbiosis,* eds. S. Scannerini, D. Smith, P. Bonfante-Fasolo, and V. Gianinazzi-Pearson, pp. 85–100. New York: Springer Verlag.

Nash, T.E. (1995) "Antigenic variation in *Giardia lamblia.*" In *Molecular Approaches to Parasitology,* eds. J.C. Boothroyd and R. Komuniecki, pp. 31–42. New York: Wiley-Liss.

Nash, T.H., ed. (1996) *Lichen Biology.* Cambridge: Cambridge University Press.

Nealson, K.H. (1991) "Luminous bacteria symbiotic with entomopathogenic nematodes." In *Symbiosis as a Source of Evolutionary Innovation,* eds. L. Margulis and E. Fester, pp. 205–218. Cambridge, MA: The MIT Press.

Nealson, K.H. and J.W. Hastings. (1992) "The luminous bacteria." In *The Prokaryotes,* 2nd ed. vol. 4, eds. A. Balows, H.G. Truper, M. Dworkin, W. Harder, and K.H. Schleifer, pp. 625–639. New York: Springer-Verlag.

Nealson, K.H., T.M. Schmidt, and B. Bleakley. (1988) "Luminescent bacteria: symbionts of nematodes and pathogens of insects." In *Cell to Cell Signals in Plant, Animal and Microbial Symbiosis,* eds. S. Scannerini, D. Smith, P. Bonfante-Fasolo, and V. Gianinazzi-Pearson, pp. 101–113. New York: Springer-Verlag.

Nee, S. and J.M. Smith. (1990) "The evolution of molecular parasites." *Parasitology* 100: S5–S18.

Nelson, D.C. and K.D. Hagen. (1995) "Physiology and biochemistry of symbiotic and free-living chemoautotrophic sulfur bacteria." *American Zoologist* 35: 91–101.

Netzly, D. and L. Butler. (1986) "Roots of *Sorghum bicolor* contain hydrophobic droplets that contain biologically active materials." *Crop Science* 26: 775–778.

Newsham, K.K., A.H. Fitter, and A.R. Watkinson. (1995) "Multifunctionality and biodiversity in arbuscular mycorrhizas." *Trends in Ecology and Evolution* 10: 407–411.

Nickle, W., ed. (1984) *Plant and Insect Nematodes.* New York: Marcel Dekker.

Niebel, A., G. Gheysen, and M. Van Montagu. (1994) "Plant-cyst nematode and plant-root-knot nematode interactions." *Parasitology Today* 10: 424–430.

Nierzwicki-Bauer, S.A. (1990) "*Azolla-Anabaena* symbiosis: use in agriculture." In *Handbook of Symbiotic Cyanobacteria,* ed. A.N. Rai, pp. 119–136. Boca Raton, FL: CRC Press.

Nikiforuk, A. (1991) *The Fourth Horseman. A Short History of Epidemics, Plagues and Other Scourges.* London: Phoenix.

Nilsson, L. (1998) "Deep flowers for long tongues." *Trends in Ecology and Evolution* 13: 259–260.

Nordbring-Hertz, B. (1988) "Ecology and recognition in the nematode-nematophagous fungus system." *Advances in Microbial Ecology* 10: 81–114.

Norris, D.M. (1979) "The mutualistic fungi of *Xyleborini* beetles." In *Insect-Fungus Symbiosis: Nutrition, Mutualism, and Commensalism,* ed. L.R. Batra, pp. 53–63. Totowa, NJ: Allanheld, Osmuun.

Norstog, K. (1987) "Cycads and the origin of insect pollination." *American Scientist* 75: 270–279.

North, R.D., C.W. Jackson, and P.E. Howse. (1997) "Evolutionary aspects of ant-fungus interactions in leaf-cutting ants." *Trends in Ecology and Evolution* 12: 384–389.

Norton, D.A. and M.A. Carpenter. (1998) "Mistletoes as parasites: host specificity and speciation." *Trends in Ecology and Evolution* 13: 101–105.

Norton, J.H., M.A. Shepherd, H.M. Long, and W.K. Fitt. (1992) "The zooxanthellal tubular system in giant clam." *Biological Bulletin* 183: 503–506.

Nye, K.E. and J.M. Parkin. (1994) *HIV and AIDS.* Oxford: BIOS Scientific Publishers.

O'Brian, M. (1996) "Heme synthesis in the *Rhizobium*-legume symbiosis: a palette for bacterial and eukaryotic pigments." *Journal of Bacteriology* 178: 2471–2478.

O'Callaghan, J. and M. Conrad. (1992) "Symbiotic interactions in the Evolve III ecosystem model." *BioSystems* 26: 199–209.

O'Gara, F., D. Dowling, and B. Boeston, eds. (1995) *The Molecular Ecology of the Rhizosphere.* Weinham: VCH Publishers.

O'Neill, S., A.A. Hoffmann, and J.H. Werren, eds. (1997) *Influential Passengers. Inherited Microorganisms and Arthropod Reproduction.* Oxford: Oxford University Press.

Ofek, I. (1995) "Non opsonic phagocytosis of microorganisms." *Annual Review of Microbiology* 49: 239–276.

Oldstone, M.B. (1989) "Viral alteration of cell function." *Scientific American* 261: 42–48.

Oldstone, M.B. (1998a) "Molecular mimicry and immune-mediated diseases." *The FASEB Journal* 12: 1255–1265.

Oldstone, M.B.A. (1998b) *Viruses, Plagues and History.* New York: Oxford University Press.

Olff, H. and M. Ritchie. (1998) "Effects of herbivores on grassland plant diversity." *Trends in Ecology and Evolution* 13: 261–265.

Olsen, G.J., C.R. Woese, and R. Overbeek. (1994) "The winds of (evolutionary) change: breathing new life into microbiology." *Journal of Bacteriology* 176: 1–6.

Osawa, S. and T. Honjo, eds. (1991) *Evolution of Life, Fossils, Molecules and Culture.* Tokyo: Springer-Verlag.

Ostergren, G. (1945) "Parasitic nature of extra fragment chromosomes." *Botaniska Notiser* 2: 157–163.

Otte, D. and J.A. Endler, eds. (1989) *Speciation and Its Consequences.* Sinauer Associates.

Pace, J., M.J. Hayman, and J.E. Galan. (1993) "Signal transduction and invasion of epithelial cells by *S. typhimurium.*" *Cell* 72: 505–514.

Pace, N.R. (1997) "A molecular view of microbial diversity and the biosphere." *Science* 276: 734–740.

Paerl, H.W. (1992) "Epi- and endobiotic interactions of cyanobacteria." In *Algae and Symbioses,* ed. W. Reisser, pp. 537–565. Bristol, UK: Biopress Ltd.

Pak, J.W. and K.W. Jeon. (1997) "A symbiont-produced protein and bacterial symbiosis in *Amoeba proteus.*" *Journal of Eukaryotic Microbiology* 44: 614–619.

Palincsar, J.S., W.R. Jones, and E.E. Palincsar. (1988) "Effects of isolation of the endosymbiont *Symbiodinium microadriaticum* (Dinophyceae) from its host *Aiptasia pallida* (Anthozoa) on cell wall ultrastructure and mitotic rate." *Transactions of American Microscopic Society* 107: 53–66.

Palmer, J.D. (1992) "A degenerate plastid genome in malaria parasite?" *Current Biology* 2: 318–320.

Palmer, J.D. (1993) "A genetic rainbow of plastids." *Nature* 363: 762–763.

Palmer, J.D. (1997) "Organelle genomes: going, going, gone!" *Science* 275: 790–791.

Palmqvist, K. (1993) "Photosynthetic CO_2-use efficiency in lichens and their isolated photobionts: the possible role of a CO_2-concentrating mechanism." *Planta* 191: 48–56.

Pamilo, P., P. Gertsch, P. Thoren, and P. Seppa. (1997) "Molecular population genetics of social insects." *Annual Review of Ecology and Systematics* 28: 1–25.

Parker, C. and C.R. Riches. (1993) *Parasitic Weeds of the World: Biology and Control.* Wallingford: CAB International.

Partridge, L. and L. Hurst. (1998) "Sex and conflict." *Science* 281: 2003–2008.

Passador, L., J.M. Cook, M.J. Gambello, L. Rust, and B.H. Iglewski. (1993) "Expression of *Pseudomonas aeruginosa* virulence genes requires cell-to-cell communication." *Science* 260: 1127–1130.

Pasteur, G. (1982) "A classificatory review of mimicry systems." *Annual Review of Ecology and Systematics* 13: 169–199.

Patience, C., D.A. Wilkinson, and R.A. Weiss. (1997) "Our retroviral heritage." *Trends in Genetics* 13: 116–120.

Patterson, D.J. (1985) "The fine structure of *Opalina ranarum* (Family Opalinidae); opalinid phylogeny and classification." *Protistologica* 21: 413–428.

Patterson, D.J. and B.L.J. Delvinquier. (1990) "The fine structure of the cortex of the protist *Protoopalina australis* (Slopalininda, Opalinidae) from *Litoria nasuta* and *Litoria inermis* (Amphibia: Anura: Hylidae) in Queensland, Australia." *Journal of Protozoology* 37: 449–455.

Patterson, D.J. and M.L. Sogin. (1992) "Eukaryotic origins and protistan diversity." In *The Origin and Evolution of the Cell,* eds. H. Hartman and K. Matsuno, pp. 13–46. Singapore: World Scientific.

Pawlowski, C. and T. Bisseling. (1996) "Rhizobial and actinorhizal symbioses: what are the shared features?" *The Plant Cell* 8: 1899–1913.

Payne, R.B. (1998) "Brood parasitism in birds: strangers in the nest." *BioScience* 48: 377–386.

Peakall, R., S.N. Handel, and A.J. Beattie. (1991) "The evidence for, and importance of, ant pollination." In *Ant-Plant Interactions,* eds. C.R. Huxley and D.F. Cutler, pp. 421–429. Oxford: Oxford University Press.

Pellmyr, O. (1992) "Evolution of insect pollination and angiosperm diversification." *Trends in Ecology and Evolution* 7: 46–49.

Penn, D. and W.K. Potts. (1998) "Chemical signals and parasite-mediated sexual selection." *Trends in Ecology and Evolution* 13: 391–396.

Pennisi, E. (1999) "Is it time to uproot the tree of life?" *Science* 284: 1305–1307.

Percey, B. (1822) "Phosophorescence of wounds." *Quarterly Journal of Science, Literature and Art.* 12: 180–181.

Pereira, M. E. A. (1995) "Cell biology of *Trypanosoma cruzi.* " In *Modern Parasite Biology,* ed. D. J. Wyler. 64–78. New York: W. H. Freeman and Company.

Perez-Urria, E., M. Rodriguez, and C. Vicente. (1989) "Algal partner regulates fungal urease in the lichen *Evernia prunastri* by producing a protein which inhibits urease synthesis." *Plant Molecular Biology* 13: 665–672.

Perotto, S. and P. Bonfante. (1997) "Bacterial associations with mycorrhizal fungi: close and distant friends in the rhizosphere." *Trends in Microbiology* 5: 496–501.

Perry, N. (1983) *Symbiosis: Close Encounters of the Natural Kind.* Poole, Dorset, UK: Blandford Press.

Pesci, E.C. and B.H. Iglewski. (1997) "The chain of command in *Pseudomonas* quorum sensing." *Trends in Microbiology* 5: 132–135.

Peters, G.A. and J.C. Meeks. (1989) "The *Azolla-Anabaena* symbiosis: basic biology." *Annual Review of Plant Physiology and Plant Molecular Biology* 40: 193–210.

Petersen, J. and J. Cupello. (1981) "Commercial development and future prospects for entomogenous nematodes." *Journal of Nematology* 13: 280–284.

Phares, C. (1987) "Plerocercoid growth factor: A homolog of human growth hormone." *Parasitology Today* 3: 346–349.

Pianka, E. (1970) "On r- and K-selection." *American Naturalist* 104: 592–597.

Pierce, S.K., R.W. Biron, and M.E. Rumpho. (1996) "Endosymbiotic chloroplasts in molluscan cells contain proteins synthesized after plastid capture." *Journal of Experimental Biology* 199: 2323–2330.

Pirozynski, K.A. and D.L. Hawksworth, eds. (1988) *Coevolution of Fungi with Plants and Animals.* London: Academic Press.

Plazinski, J. (1990) "The *Azolla-Anabaena* symbiosis." In *Molecular Biology of Symbiotic Nitrogen Fixation,* ed. P.M. Gresshoff, pp. 51–75. Boca Raton, FL: CRC Press.

Poinar, G. (1979) *Nematodes for Biological Control of Insects.* Boca Raton, FL: CRC Press.

Pond, F.R., I. Gibson, Jorge Lalucat, and R.L. Quackenbush. (1989) "R-body-producing bacteria." *Microbiological Reviews* 53: 25–67.

Postgate, J.R. (1992) *Microbes and Man,* 3rd ed. Cambridge, Cambridge University Press.

Postgate, J.R. (1994) *The Outer Reaches of Life.* Cambridge: Cambridge University Press.

Poulin, R. (1996) "The evolution of life history strategies in parasitic animals." *Advances in Parasitology* 37: 107–134.

Poulin, R. and A.S. Grutter. (1996) "Cleaning symbioses: proximate and adaptive explanations." *BioScience* 46: 512–517.

Powell, J.A. (1992) "Interrelationships of *Yucca* and *Yucca* moths." *Trends in Ecology and Evolution* 7: 10–15.

Press, M. and J. Graves, eds. (1995) *Parasitic Plants.* London: Chapman and Hall.

Preston, G.M., B. Haubald, and P.B. Rainey. (1998) "Bacterial genomics and adaptation to life on plants: implications for the evolution of pathogenicity and symbiosis." *Current Opinion in Microbiology* 1: 589–597.

Price, P. (1980) *Evolutionary Biology of Parasites.* Princeton, NJ: Princeton University Press.

Price, P. (1991) "The web of life: Development over 3.8 billion years of trophic relationships." In *Symbiosis as a Source of Evolutionary Innovation: Speciation and morphogenesis,* eds. L. Margulis and R. Fester, pp. 262–272. Cambridge, MA: MIT Press.

Price, P.W., T.M. Lewinsohn, G.W. Fernandes, and W.W. Benson, eds. (1991) *Plant-Animal Interactions: Evolutionary Ecology in Tropical and Temperate Regions.* New York: John Wiley and Sons.

Proctor, L.M. and J.A. Fuhrman. (1990) "Viral mortality of marine bacteria and cyanobacteria." *Nature* 343: 60–62.

Proctor, M., P. Yeo, and A. Lack. (1996) *The Natural History of Pollination.* Portland, OR: Timber Press.

Prusiner, S.B. (1991) "Molecular biology of prion diseases." *Science* 252: 1515–1522.

Prusiner, S.B. (1994) "Biology and genetics of prion diseases." *Annual Review of Microbiology* 48: 655–686.

Prusiner, S.B. (1996) "Prions, prions, prions." *Current Topics in Microbiology and Immunology* 207: 1–163.

Prusiner, S.B. (1998) "Prions." *Proceedings National Academy of Sciences, USA* 95: 13363–13383.

Prusiner, S.B. and M.R. Scott. (1997) "Genetics of prions." *Annual Review of Genetics* 31: 139–175.

Pueppke, S.G. (1996) "The genetic and biochemical basis for nodulation of legumes by rhizobia." *Critical Reviews in Biotechnology* 16: 1–51.

Rabeneck, L. and D.F. Ranshoff. (1991) "Is *Helicobacter pylori* a cause of duodenal ulcer: a methodological critique of current evidence." *American Journal of Medicine* 91: 566–572.

Radetsky, P. (1994) *The Invisible Invaders: Viruses and the Scientists Who Pursue Them.* Boston: Little Brown.

Rae, A.L., Bonfante-Fasolo, and N.J. Brewin. (1992) "Structure and growth of infection threads in the legume symbiosis with *Rhizobium leguminosarum." The Plant Journal* 2: 385–395.

Rahat, M. (1992) "Algae/hydra symbioses." In *Algae and Symbioses,* ed. W. Reisser, pp. 41–62. Bristol, UK: Biopress Ltd.

Rai, A.N., N., ed. (1990) *CRC Handbook of Symbiotic Cyanobacteria.* Boca Raton: FL: CRC Press.

Rai, A.N. (1990) "Cyanobacterial-fungal symbioses: the cyanolichens." In *CRC Handbook of Symbiotic Cyanobacteria,* ed. A.N. Rai. Boca Raton, FL: CRC Press.

Rands, M.L., A.E. Douglas, B.C. Loughman, and C.R. Hawes. (1992) "The pH of the perisymbiont space in the green hydra-*Chlorella* symbiosis: an immunocytochemical investigation." *Protoplasma* 170: 90–93.

Rands, M.L., B.C. Loughman, and A.E. Douglas. (1993) "The symbiotic interface in an alga-invertebrate symbiosis." *Proceedings of the Royal Society of London B* 253: 161–165.

Rautian, M.S., I.I. Skoblo, N.A. Lebedeva, and D.V. Ossipov. (1993) "Genetics of symbiotic interactions between *Paramecium bursaria* and the intranuclear bacterium *Holospora acuminata*; natural genetic variability, infectivity and susceptibility." *Acta Protozoologica* 32: 165–173.

Ravdin, J.I. (1990) "Cell biology of *Entamoeba histolytica* and immunology of amebiasis." In *Modern Parasite Biology*, ed. D.J. Wyler, pp. 126–150. New York: W.H. Freeman.

Read, A.F. (1994) "The evolution of virulence." *Trends in Microbiology* 2: 73–76.

Redlin, S.C. and L.M. Carris, eds. (1996) *Endophytic Fungi in Grasses and Woody Plants. Systematics, Ecology and Evolution.* St. Paul, MN: American Phytopathological Society Press.

Reeve, J.N. (1992) "Molecular biology of methanogens." *Annual Review of Microbiology* 46: 165–191.

Reinheimer, H. (1913) *Evolution by Cooperation. A Study of Bio-Economics.* London: Kegal Paul, Trench and Trubner.

Reinhold-Hurek, B. and T. Hurek. (1998) "Life in grasses: diazotrophic endophytes." *Trends in Microbiology* 6: 139–144.

Reisser, W. (1992a) "Basic mechanisms of signal exchange, recognition, specificity and regulation in endosymbiotic systems." In *Algae and Symbioses,* ed. W. Reisser, pp. 657–674. Bristol, UK: Biopress Ltd.

Reisser, W., ed. (1992b) *Algae and Symbioses: Plants, Animals, Fungi, Viruses Interactions Explored.* Bristol, UK: Biopress Ltd.

Reisser, W. (1993) "Green ciliates: principles of symbiosis formation between autotrophic and heterotrophic partners." In *Origins of Plastids,* ed. R.A. Lewin, pp. 27–43. New York: Chapman and Hall.

Relman, D.A. (1999) "The search for unrecognized pathogens." *Science* 284: 1308–1310.

Relman, D.A., T.M. Schmidt, R.P. MacDermott, and S. Falkow. (1992) "Using 16S rRNA to find the cause of Whipple's disease." *New England Journal of Medicine* 327: 293–301.

Rennie, J. (1992) "Living together." *Scientific American* 266: 122–133.

Richardson, D.H.S. (1992) *Pollution Monitoring with Lichens.* Naturalist's Handbooks 19. Slough: Richmond Publishing.

Richardson, L.L. (1998) "Coral diseases: what is really known?" *Trends in Ecology and Evolution* 13: 438–443.

Rigaud, T. (1999) "Further *Wolbachia* endosymbiont diversity: a tree hiding in the forest?" *Trends in Ecology and Evolution* 14: 212–213.

Rindos, D. (1984) *The Origin of Agriculture. An Evolutionary Perspective.* Orlando, FL: Academic Press.

Riopel, J. and M. Timko. (1995) "Haustorial initiation and differentiation." In *Parasitic Plants,* eds. M. Press and J. Graves, pp. 39–79. London: Chapman and Hill.

Rivera, M.C. and J.A. Lake. (1992) "Evidence that eukaryotes and eocyte prokaryotes are immediate relatives." *Science* 257: 74–76.

Roberts, C.M. (1993) "Coral reefs: health, hazards and history." *Trends in Ecology and Evolution* 8: 425–427.

Roberts, D., P. Sharp, G. Alderson, and M.A. Collins, eds. (1996) *Evolution of Microbial Life.* 54th Symposium of the Society for General Microbiology. Cambridge: Cambridge University Press.

Roberts, L.S. and J. Janovy. (1996) *Foundations of Parasitology,* 5th ed. Dubuque, IA: W.C. Brown.

Robledo, D.R., P.A. Sosa, G. Garcia-Reina, and D.G. Muller. (1994) "Photosynthetic performance of healthy and virus-infected *Feldmannia irregularis* and *F. simplex* (Phaeophyceae)." *European Journal of Phycology* 29: 247–251.

Roger, A.J., C.G. Clark, and W.F. Doolittle. (1996) "A possible mitochondrial gene in the early- branching amitochondriate protist *Trichomonas vaginalis*." *Proceedings of the National Academy of Sciences, USA* 93: 14618–14622.

Roger, A.J., S.G. Svard, J. Tovar, C.G. Clark, M.W. Smith, F.D. Gillin, and M.L. Sogin. (1998) "A mitochondrial-like chaperonin 60 gene in *Giardia lamblia*: evidence that diplomonads once harbored an endosymbiont related to the progenitor of mitochondria." *Proceedings of the National Academy of Sciences, USA* 95: 229–234.

Roizman, B., ed. (1996) *Infectious Diseases in an Age of Change: the Impact of Human Ecological Behavior on Disease Transmission.* Washington, DC: National Academy Press.

Ronald, P.C. (1997) "Making rice disease-resistant." *Scientific American* 278: 100–105.

Roth, J.A., C.A. Bolin, K.A. Brogden, F.C. Minion, and M.J. Wannemmuehler, eds. (1995) *Virulence Mechanisms of Bacterial Pathogenesis.* St. Paul, MN: American Society for Microbiology Press.

Roth, L.E., K. Jeon, and G. Stacey. (1988) "Homology in endosymbiotic systems: the term 'symbiosome'." In *Molecular Genetics of Plant-Microbe Interactions,* eds. R. Palacios and D.P.S. Verma, pp. 220–225. St. Paul, MN: American Phytopathological Society Press.

Rothstein, S.I. and S.K. Robinson. (1994) "Conservation and coevolutionary implications of brood parasitism by cowbirds." *Trends in Ecology and Evolution* 9: 162–164.

Rothstein, S.I. and S.K. Robinson, eds. (1998) *Parasitic Birds and Their Hosts.* New York: Oxford University Press.

Rothwell, T. (1989) "Immune expulsion of parasitic nematodes from the alimentary tract." *International Journal of Parasitology* 29: 139–168.

Rottem, S. and Y. Naot. (1998) "Subversion and exploitation of host cells by mycoplasmas." *Trends in Microbiology* 6: 436–440.

Rowan, R. (1998) "Diversity and ecology of zooxanthellae on coral reefs." *Journal of Phycology* 34: 407–417.

Rowan, R. and D.A. Powers. (1991) "A molecular genetic classification of zooxanthellae and the evolution of animal-algal symbioses." *Science* 251: 1348–1351.

Rowan, R. and D.A. Powers (1992) "Ribosomal RNA sequences and the diversity of symbiotic dinoflagellates (zooxanthellae)." *Proceedings of the National Academy of Sciences, USA* 89: 3639–3643.

Rowan, R., N. Knowlton, A. Baker, and J. Jara. (1997) "Landscape ecology of algal symbionts creates variation in episodes of coral bleaching." *Nature* 388: 265–269.

Roy, B.A. (1993) "Floral mimicry by a plant pathogen." *Nature* 362: 56–58.

Ruby, E.G. (1996) "Lessons from a cooperative, bacterial-animal association: the *Vibrio fischeri-Euprymna scolopes* light organ symbiosis." *Annual Review of Microbiology* 50: 591–624.

Ruby, E.G. and M.J. McFall-Ngai. (1992) "A squid that glows in the night: development of an animal-bacterial mutualism." *Journal of Bacteriology* 174: 4865–4870.

Ruehle, J.L. and D.H. Marx. (1979) "Fiber, food, fuel, and fungal symbionts." *Science* 206: 419–422.

Russell, D. (1998) "Underwater epidemic." *The Amicus Journal* 20: 28–33.

Rutzler, K. ed. (1990) *New Perspectives in Sponge Biology.* Washington, DC: Smithsonian Institution Press.

Saffo, M.B. (1992a) "Coming to terms with a field: words and concepts in symbiosis." *Symbiosis* 14: 17–31.

Saffo, M.B. (1992b) "Invertebrates in endosymbiotic associations." *American Zoologist* 32: 557–565.

Safir, G.R., ed. (1987) *Ecophysiology of VA Mycorrhizal Plants.* Boca Raton, FL: CRC Press.

Sagan, D. and L. Margulis. (1993) *Garden of Microbial Delights.* Dubuque: Kendall Hunt Publishing.

Saikkonen, K., S.H. Faeth, M. Helander, and T.J. Sullivan. (1998) "Fungal endophytes: a continuum of interactions with host plants." *Annual Review of Ecology and Systematics* 29: 319–343.

Salyers, A.A. and D. Whitt. (1994) *Bacterial Pathogenesis. A Molecular Approach.* St. Paul, MN: American Society for Microbiology Press.

Samson, R.A. (1988) *Atlas of Entomopathogenic Fungi.* Berlin: Springer-Verlag.

Sanderson, B. and B. Ogilvie. (1971) "A study of acetylcholinesterase throughout the life cycle of *Nippostrongylus brasilinesis.*" *Parasitology* 62: 367–373.

Sansonetti, P. (1992) "Pathogenesis of shigellosis." *Current Topics in Microbiology and Immunology* 180: 1–143.

Sapp, J. (1987) *Beyond the Gene, Cytoplasmic Inheritance and the Struggle for Authority in Genetics.* New York: Oxford University Press.

Sapp, J. (1994a) "Symbiosis and disciplinary demarcations; the boundaries of the organism." *Symbiosis* 17: 91–115.

Sapp, J. (1994b) *Evolution by Association; A History of Symbiosis.* New York: Oxford University Press.

Sapp, J. (1998) "Cytoplasmic heretics." *Perspectives in Biology and Medicine* 41: 224–242.

Sarkar, S. (1992) "Sex, disease, and evolution: variations on a theme from J.B.S. Haldane." *BioScience* 42: 448–454.

Sarkisov, A.M. (1954) *Mycotoxicoses.* Moscow: State Publishing House for Agricultural Literature.

Sasa, M. (1976) *Human Filariasis: A Global Survey of Epidemiology and Control.* Baltimore: University Park Press.

Sato, S., M. Ishida, and H. Ishikawa, eds. (1993) *Endocytobiology,* vol. 5. Tubingen: Tubingen University Press.

Sayre, R.M. and D.E. Walter. (1991) "Factors affecting the efficacy of natural enemies of nematodes." *Annual Review of Phytopathology* 29: 149–166.

Scannerini, S., D. Smith, P. Bonfante-Fasolo, and V. Gianinazzi-Pearson, eds. (1988) *Cell-to-Cell signals in Plant, Animal, and Microbial Symbiosis,* vol. 17. Berlin: Springer-Verlag.

Schafer, W., D. Straney, L. Ciuffetti, H.D. Van Etten, and O.C. Yoder. (1989) "One enzyme makes a fungal pathogen, but not a saprophyte, virulent on a new host plant." *Science* 246: 247–249.

Schardl, C.L. (1996) "*Epichloe* species: fungal symbionts of grasses." *Annual Review of Phytopathology* 34: 109–130.

Scheidegger, C. (1994) "Low-temperature scanning electron microscopy: the localization of free and perturbed water and its role in the morphology of the lichen symbionts." *Cryptogamic Botany* 4: 290–299.

Schenk, H.E.A. (1992) "Cyanobacterial symbioses." In *The Prokaryotes,* 2nd ed., vol. 4, eds. A. Balows, H.G. Truper, M. Dworkin, W. Harder, and K.H. Schleifer, 3819–3854. New York: Springer-Verlag.

Schenk, H.E.A. (1993) "Is endocytobiology an independent science?" *Endocytobiosis and Cell Research* 10: 229–240.

Schlegel, M. (1994) "Molecular phylogeny of eukaryotes." *Trends in Ecology and Evolution* 9: 330–334.

Schofield, C.J. (1985) "Control of Chagas' disease vectors." *British Medical Bulletin* 41: 187–194.

Schofield, C.J. (1988) "Complement mediated evolution?" *Parasitology Today* 4: 89–90.

Scholtissek, C. (1996) "Molecular evolution of influenza viruses." *Virus Genes* 11: 209–215.

Schumann, G.L. (1991) *Plant Diseases: Their Biology and Social Impact.* St. Paul, MN: American Phytopathogical Society Press.

Schwemmler, W., ed. (1989) *Handbook of Insect Endocytobiosis: Morphology, Physiology, Genetics, Evolution.* Boca Raton, FL: CRC Press.

Schwemmler, W. (1991) "Symbiogenesis in insects as a model for cell differentiation, morphogenesis, and speciation." In *Symbiosis as a Source of Evolutionary Innovation,* eds. L. Margulis and R. Fester, pp. 178–204. Cambridge; MIT Press.

Schwemmler, W. (1993) "Ecophysiological significance of insect symbiogenesis." *Endocytobiosis and Cell Research* 9: 245–254.

Seckbach, J., ed. (1994) *Evolutionary Pathways and Enigmatic Algae: Cyanidium caldarium (Rhodophyta) and Related Cells.* Dordrecht: Kluwer.

Secord, D. and P. Kareiva. (1996) "Perils and pitfalls in the host specificity paradigm." *BioScience* 46: 448–453.

Seifert, R.P. (1981) "Application of mycological data base to principles and concepts of population and community ecology." In *The Fungal Community. Its Organization and Role in the Ecosystem,* eds. D.T. Wicklow and G.C. Carroll, pp. 11–23. New York: Marcel Dekker.

Selander, R.K., A.G. Clark, and T.S. Whittam, eds. (1991) *Evolution at the Molecular Level.* Sunderland, MA: Sinauer Associates.

Selosse, M.A. and F. LeTacon. (1998) "The land flora: a phototroph-fungus partnership?" *Trends in Ecology and Evolution* 13: 15–20.

Sequeira, L. (1984) "Cross protection and induced resistance. Their potential for plant disease control." *Trends in Biotechnology* 2: 25–29.

Service, R.S. (1997) "Microbiologists explore life's rich hidden kingdoms." *Science* 275: 1740–1742.

Shad, G. and K. Warren, eds. (1990) *Hookworm Disease: Current Status and New Directions.* London: Taylor and Francis.

Shadan, F.F. and L.P. Villarreal. (1996) "The evolution of small DNA viruses of eukaryotes: past and present considerations." *Virus Genes* 11: 239–257.

Shapiro, J. (1997) "Multicellularity: the rule, not the exception. Lessons from *Escherichia coli* colonies." In *Bacteria as Multicellular Organisms,* eds. J. Shapiro and M. Dworkin, pp. 14–49. New York: Oxford University Press.

Shapiro, J.A. (1998) "Thinking about bacterial populations as multicellular organisms." *Annual Review of Microbiology* 52: 81–104.

Shapiro, J. and M. Dworkin, eds. (1997) *Bacteria as Multicellular Organisms.* New York: Oxford University Press.

Shenk, T.E. (1993) "Virus and cell: determinants of tissue tropism." In *Emerging Viruses,* ed. S.S. Morse, pp. 79–90. New York: Oxford University Press.

Sherwood-Pike, M. (1991) "Fossils as keys to evolution in fungi." *BioSystems* 25: 121–129.

Sidhu, G. (1984) "Genetics of plant and animal parasitic systems." *BioScience* 34: 368–373.

Simonet, P., P. Normand, A.M. Hirsch, and A.D.L. Akkermans. (1990) "The Genetics of the *Frankia*-actinorhizal symbiosis." In *Molecular Biology of Symbiotic Nitrogen Fixation,* ed. P.M. Gresshoff, pp. 78–109. Boca Raton, FL: CRC Press.

Sitte, P. and S. Eschbach. (1992) "Cytosymbiosis and its significance in cell evolution." *Progress in Botany* 53: 29–43.

Sitte, P., S. Eschbach, and M. Maerz. (1992) "The role of symbiosis in algal evolution." In *Algae and Symbioses,* ed. W. Reisser, pp. 711–733. Bristol, UK: Biopress Ltd.

Smith, D.C. (1991) "Why do so few animals form endosymbiotic associations with photosynthetic microbes?" *Philosophical Transactions of the Royal Society of London B* 333: 225–230.

Smith, D.C. and A.E. Douglas. (1987) *The Biology of Symbiosis.* London: Edward Arnold.

Smith, E.C. and G. Griffiths. (1996) "The occurrence of the chloroplast pyrenoid is correlated with the activity of a CO_2-concentrating mechanism and carbon isotope discrimination in lichens and bryophytes." *Planta* 198: 6–16.

Smith, E.F. (1911) "Anton De Bary." *Phytopathology* 1: 1–2.

Smith, F.A. and S.E. Smith. (1996) "Mutualism and parasitism; diversity in function and structure in the "arbuscular" (VA) mycorrhizal symbiosis." *Advances in Botanical Research* 22: 1–43.

Smith, G.A. and D.A. Portnoy. (1997) "How the *Listeria monocytogenes* ActA protein converts actin polymerization into a motile force." *Trends in Microbiology* 5: 272–276.

Smith, G.L. (1994) "Virus strategies for evasion of the host response to infection." *Trends in Microbiology* 2: 81–88.

Smith, M.L., J.N. Bruhn, and J.A. Anderson. (1992) "The fungus *Armillaria bulbosa* is among the largest and oldest living organisms." *Nature* 356: 428–431.

Smith, M.W., D.F. Feng, and R.F. Doolittle. (1992) "Evolution by acquisition: the case for horizontal gene transfers." *Trends in Biochemical Sciences* 17: 489–493.

Smith, N. (1968) "The advantage of being parasitized." *Nature* 219: 690–694.

Smith, S.E. and D.J. Read. (1997) *Mycorrhizal Symbiosis.* San Diego, CA: Academic Press.

Smith Trail, D. (1980) "Behavioral interactions between parasites and hosts: host suicide and the evolution of complex life cycles." *American Naturalist* 116: 77–91.

Smyth, J.D. (1994) *Introduction to Animal Parasitology,* 3rd ed. Cambridge: Cambridge University Press.

Smyth, J.D., and D.P. McManus. (1989) *The Physiology and Biochemistry of Cestodes.* Cambridge: Cambridge University Press.

Smyth, J.D. and M.M. Smyth. (1980) *Frogs as Host-Parasite Systems. I. An Introduction to Parasitology through the Parasites of Rana temporaria, R. esculenta and R. pipiens.* London: Macmillan.

Sogin, M.L. (1989) "Evolution of eukaryotic microorganisms and their small subunit ribosomal RNAs." *American Zoologist* 29: 487–499.

Sogin, M.L. (1991) "Early evolution and the origin of eukaryotes." *Current Opinion in Genetics and Development* 1: 457–463.

Sogin, M.L. (1997) "History assignment: when was the mitochondrion founded?" *Current Opinion in Genetics and Development* 7: 792–799.

Sogin, M.L., J.H. Gunderson, H.J. Elwood, R.A. Alonso, and D.A. Peattie. (1989) "Phylogenetic meaning of the kingdom concept: an unusual ribosomal RNA from *Giardia lamblia.*" *Science* 243: 75–77.

Sogin, M.L., H.G. Morrison, G. Hinkle, and J.D. Silberman. (1996) "Ancestral relationships of the major eukaryotic lineages." *Microbiologica* 12: 17–28.

Sonea, S. (1988) "The global organism. A new view of bacteria." *Annals of the New York Academy of Sciences* 28: 38–45.

Spaink, H.P. (1995) "The molecular basis of infection and nodulation by Rhizobia: the ins and outs of sympathogenesis." *Annual Review of Phytopathology* 33: 345–368.

Spencer, D.M., ed. (1978) *The Powdery Mildews.* London: Academic Press.

Spencer-Phillips, P.T.N. (1997) "Function of fungal haustoria in epiphytic and endophytic infections." *Advances in Botanical Research* 24: 309–333.

Sprent, J.I. (1989) "Which steps are essential for the formation of functional legume nodules?" *New Phytologist* 111: 129–153.

Sprent, J.I. (1997) "Coevolution of legume-rhizobial symbioses." In *Biological Fixation of Nitrogen for Ecology and Sustainable Agriculture.* eds. A. Legocki, H. Bothe, and A. Puhler, pp. 313–316. Berlin: Springer-Verlag.

Sprent, J.I. and P. Sprent. (1990) *Nitrogen Fixing Organisms: Pure and Applied Aspects.* London: Chapman and Hall.

Stacey, G., R.H. Burris, and H.J. Evans, eds. (1992) *Biological Nitrogen Fixation.* New York: Chapman and Hall.

Staddon, P.L. and A.H. Fitter. (1998) "Does elevated atmospheric carbon dioxide affect arbuscular mycorrhizas?" *Trends in Ecology and Evolution* 13: 455–458.

Stafford, H.A. (1997) "Roles of flavonoids in symbiotic and defense functions in legume roots." *Botanical Review* 63: 27–39.

Starr, M.P. (1975) "A generalized scheme for classifying organismic associations." In *Symbiosis,* eds. D.H. Jennings and D.L. Lee, pp. 1–20. Cambridge: Cambridge University Press.

Staskawicz, B., F. Ausubel, B. Baker, J. Ellis, and J. Jones. (1995) "Molecular genetics of plant disease resistance." *Science* 268: 661–667.

Stern, D.B. and J.D. Palmer. (1984) "Extensive and widespread homologies between mitochondrial DNA and chloroplast DNA in plants." *Proceedings of the National Academy of Sciences, USA* 81: 1946–1950.

Stevens, L. and R. Fialho. (1999) "Virulence of symbiotic bacteria (*Wolbachia*) in their arthropod hosts." *Annual Review of Ecology and Systematics.* In press.

Stevenson, R.J., M.L. Bothwell, and R.L. Lowe. (1996) *Algal Ecology, Freshwater Benthic Ecosystems.* San Diego, CA: Academic Press.

Stiller, J.W. and B.D. Hall. (1997) "The origin of red algae: implications for plastid evolution." *Proceedings of the National Academy of Sciences, USA* 94: 4520–4525.

Stocker-Worgotter, E. (1995) "Experimental cultivation of lichens and lichen symbionts." *Canadian Journal of Botany* (suppl. 1) vol. 13: 73 S579–S589.

Stolp, H. (1979) "Interaction between *Bdellovibrio* and its host cell." *Proceedings of the Royal Society of London. B. Biological Sciences* 204: 211–217.

Stouthamer, R., R.F. Luck, and W.D. Hamilton. (1990) "Antibiotics cause parthenogenetic *Trichogramma* to revert to sex." *Proceedings of the National Academy of Sciences, USA* 90: 2424–2427.

Strauss, E. (1999) "A symphony of bacterial voices." *Science* 284: 1302–1304.

Stuart, K. (1995) "RNA editing: an overview, status report and personal perspective." In *Molecular Approaches to Parasitology,* eds. J.C. Boothroyd and E. Komuniecki, pp. 243–254. New York: Wiley-Liss.

Stuart, K. and J.E. Feagin. (1992) "Mitochondrial DNA of kinetoplastids." *International Review of Cytology* 141: 65–88.

Surico, G. and N.S. Iacobellis. (1992) "Phytohormones and olive knot disease." In *Molecular Signals in Plant-Microbe Communications,* ed. D.S. Verma, pp. 209–227. Boca Raton, FL: CRC Press.

Suttle, C.A., A.M. Chan, and M.T. Cottrell. (1990) "Infection of phytoplankton by viruses and reduction of primary production." *Nature* 347: 467–469.

Syvanen, M. (1994) "Horizontal gene transfer: evidence and possible consequences." *Annual Review of Genetics* 28: 237–261.

Takemoto, J.Y. (1992) "Bacterial phytotoxin syringomycin and its interaction with host membranes." In *Molecular Signals in Plant Microbe Communications,* ed. D.S. Verma, pp. 247–260. Boca Raton, FL: CRC Press.

Talbot, N.J. (1995) "Having a blast: exploring the pathogenicity of *Magnaporthe grisea.*" *Trends in Microbiology* 3: 9–16.

Tannock, G.W. (1995) *Normal Microflora: An Introduction to Microbes Inhabiting the Human Body.* London: Chapman and Hall.

Tavares, I.I. (1985) *"Laboulbeniales."* *Mycological Memoirs* 9: 1–627.

Taylor, F.J.R. (1987) *The Biology of Dinoflagellates.* Oxford: Blackwell.

Taylor, F.J.R., and P.J. Harrison. (1983) "Ecological aspects of intracellular symbiosis." In *Endocytobiology.* vol. 2. eds. H.E.A. Schenk and W. Schwemmler, pp. 827–842. Berlin: Walter de Gruyter.

Temin, H.M. (1989) "Retrons in bacteria." *Nature* 339: 254–255.

Terborgh, J. (1989) *Where Have All the Birds Gone?* Princeton, NJ: Princeton University Press.

Thaxter, R. (1896–1931) "Contributions towards a Monograph of the Laboulbeniaceae. Part I-V." *Memoirs of the American Academy Arts Science*

Theriot, J.A. (1995) "The cell biology of infection by intracellular bacterial pathogens." *Annual Review of Cell and Developmental Biology* 11: 213–239.

Thingstad, T.F., M. Heldal, G. Bratbak, and I. Dundas. (1993) "Are viruses important partners in pelagic food webs?" *Trends in Ecology and Evolution* 8: 209–213.

Thompson, J. (1981) "Reversed animal-plant interactions: the evolution of insectivorous and ant-fed plants." *Biological Journal of Linnean Society* 16: 147–155.

Thompson, J.N. (1989) "Concepts of coevolution." *Trends in Ecology and Evolution* 4: 179–183.

Thompson, J.N. (1994) *The Coevolutionary Process.* Chicago: University of Chicago Press.

Thompson, R.C.A., J.A. Reynoldson, and A.H.W. Mendis. (1993) "*Giardia* and giardiasis." *Advances in Parasitology* 32: 71–160.

Threadgold, L. (1984) "Parasitic Platyhelminthes." In *Biology of the Integument,* eds. J. Bereiter-Hahn, A. Maltoltsky, and A. Richards, pp. 132–191. Berlin: Springer-Verlag.

Tilney, L.G., D.J. DeRosier, and M.S. Tilney. (1992) "How *Listeria* exploits host cell actin to form its own cytoskeleton." *Journal of Cell Biology* 118: 71–93.

Tiollais, P. and M.A. Buenidia. (1991) "Hepatitis B virus." *Scientific American* 264: 116–123.

Toft, C.A., A. Aeschlimann, and L. Bolis, eds. (1991) *Parasite-Host Associations: Coexistence or Conflict?* Oxford: Oxford University Press.

Toft, C.A. and A.J. Karter. (1990) "Parasite-host coevolution." *Trends in Ecology and Evolution* 5: 326–329.

Topoff, H. (1990) "Slave-making ants." *American Scientist* 78: 520–528.

Topoff, H. (1994) "The ant who would be queen." *Natural History* 103: 41–47.

Trager, W. (1986) *Living Together: The Biology of Animal Parasitism.* New York: Plenum Press.

Trench, R.K. (1993) "Microalgal-invertebrate symbioses: a review." *Endocytobiosis and Cell Research* 9: 135–175.

Treseder, K.K., D.W. Davidson, and J.R. Ehleringer. (1995) "Absorption of ant-provided carbon dioxide and nitrogen by a tropical epiphyte." *Nature* 375: 137–139.

Trinchant, J.C. and J. Rigaud. (1987) "Acetylene reduction by bacteroids isolated from stem nodules of *Sesbania ristrata*. Specific role of lactate as an energy-yielding substrate." *Journal of General Microbiology* 29: 489–496.

Trivers, R. (1971) "The evolution of reciprocal altruism." *The Quarterly Review of Biology* 46: 35–57.

Troyer, K. (1984) "Microbes, herbivory, and the evolution of social behavior." *Journal of Theoretical Biology* 106: 157–169.

Tunlid, A., H-B. Jansson, and B. Nordbring-Hertz. (1992) "Fungal attachment to nematodes." *Mycological Research* 96: 401–412.

Tunnicliffe, V. (1991) "The biology of hydrothermal vents: ecology and evolution." *Annual Review of Oceanography and Marine Biology* 29: 319–407.

Tunnicliffe, V. (1992) "The nature and origin of the modern hydrothermal vent fauna." *Palaios* 7: 338–350.

Turner, P.E., V.S. Cooper, and R.E. Lenski. (1998) "Tradeoff between horizontal and vertical modes of transmission in bacterial plasmids." *Evolution* 52: 315–329.

Udvardi, M.K. and D.A. Day. (1997) "Metabolite transport across symbiotic membranes of legume nodules." *Annual Review of Plant Physiology and Plant Molecular Biology* 48: 493–523.

Upcroft, J. and P. Upcroft. (1998) "My favorite cell: *Giardia*." *BioEssays* 20: 256–263.

Valdivia, R.H. and S. Falkow. (1997) "Probing bacterial gene expression within host cells." *Trends in Microbiology* 5: 360–362.

Valent, B.S. (1999) "Fungal plant pathology from the fungus' point of view." *Annual Review of Genetics* 33: In press.

Valentin, K., R.A. Cattolico, and K. Zetsche. (1992) "Phylogenetic origin of the plastids." In *Origins of Plastids,* ed. R.A. Lewin, pp. 193–221. New York: Chapman and Hall.

Van Beneden, P.J. (1876) *Animal Parasites and Messmates.* New York: Appleton.

Van Dover, C.L. (1996) *The Octopus's Garden: Hydrothermal Vents and Other Mysteries of the Deep Sea.* Reading, MA: Addison Wesley.

Van Etten, J.L. and R.H. Meints. (1999) "Giant viruses infecting algae." *Annual Review of Microbiology* 53: 447–494.

Van Etten, J.L., L.C. Lane, and R.H. Meints. (1991) "Viruses and viruslike particles of eukaryotic algae." *Microbiological Reviews* 55: 586–620.

Van Rhijn, P. and J. Vanderleyden. (1995) "The *Rhizobium*-plant symbiosis." *Microbiological Reviews* 59: 124–142.

Van Rhijn, P., R.B. Goldberg, and A.M. Hirsch. (1998) "*Lotus corniculatus* nodulation specificity is changed by the presence of a soybean lectin gene." *Plant Cell* 10: 1233–1249.

Van Soest, P.J. (1994) *Nutritional Ecology of the Ruminant,* 2nd ed. Ithaca, NY: Cornell University Press.

Varma, A. (1995) "Arbuscular mycorrhizal fungi: the state of art." *Critical Reviews in Biotechnology* 15: 179–199.

Varma, A. and B. Hock, eds. (1995) *Mycorrhiza: Structure, Function, Molecular Biology, and Biotechnology.* New York: Springer-Verlag.

Varmus, H. (1988) "Retroviruses." *Science* 240: 1427–1435.

Varon, M. (1974) "The bdellophage three-membered parasitic system." *Critical Review of Microbiology* 3: 221–241.

Vasil, I.K. (1994) "Molecular improvement of cereals." *Plant Molecular Biology* 25: 925–937.

Vellai, T., K. Takacs, and G. Vida (1998) "A new aspect to the origin and evolution of eukaryotes." *Journal Molecular Evolution* 46: 499–507.

Verma, D.P.S. (1992) "Signals in root nodule organogenesis and endocytosis of *Rhizobium*." *The Plant Cell* 4: 373–382.

Verma, D.P.S. and A.J. Delauney. (1989) "Root nodule symbiosis: nodulins and nodulin genes." In *Temporal and Spatial Regulation of Plant Genes,* eds. D.P.S. Verma and R.B. Goldberg, pp. 169–199. Vienna: Springer-Verlag.

Verma, D.P.S., C.A. Hu, and M. Zhang. (1992) "Plant nodule development: origin, function and regulation of nodulin genes." *Physiologia Plantarum* 85: 253–265.

Vermeij, G.J. (1994) "The evolutionary inter-actions among species: selection, escala-tion, and coevolution." *Annual Review of Ecology and Systematics* 25: 219–236.

Viale, A.M. and A.K. Arakaki. (1994) "The chaperone connection to the origins of the eukaryotic organelles." *FEBS Letters* 341: 146–151.

Vickerman, K. (1971) "Morphological and physiological considerations of extracellu-lar blood protozoa." In *Ecology and Physiology of Parasites,* ed. A.M. Fallis, pp. 58–91. Toronto: University of Toronto Press.

Virchow, Rudolf. (1860) *Cellular Pathology.* (translated by Frank Chance from the 2nd ed.) New York: R.M. De Witt.

Vlassak, M.K. and J. Vanderleyden. (1997) "Factors influencing nodule occupancy by inoculant rhizobia." *Critical Reviews in Plant Sciences* 16: 163–229.

Vogel, G. (1997) "Parasites shed light on cel-lular evolution." *Science* 275: 1422.

Volk, T. (1997) *Gaia's Body.* New York: Springer-Verlag.

Vollrath, F. (1998) "Dwarf males." *Trends in Ecology and Evolution* 13: 159–163.

Volterra, V. (1926) "Variation and fluctuations of numbers of individual species living to-gether." Reprinted 1931, In *Animal Ecol-ogy,* R.N. Chapman, pp. 409–448. New York: McGraw Hill.

Voyles, B.A. (1993) *The Biology of Viruses.* St. Louis, MO: Mosby.

Vulliamy, T., P. Mason, and L. Luzzatto. (1992) "The molecular basis of glucose-6-phosphate dehydrogenase deficiency." *Trends in Genetics* 8: 138–143.

Wagner, G.M. (1997) "*Azolla*: a review of its biology and utilization." *Botanical Review* 63: 1–26.

Wakelin, D. (1996) *Immunity to Parasites. How Parasitic Infections Are Controlled,* 2nd ed. Cambridge: Cambridge University Press.

Walden, R. and R. Wingender. (1995) "Gene transfer and plant regeneration tech-niques." *Trends in Biotechnology* 13: 324–331.

Walker, C. (1995) "AM or VAM: What's in a word?" In *Mycorrhiza,* eds. A. Varma and B. Hock, pp. 25–26. Berlin: Springer-Verlag.

Wallin, I.E. (1927) *Symbionticism and the Origin of Species.* Baltimore: Williams and Wilkins.

Walter, D.E. (1996) "Living on leaves: mites, tomenta, and leaf domatia." *Annual Re-view of Entomology* 41: 101–114.

Wang, A.L. and C.C. Wang. (1991) "Viruses of parasitic protozoa." *Parasitology Today* 7: 76–80.

Wang, C.C. (1989) "Nucleic acid metabolism in *Trichomonas vaginalis.*" In *Tricho-monads Parasites in Humans,* ed. B.M. Honigberg, pp. 84–90. New York: Springer-Verlag.

Warren, K. (1988) "The global impact of para-sitic diseases." In *The Biology of Para-sitism,* eds. P. Englund and A. Sher, pp. 3–12. New York: Alan R. Liss.

Wasmann, C.C., W. Loefferhardt, and J.J. Bohnert. (1987) "Cyanelles: organization and molecular biology." In *The Cyano-bacteria: A Comprehensive Review,* eds. P. Fay and C. van Baalen. Amsterdam: Elsevier.

Watt, W.B. (1999) "Molecular functional stud-ies of natural genetic variation: compari-son between prokaryotes and eukaryotes." *Annual Review of Genetics.* In press.

Webbe, G. (1981) "Schistosomiasis: some ad-vances." *British Medical Journal* 283: 1–8.

Weber, N.A. (1966) "Fungus-growing ants." *Science* 153: 587–604.

Webster, R.G., W.J. Bean, O.T. Gorman, T.M. Chambers, and Y. Kawaoka. (1992) "Evo-lution and ecology of influenza A viruses." *Microbiological Reviews* 56: 152–179.

Wei, X. (1995) "Viral dynamics in HIV-1 in-fection." *Nature* 373: 117–122.

Weinberg, E.D. (1997) "The *Lactobacillus* anomaly: total iron abstinence." *Perspec-tives in Biology and Medicine* 40: 578–583.

Weiss, R.A. (1993) "How does HIV cause AIDS?" *Science* 260: 1273–1279.

Wells, S.M., C. Sheppard, and C. Jenkins, eds. (1988) *Coral Reefs of the World.* Nairobi, Kenya: United Nations Environment Pro-gramme.

Werner, D. (1992) *Symbiosis of Plants and Microbes.* London: Chapman and Hall.

Werren, J.H. (1997) "Biology of *Wolbachia.*" *Annual Review of Entomology* 42: 587–609.

Werren, J.H. and L.W. Beukeboom. (1998) "Sex determination, sex ratios, and genetic conflict." *Annual Review of Ecology and Systematics* 29: 233–261.

Werren, J.H., W. Zhang, and L.R. Guo. (1995) "Evolution and phylogeny of *Wolbachia*: reproductive parasites of arthropods." *Proceedings of the Royal Society of London B. Biological Sciences* 261: 55–63.

Wessels, J.G.H. (1997) "Hydrophobins: pro-teins that change the nature of fungal sur-face." *Advances in Microbial Physiology* 38: 1–36.

Westneat, D.F. and T.R. Birkhead. (1998) "Alternative hypotheses linking the immune system and mate choice for good genes." *Proceedings of the Royal Society of London B* 265: 1065–1073.

Wheeler, W.M. (1911) "The ant-colony as an organism." *Journal of Morphology* 22: 307–325.

White, D.C. (1995) *The Physiology and Biochemistry of Prokaryotes.* New York: Oxford University Press.

White, D.O. and F.J. Fenner. (1994) *Medical Virology,* 4th ed. Orlando, FL: Academic Press.

Whitefield, P. (1993) *From So Simple a Beginning. The Book of Evolution.* New York: Macmillan.

Whitfield, J.B. (1990) "Parasitoids, polydnaviruses and endosymbiosis." *Parasitology Today* 6: 381–384.

Whitfield, J.B. (1998) "Phylogeny and evolution of host-parasitoid interactions in hymenoptera." *Annual Review of Entomology* 43: 129–151.

Whittaker, R.H. (1969) "New concepts of kingdoms." *Science* 163: 150–160.

Wilding, N., N.M. Collins, P.M. Hammond, and J.F. Webber, eds. (1989) *Insect-Fungus Interactions.* London: Academic Press.

Wilkinson, C.R. (1987) "Significance of microbial symbionts in sponge evolution and ecology." *Symbiosis* 4: 135–146.

Wilkinson, C.R. (1992) "Symbiotic interactions between marine sponges and algae." In *Algae and Symbioses,* ed. W. Reisser, pp. 111–151. Bristol, UK: Biopress Ltd.

Wilkinson, T. (1998) "*Wolbachia* comes of age." *Trends in Ecology and Evolution* 13: 213–214.

Williams, A.G. and G.S. Coleman. (1991) *The Rumen Protoza.* Berlin: Springer-Verlag.

Williams, A.G. and D. Lloyd. (1993) "Biological activities of symbiotic and parasitic protozoa and fungi in low-oxygen environments." *Advances in Microbial Ecology* 13: 211–262.

Williams, D.M. and T.M. Embley. (1996) "Microbial biodiversity: domains and kingdoms." *Annual Review of Ecology and Systematics* 27: 569–595.

Williams. G.R. (1996) *The Molecular Biology of Gaia.* New York: Columbia University Press.

Williamson, G. (1982) "Plant mimicry—evolutionary constraints." *Biological Journal of Linnean Society* 18: 44–58.

Wills, C. (1996) *Yellow Fever Black Goddess: The Coevolution of People and Plagues.* Reading, MA: Addison-Wesley.

Wilson, D.S. and E. Sober. (1989) "Reviving the superorganism." *Journal of Theoretical Biology* 136: 337–356.

Wilson, E.O. (1971) *The Insect Societies.* Cambridge, MA: Harvard University Press.

Wilson, E.O. (1975) *Sociobiology: The New Synthesis.* Cambridge, MA: Harvard University Press.

Wilson, E.O. (1996) *In Search of Nature.* Washington, DC: Island/Shearwater.

Wilson, E.O. (1998) *Consilience. The Unity of Knowledge.* New York: Alfred A. Knopf.

Wilson, R.J.M., M.J. Gardner, D.H. Williamson, and J.E. Feagin. (1991) "Have malaria parasites three genomes?" *Parasitology Today* 7: 134–136.

Wilson, T. and J.W. Hastings. (1998) "Bioluminescence." *Annual Review of Cell and Developmental Biology* 14: 197–230.

Winans, S.C. (1992) "Two-way chemical signaling in *Agrobacterium*-plant interactions." *Microbiological Reviews* 56: 12–31.

Winans, S.C. (1998) "Command, control and communication in bacterial pathogenesis." *Trends in Microbiology* 6: 382–383.

Windsor, D. (1997) "Equal rights for parasites." *Perspectives in Biology and Medicine* 40: 222–229.

Woese, C.R. (1977) "Endosymbionts and mitochondrial origins." *Journal of Molecular Evolution* 10: 93–96.

Woese, C.R. (1993) "The archaea: their history and significance." In *The Biochemistry of Archaea (Archaebacteria),* eds. M. Kates, D.J. Kushner, and A.T. Matheson, pp. 7–29. Amsterdam: Elsevier.

Woese, C.R. (1994) "There must be a prokaryote somewhere: microbiology's search for itself." *Microbiological Reviews* 56: 1–9.

Woese, C.R. (1998) "The universal ancestor." *Proceedings of the National Academy of Sciences USA* 95: 6854–6859.

Woese, C.R. and G.E. Fox (1977) "Phylogenetic structure of the prokaryotic domain: the primary kingdoms." *Proceedings of the National Academy of Sciences, USA* 74: 5088–5090.

Woese, C.R., O. Kandler, and M.L. Wheelis. (1990) "Towards a natural system of organisms: proposal for the domains: archaea, bacteria and eucarya." *Proceedings of the National Academy of Sciences, USA* 87: 4576–4579.

Wood, H.A., and R.R. Granados. (1991) "Genetically engineered baculoviruses as agents for pest control." *Annual Review of Microbiology* 45: 69–87.

Wood, R. (1998) "The ecological evolution of reefs." *Annual Review of Ecology and Systematics* 29: 179–206.

Wood, T.G. and R.J. Thomas. (1989) "The mutualistic association between Macrotermitinae and Termitomyces." In *Insect-Fungus Interactions*, eds. N. Wilding, N.M. Collins, P.M. Hammond, and J.F. Webber, pp. 69–92. London: Academic Press.

Wyler, D.J. (1998) *Modern Parasite Biology*. New York: W.H. Freeman.

Yamamoto, H. (1995) "Pathogenesis and host-parasite specificity in rusts." In *Pathogenesis and Host Specificity in Plant Diseases*, vol. 2, eds. K. Kohmoto, U.S. Singh, and R.P. Singh, pp. 203–215. New York: Elsevier.

Yanaki, M. and K. Yamasato. (1993) "Phylogenetic analysis of the family Rhizobiaceae and related bacteria by sequencing of 16S rRNA gene using PCR and DNA sequencer." *FEMS Microbiological Letter* 107: 115–120.

Yang He, S. (1998) "Type III Protein secretion systems in plant and animal pathogenic bacteria." *Annual Review of Phytopathology* 36: 363–392.

Yellowlees, D., T. Rees, and W. Fitt. (1993) "The giant clam as a model animal for study of marine algal-invertebrate associations." In *Biology and Mariculture of Giant Clams*, ACIAR Monograph no. 47, ed. W. Fitt. Canberra.

Yen, J.H. and A. R. Barr. (1973) "The etiological agent of cytoplasmic incompatibility in *Culex pipens*." *Journal of Invertebrate Pathology* 22: 242–250.

Yoshimura, I., T. Kurokawa, Y. Yamamoto, and Y. Kinoshita. (1993) "Development of lichen thalli *in vitro*." *Bryologist* 96: 412–421.

Young, J.P.W. and A.W.B. Johnston. (1989) "The evolution of specificity in the legume-*Rhizobium* symbiosis." *Trends in Ecology and Evolution* 4: 341–349.

Zacheo, G. and T. Bleve-Zacheo. (1995) "Plant-nematode interactions: histological, physiological and biochemical interactions." In *Pathogenesis and Host Specificity in Plant Diseases*, eds. K. Kohmoto, U.S. Singh, and R.P. Singh, pp. 321–353. New York: Elsevier.

Zamora, R., J. Gomez, and J. Hodar. (1998) "Fitness responses of a carnivorous plant in contrasting ecological scenarios." *Ecology* 79: 1630–1644.

Zavars, E. and L. Roberts. (1985) "Developmental physiology of cestodes: cyclic nucleotide and the identity of putative crowding factors in *Hymenolepis diminuta*." *Journal of Parasitology* 71: 96–105.

Zeyl, C. and G. Bell. (1995) "Symbiotic DNA in eukaryotic genomes." *Trends in Ecology and Evolution* 11: 10–15.

Zhao, Z-S., F. Granucci, L. Yeh, P.A. Schaffer and H. Cantor. (1998) "Molecular mimicry by herpes simplex virus-type 1: autoimmune disease after viral infection." *Science* 279: 1344–1347.

Zimmer, C. (1991) "Hypersea invasion." *Discover Magazine* (October) 77–87.

Zuk, M. (1996) "Disease, endocrine-immune interactions, and sexual selection." *Ecology* 77: 1037–1042.

Zuk, M. and G.R. Kolluru. (1998) "Exploitation of sexual signals by predators and parasitoids." *The Quarterly Review of Biology* 73: 415–438.

Zupan, J. and P. Zambryski. (1997) "The *Agrobacterium* DNA transfer complex." *Critical Reviews in Plant Science* 16: 279–295.

Zychlinsky, A., J.J. Perdomo, and P.J. Sansonetti. (1994) "Molecular and cellular mechanisms of tissue invasion by *Shigella flexneri*." In *Microbial Pathogenesis and Immune Response*, eds. E.W. Ades, R.F. Rest, and S.A. Morse, pp. 197–208. New York: The New York Academy of Sciences.

INDEX